Real analysis for graduate students

Second edition

Richard F. Bass

ii

To the memory of my parents

Jay and Betty Bass

iv

Contents

20 Topology   197

21 Probability   247

Preface

Nearly every Ph.D. student in mathematics needs to pass a preliminary or qualifying examination in real analysis. The purpose of this book is to teach the material necessary to pass such an examination.

I had three main goals in writing this text:
  (1) present a very clear exposition;
  (2) provide a large collection of useful exercises;
  (3) make the text affordable.

Let me discuss each of these in more detail.

(1) There are a large number of real analysis texts already in existence. Why write another? In my opinion, none of the existing texts are ideally suited to the beginning graduate student who needs to pass a "prelim" or "qual." They are either too hard, too advanced, too encyclopedic, omit too many important topics, or take a nonstandard approach to some of the basic theorems.

Students who are starting their graduate mathematics education are often still developing their mathematical sophistication and find that the more details that are provided, the better (within reason). I have tried to make the writing as clear as possible and to provide the details. For the sake of clarity, I present the theorems and results as they will be tested, not in the absolutely most general abstract context. On the other hand, a look at the index will show that no topics that might appear on a preliminary or qualifying examination are omitted.

All the proofs are "plain vanilla." I avoid any clever tricks, sneaky proofs, unusual approaches, and the like. These are the proofs and methods that most experts grew up on.

(2) There are over 400 exercises. I tried to make them interesting and useful and to avoid problems that are overly technical. Many are routine, many are of moderate difficulty, but some are quite challenging. A substantial number are taken from preliminary examinations given in the past at major universities.

I thought long and hard as to whether to provide hints to the exercises. When I teach the real analysis course, I give hints to the harder questions. But some instructors will want a more challenging course than I give and some a less challenging one. I leave it to the individual instructor to decide how many hints to give.

(3) I have on my bookshelf several books that I bought in the early 1970's that have the prices stamped in them: \$10-\$12. These very same books now sell for \$100-\$200. The cost of living has gone up in the last 40 years, but only by a factor of 5 or 6, not a factor of 10. Why do publishers make textbooks so expensive? This is particularly troublesome when one considers that nowadays authors do their own typesetting and frequently their own page layout.

My aim was to make the soft cover version of this text cost less than \$20 and to provide a version in .pdf format for free. To do that, I am self-publishing the text.

At this point I should tell you a little bit about the subject matter of real analysis. For an interval contained in the real line or a nice region in the plane, the length of the interval or the area of the region give an idea of the size. We want to extend the notion of size to as large a class of sets as possible. Doing this for subsets of the real line gives rise to Lebesgue measure. Chapters 2–4 discuss classes of sets, the definition of measures, and the construction of measures, of which one example is Lebesgue measure on the line. (Chapter 1 is a summary of the notation that is used and the background material that is required.)

Once we have measures, we proceed to the Lebesgue integral. We talk about measurable functions, define the Lebesgue integral, prove the monotone and dominated convergence theorems, look at some simple properties of the Lebesgue integral, compare it to the Riemann integral, and discuss some of the various ways a sequence of functions can converge. This material is the subject of Chapters 5–10.

Closely tied with measures and integration are the subjects of product measures, signed measures, the Radon-Nikodym theorem, the differentiation of functions on the line, and L^p spaces. These are covered in Chapters 11–15.

Many courses in real analysis stop at this point. Others also include some or all of the following topics: the Fourier transform, the Riesz representation theorem, Banach spaces, and Hilbert spaces. We present these in Chapters 16–19.

Topology and probability are courses in their own right, but they are something every analyst should know. The basics are given in Chapters 20 and 21, resp.

Chapters 22–26 include a number of topics that are sometimes included in examinations at some universities. These topics are harmonic functions, Sobolev spaces, singular integrals, spectral theory, and distributions.

The first edition of this text, which was titled *Real analysis for graduate students: measure and integration theory*, stopped at Chapter 19. The main comments I received on the first edition were that I should cover additional topics. Thus, the second edition includes Chapters 20 to 26. This increased the length from around 200 pages to around 400 pages.

The prerequisites to this text are a solid background in undergraduate mathematics. An acquaintance with metric spaces is assumed, but no other topology. A summary of what you need to know is in Chapter 1. All the necessary background material can be learned from many sources; one good place is the book [7].

At some universities preliminary or qualifying examinations in real analysis are combined with those in undergraduate analysis or complex analysis. If that is the case at your university, you will have to supplement this book with texts in those subjects.

Further reading is always useful. I have found the books [4], [6], and [8] helpful.

I would like to thank A. Baldenko, I. Ben-Ari, K. Bharath, K. Burdzy, D. Ferrone, E. Giné, M. Gordina, E. Hsu, G. Lawler, L. Lu, K. Marinelli, J. Pitman, M. Poehlitz, H. Ren,and L. Rogers for some very useful suggestions. I would particularly like to thank my colleague Sasha Teplyaev for using the text in his class and suggesting innumerable improvements.

If you have comments, suggestions, corrections, etc., I would be glad to hear from you: **r.bass@uconn.edu**. I cannot, however, provide hints or solutions to the exercises.

Good luck with your exam!

Chapter 1

Preliminaries

In this short chapter we summarize some of the notation and terminology we will use and recall a few definitions and results from undergraduate mathematics.

1.1 Notation and terminology

We use A^c, read "A complement," for the set of points not in A. To avoid some of the paradoxes of set theory, we assume all our sets are subsets of some given set X, and to be precise, define

$$A^c = \{x \in X : x \notin A\}.$$

We write

$$A - B = A \cap B^c$$

(it is common to also see $A \setminus B$) and

$$A \triangle B = (A - B) \cup (B - A).$$

The set $A \triangle B$ is called the *symmetric difference* of A and B and is the set of points that are in one of the sets but not the other. If I is some non-empty index set, a collection of subsets $\{A_\alpha\}_{\alpha \in I}$ is disjoint if $A_\alpha \cap A_\beta = \emptyset$ whenever $\alpha \neq \beta$.

We write $A_i \uparrow$ if $A_1 \subset A_2 \subset \cdots$ and write $A_i \uparrow A$ if in addition $A = \cup_{i=1}^\infty A_i$. Similarly $A_i \downarrow$ means $A_1 \supset A_2 \supset \cdots$ and $A_i \downarrow A$ means that in addition $A = \cap_{i=1}^\infty A_i$.

1

We use $\mathbb{Q}$ to denote the set of rational numbers, $\mathbb{R}$ the set of real numbers, and $\mathbb{C}$ the set of complex numbers. We use

$$x \vee y = \max(x, y) \qquad \text{and} \qquad x \wedge y = \min(x, y).$$

We can write a real number x in terms of its positive and negative parts: $x = x^+ - x^-$, where

$$x^+ = x \vee 0 \qquad \text{and} \qquad x^- = (-x) \vee 0.$$

If z is a complex number, then $\bar{z}$ is the complex conjugate of z. The composition of two functions is defined by $f \circ g(x) = f(g(x))$.

If f is a function whose domain is the reals or a subset of the reals, then $\lim_{y \to x+} f(y)$ and $\lim_{y \to x-} f(y)$ are the right and left hand limits of f at x, resp.

We say a function $f : \mathbb{R} \to \mathbb{R}$ is *increasing* if $x < y$ implies $f(x) \leq f(y)$ and f is *strictly increasing* if $x < y$ implies $f(x) < f(y)$. (Some authors use "nondecreasing" for the former and "increasing" for the latter.) We define *decreasing* and *strictly decreasing* similarly. A function is *monotone* if f is either increasing or decreasing.

Given a sequence $\{a_n\}$ of real numbers,

$$\limsup_{n \to \infty} a_n = \inf_n \sup_{m \geq n} a_m,$$

$$\liminf_{n \to \infty} a_n = \sup_n \inf_{m \geq n} a_m.$$

For example, if

$$a_n = \begin{cases} 1, & n \text{ even}; \\ -1/n, & n \text{ odd}, \end{cases}$$

then $\limsup_{n \to \infty} a_n = 1$ and $\liminf_{n \to \infty} a_n = 0$. The sequence $\{a_n\}$ has a limit if and only if $\limsup_{n \to \infty} a_n = \liminf_{n \to \infty} a_n$ and both are finite. We use analogous definitions when we take a limit along the real numbers. For example,

$$\limsup_{y \to x} f(y) = \inf_{\delta > 0} \sup_{|y - x| < \delta} f(y).$$

1.2   Some undergraduate mathematics

We recall some definitions and facts from undergraduate topology, algebra, and analysis. The proofs and more details can be found in many places. A good source is [7]. Some of the results from topology can also be found in Chapter 20.

A set X is a *metric space* if there exists a function $d : X \times X \to \mathbb{R}$, called the metric, such that
(1) $d(x, y) = d(y, x)$ for all $x, y \in X$;
(2) $d(x, y) \geq 0$ for all $x, y \in X$ and $d(x, y) = 0$ if and only if $x = y$;
(3) $d(x, z) \leq d(x, y) + d(y, z)$ for all $x, y, z \in X$.

Condition (3) is called the triangle inequality.

Given a metric space X, let

$$B(x, r) = \{y \in X : d(x, y) < r\}$$

be the *open ball* of radius r centered at x. If $A \subset X$, the *interior* of A, denoted A^o, is the set of x such that there exists $r_x > 0$ with $B(x, r_x) \subset A$. The *closure* of A, denoted $\overline{A}$, is the set of $x \in X$ such that every open ball centered at x contains at least one point of A. A set A is *open* if $A = A^o$, *closed* if $A = \overline{A}$. If $f : X \to \mathbb{R}$, the *support* of f is the closure of the set $\{x : f(x) \neq 0\}$. f is continuous at a point x if given $\varepsilon > 0$, there exists $\delta > 0$ such that $|f(x) - f(y)| < \varepsilon$ whenever $d(x, y) < \delta$. f is continuous if it is continuous at every point of its domain. One property of continuous functions is that $f^{-1}(F)$ is closed and $f^{-1}(G)$ is open if f is continuous, F is closed, and G is open.

A sequence $\{x_n\} \subset X$ converges to a point $x \in X$ if for each $\varepsilon > 0$ there exists N such that $d(x_n, x) < \varepsilon$ whenever $n \geq N$. A sequence is a *Cauchy sequence* if for each $\varepsilon > 0$ there exists N such that $d(x_m, x_n) < \varepsilon$ whenever $m, n \geq N$. If every Cauchy sequence in X converges to a point in X, we say X is *complete*.

An *open cover* of a subset K of X is a non-empty collection $\{G_\alpha\}_{\alpha \in I}$ of open sets such that $K \subset \cup_{\alpha \in I} G_\alpha$. The index set I can be finite or infinite. A set K is *compact* if every open cover contains a finite subcover, i.e., there exists $G_1, \ldots, G_n \in \{G_\alpha\}_{\alpha \in I}$ such that $K \subset \cup_{i=1}^n G_i$.

We have the following two facts about compact sets.

Proposition 1.1 *If K is compact, $F \subset K$, and F is closed, then F is compact.*

Proposition 1.2 *If K is compact and f is continuous on K, then there exist x_1 and x_2 such that $f(x_1) = \inf_{x \in K} f(x)$ and $f(x_2) =$*

$\sup_{x \in K} f(x)$. *In other words, f takes on its maximum and minimum values.*

Remark 1.3 If $x \neq y$, let $r = d(x, y)$ and note that $d(x, r/2)$ and $d(y, r/2)$ are disjoint open sets containing x and y, resp. Therefore metric spaces are also what are called Hausdorff spaces.

Let F be either $\mathbb{R}$ or $\mathbb{C}$. X is a *vector space* or *linear space* if there exist two operations, addition $(+)$ and scalar multiplication, such that

(1) $x + y = y + x$ for all $x, y \in X$;

(2) $(x + y) + z = x + (y + z)$ for all $x, y, z \in X$;

(3) there exists an element $0 \in X$ such that $0 + x = x$ for all $x \in X$;

(4) for each x in X there exists an element $-x \in X$ such that $x + (-x) = 0$;

(5) $c(x + y) = cx + cy$ for all $x, y \in X$, $c \in F$;

(6) $(c + d)x = cx + dx$ for all $x \in X$, $c, d \in F$;

(7) $c(dx) = (cd)x$ for all $x \in X$, $c, d \in F$;

(8) $1x = x$ for all $x \in X$.

We use the usual notation, e.g., $x - y = x + (-y)$.

X is a *normed linear space* if there exists a map $x \to \|x\|$ such that

(1) $\|x\| \geq 0$ for all $x \in X$ and $\|x\| = 0$ if and only if $x = 0$;

(2) $\|cx\| = |c| \, \|x\|$ for all $c \in F$ and $x \in X$;

(3) $\|x + y\| \leq \|x\| + \|y\|$ for all $x, y \in X$.

Given a normed linear space X, we can make X into a metric space by setting $d(x, y) = \|x - y\|$.

A set X has an *equivalence relationship* "$\sim$" if

(1) $x \sim x$ for all $x \in X$;

(2) if $x \sim y$, then $y \sim x$;

(3) if $x \sim y$ and $y \sim z$, then $x \sim z$.

Given an equivalence relationship, X can be written as the union of disjoint equivalence classes. x and y are in the same equivalence class if and only if $x \sim y$. For an example, let $X = \mathbb{R}$ and say $x \sim y$ if $x - y$ is a rational number.

A set X has a *partial order* "$\leq$" if

(1) $x \leq x$ for all $x \in X$;

(2) if $x \leq y$ and $y \leq x$, then $x = y$.

Note that given $x, y \in X$, it is not necessarily true that $x \leq y$ or $y \leq x$. For an example, let Y be a set, let X be the collection of all subsets of Y, and say $A \leq B$ if $A, B \in X$ and $A \subset B$.

We need the following three facts about the real line.

Proposition 1.4 *Suppose $K \subset \mathbb{R}$, K is closed, and K is contained in a finite interval. Then K is compact.*

Proposition 1.5 *Suppose $G \subset \mathbb{R}$ is open. Then G can be written as the countable union of disjoint open intervals.*

Proposition 1.6 *Suppose $f : \mathbb{R} \to \mathbb{R}$ is an increasing function. Then both $\lim_{y \to x+} f(y)$ and $\lim_{y \to x-} f(y)$ exist for every x. Moreover the set of x where f is not continuous is countable.*

For an application of Hilbert space techniques to Fourier series, which is the last section of Chapter 19, we will use the *Stone-Weierstrass theorem*. The particular version we will use is the following.

Theorem 1.7 *Let X be a compact metric space and let $\mathcal{A}$ be a collection of continuous complex-valued functions on X with the following properties:*
(1) If $f, g \in \mathcal{A}$ and $c \in \mathbb{C}$, then $f + g$, fg, and cf are in $\mathcal{A}$;
(2) If $f \in \mathcal{A}$, then $\overline{f} \in \mathcal{A}$, where $\overline{f}$ is the complex conjugate of f;
(3) If $x \in X$, there exists $f \in \mathcal{A}$ such that $f(x) \neq 0$;
(4) If $x, y \in X$ with $x \neq y$, there exists $f \in \mathcal{A}$ such that $f(x) \neq f(y)$.
Then the closure of $\mathcal{A}$ with respect to the supremum norm is the collection of continuous complex-valued functions on X.

The conclusion can be rephrased as saying that given f continuous on X and $\varepsilon > 0$, there exists $g \in \mathcal{A}$ such that

$$\sup_{x \in X} |f(x) - g(x)| < \varepsilon.$$

When (3) holds, $\mathcal{A}$ is said to *vanish at no point* of X. When (4) holds, $\mathcal{A}$ is said to *separate points*. In (3) and (4), the function f depends on x and on x and y, resp.

For a proof of Theorem 1.7 see [4], [7], or Section 20.12.

Chapter 2

Families of sets

2.1 Algebras and σ-algebras

When we turn to constructing measures in Chapter 4, we will see that we cannot in general define the measure of an arbitrary set. We will have to restrict the class of sets we consider. The class of sets that we will want to use are σ-algebras (read "sigma algebras").

Let X be a set.

Definition 2.1 An *algebra* is a collection $\mathcal{A}$ of subsets of X such that
(1) $\emptyset \in \mathcal{A}$ and $X \in \mathcal{A}$;
(2) if $A \in \mathcal{A}$, then $A^c \in \mathcal{A}$;
(3) if $A_1, \ldots, A_n \in \mathcal{A}$, then $\cup_{i=1}^n A_i$ and $\cap_{i=1}^n A_i$ are in $\mathcal{A}$.

$\mathcal{A}$ is a *σ-algebra* if in addition
(4) whenever $A_1, A_2, \ldots$ are in $\mathcal{A}$, then $\cup_{i=1}^\infty A_i$ and $\cap_{i=1}^\infty A_i$ are in $\mathcal{A}$.

In (4) we allow countable unions and intersections only; we do not allow uncountable unions and intersections. Since $\cap_{i=1}^\infty A_i = (\cup_{i=1}^\infty A_i^c)^c$, the requirement that $\cap_{i=1}^\infty A_i$ be in $\mathcal{A}$ is redundant.

The pair $(X, \mathcal{A})$ is called a *measurable space*. A set A is *measurable* or *$\mathcal{A}$ measurable* if $A \in \mathcal{A}$.

Example 2.2 Let $X = \mathbb{R}$, the set of real numbers, and let $\mathcal{A}$ be the collection of all subsets of $\mathbb{R}$. Then $\mathcal{A}$ is a σ-algebra.

Example 2.3 Let $X = \mathbb{R}$ and let

$$\mathcal{A} = \{A \subset \mathbb{R} : A \text{ is countable or } A^c \text{ is countable}\}.$$

Verifying parts (1) and (2) of the definition is easy. Suppose $A_1, A_2, \ldots$ are each in $\mathcal{A}$. If each of the A_i are countable, then $\cup_i A_i$ is countable, and so is in $\mathcal{A}$. If $A_{i_0}^c$ is countable for some i_0, then

$$(\cup_i A_i)^c = \cap_i A_i^c \subset A_{i_0}^c$$

is countable, and again $\cup_i A_i$ is in $\mathcal{A}$. Since $\cap A_i = (\cup_i A_i^c)^c$, then the countable intersection of sets in $\mathcal{A}$ is again in $\mathcal{A}$.

Example 2.4 Let $X = [0,1]$ and let $\mathcal{A} = \{\emptyset, X, [0, \frac{1}{2}], (\frac{1}{2}, 1]\}$. Then $\mathcal{A}$ is a σ-algebra.

Example 2.5 Let $X = \{1,2,3\}$ and let $\mathcal{A} = \{X, \emptyset, \{1\}, \{2,3\}\}$. Then $\mathcal{A}$ is a σ-algebra.

Example 2.6 Let $X = [0,1]$, and let $B_1, \ldots, B_8$ be subsets of X which are pairwise disjoint and whose union is all of X. Let $\mathcal{A}$ be the collection of all finite unions of the B_i's as well as the empty set. (Thus $\mathcal{A}$ consists of 2^8 elements.) Then $\mathcal{A}$ is a σ-algebra.

Lemma 2.7 *If $\mathcal{A}_\alpha$ is a σ-algebra for each α in some non-empty index set I, then $\cap_{\alpha \in I} \mathcal{A}_\alpha$ is a σ-algebra.*

Proof. This follows immediately from the definition.   □

If we have a collection $\mathcal{C}$ of subsets of X, define

$$\sigma(\mathcal{C}) = \cap\{\mathcal{A}_\alpha : \mathcal{A}_\alpha \text{ is a } \sigma\text{-algebra}, \mathcal{C} \subset \mathcal{A}_\alpha\},$$

the intersection of all σ-algebras containing $\mathcal{C}$. Since there is at least one σ-algebra containing $\mathcal{C}$, namely, the one consisting of all subsets of X, we are never taking the intersection over an empty class of σ-algebras. In view of Lemma 2.7, $\sigma(\mathcal{C})$ is a σ-algebra. We call $\sigma(\mathcal{C})$ the *σ-algebra generated by the collection $\mathcal{C}$*, or say that

$\mathcal{C}$ *generates the* σ-*algebra* $\sigma(\mathcal{C})$. It is clear that if $\mathcal{C}_1 \subset \mathcal{C}_2$, then $\sigma(\mathcal{C}_1) \subset \sigma(\mathcal{C}_2)$. Since $\sigma(\mathcal{C})$ is a σ-algebra, then $\sigma(\sigma(\mathcal{C})) = \sigma(\mathcal{C})$.

If X has some additional structure, say, it is a metric space, then we can talk about open sets. If $\mathcal{G}$ is the collection of open subsets of X, then we call $\sigma(\mathcal{G})$ the *Borel* σ-*algebra* on X, and this is often denoted $\mathcal{B}$. Elements of $\mathcal{B}$ are called *Borel sets* and are said to be *Borel measurable*. We will see later that when X is the real line, $\mathcal{B}$ is *not* equal to the collection of all subsets of X.

We end this section with the following proposition.

Proposition 2.8 *If* $X = \mathbb{R}$, *then the Borel* σ-*algebra* $\mathcal{B}$ *is generated by each of the following collection of sets:*
(1) $\mathcal{C}_1 = \{(a, b) : a, b \in \mathbb{R}\}$;
(2) $\mathcal{C}_2 = \{[a, b] : a, b \in \mathbb{R}\}$;
(3) $\mathcal{C}_3 = \{(a, b] : a, b \in \mathbb{R}\}$;
(4) $\mathcal{C}_4 = \{(a, \infty) : a \in \mathbb{R}\}$.

Proof. (1) Let $\mathcal{G}$ be the collection of open sets. By definition, $\sigma(\mathcal{G})$ is the Borel σ-algebra. Since every element of $\mathcal{C}_1$ is open, then $\mathcal{C}_1 \subset \mathcal{G}$, and consequently $\sigma(\mathcal{C}_1) \subset \sigma(\mathcal{G}) = \mathcal{B}$.

To get the reverse inclusion, if G is open, it is the countable union of open intervals by Proposition 1.5. Every finite open interval is in $\mathcal{C}_1$. Since $(a, \infty) = \cup_{n=1}^{\infty}(a, a + n)$, then $(a, \infty) \in \sigma(\mathcal{C}_1)$ if $a \in \mathbb{R}$ and similarly $(-\infty, a) \in \sigma(\mathcal{C}_1)$ if $a \in \mathbb{R}$. Hence if G is open, then $G \in \sigma(\mathcal{C}_1)$. This says $\mathcal{G} \subset \sigma(\mathcal{C}_1)$, and then $\mathcal{B} = \sigma(\mathcal{G}) \subset \sigma(\sigma(\mathcal{C}_1)) = \sigma(\mathcal{C}_1)$.

(2) If $[a, b] \in \mathcal{C}_2$, then $[a, b] = \cap_{n=1}^{\infty}(a - \frac{1}{n}, b + \frac{1}{n}) \in \sigma(\mathcal{G})$. Therefore $\mathcal{C}_2 \subset \sigma(\mathcal{G})$, and hence $\sigma(\mathcal{C}_2) \subset \sigma(\sigma(\mathcal{G})) = \sigma(\mathcal{G}) = \mathcal{B}$.

If $(a, b) \in \mathcal{C}_1$, choose $n_0 \geq 2/(b - a)$ and note

$$(a, b) = \cup_{n=n_0}^{\infty}\left[a + \frac{1}{n}, b - \frac{1}{n}\right] \in \sigma(\mathcal{C}_2).$$

Therefore $\mathcal{C}_1 \subset \sigma(\mathcal{C}_2)$, from which it follows that $\mathcal{B} = \sigma(\mathcal{C}_1) \subset \sigma(\sigma(\mathcal{C}_2)) = \sigma(\mathcal{C}_2)$.

(3) Using $(a, b] = \cap_{n=1}^{\infty}(a, b + \frac{1}{n})$, we see that $\mathcal{C}_3 \subset \sigma(\mathcal{C}_1)$, and as above we conclude that $\sigma(\mathcal{C}_3) \subset \sigma(\mathcal{C}_1) = \mathcal{B}$. Using $(a, b) = \cup_{n=n_0}^{\infty}(a, b - \frac{1}{n}]$, provided n_0 is taken large enough, $\mathcal{C}_1 \subset \sigma(\mathcal{C}_3)$, and as above we argue that $\mathcal{B} = \sigma(\mathcal{C}_1) \subset \sigma(\mathcal{C}_3)$.

(4) Because $(a, b] = (a, \infty) - (b, \infty)$, then $\mathcal{C}_3 \subset \sigma(\mathcal{C}_4)$. Since $(a, \infty) = \cup_{n=1}^{\infty} (a, a + n]$, then $\mathcal{C}_4 \subset \sigma(\mathcal{C}_3)$. As above, this is enough to imply that $\sigma(\mathcal{C}_4) = \mathcal{B}$. $\square$

2.2   The monotone class theorem

This section will be used in Chapter 11.

Definition 2.9 A *monotone class* is a collection of subsets $\mathcal{M}$ of X such that
(1) if $A_i \uparrow A$ and each $A_i \in \mathcal{M}$, then $A \in \mathcal{M}$;
(2) if $A_i \downarrow A$ and each $A_i \in \mathcal{M}$, then $A \in \mathcal{M}$.

The intersection of monotone classes is a monotone class, and the intersection of all monotone classes containing a given collection of sets is the smallest monotone class containing that collection.

The next theorem, the *monotone class theorem*, is rather technical, but very useful.

Theorem 2.10 *Suppose $\mathcal{A}_0$ is an algebra, $\mathcal{A}$ is the smallest σ-algebra containing $\mathcal{A}_0$, and $\mathcal{M}$ is the smallest monotone class containing $\mathcal{A}_0$. Then $\mathcal{M} = \mathcal{A}$.*

Proof. A σ-algebra is clearly a monotone class, so $\mathcal{M} \subset \mathcal{A}$. We must show $\mathcal{A} \subset \mathcal{M}$.

Let $\mathcal{N}_1 = \{A \in \mathcal{M} : A^c \in \mathcal{M}\}$. Note $\mathcal{N}_1$ is contained in $\mathcal{M}$ and contains $\mathcal{A}_0$. If $A_i \uparrow A$ and each $A_i \in \mathcal{N}_1$, then each $A_i^c \in \mathcal{M}$ and $A_i^c \downarrow A^c$. Since $\mathcal{M}$ is a monotone class, $A^c \in \mathcal{M}$, and so $A \in \mathcal{N}_1$. Similarly, if $A_i \downarrow A$ and each $A_i \in \mathcal{N}_1$, then $A \in \mathcal{N}_1$. Therefore $\mathcal{N}_1$ is a monotone class. Hence $\mathcal{N}_1 = \mathcal{M}$, and we conclude $\mathcal{M}$ is closed under the operation of taking complements.

Let $\mathcal{N}_2 = \{A \in \mathcal{M} : A \cap B \in \mathcal{M} \text{ for all } B \in \mathcal{A}_0\}$. Note the following: $\mathcal{N}_2$ is contained in $\mathcal{M}$ and $\mathcal{N}_2$ contains $\mathcal{A}_0$ because $\mathcal{A}_0$ is an algebra. If $A_i \uparrow A$, each $A_i \in \mathcal{N}_2$, and $B \in \mathcal{A}_0$, then $A \cap B = \cup_{i=1}^{\infty} (A_i \cap B)$. Because $\mathcal{M}$ is a monotone class, $A \cap B \in \mathcal{M}$, which implies $A \in \mathcal{N}_2$. We use a similar argument when $A_i \downarrow A$.

Therefore $\mathcal{N}_2$ is a monotone class, and we conclude $\mathcal{N}_2 = \mathcal{M}$. In other words, if $B \in \mathcal{A}_0$ and $A \in \mathcal{M}$, then $A \cap B \in \mathcal{M}$.

Let $\mathcal{N}_3 = \{A \in \mathcal{M} : A \cap B \in \mathcal{M} \text{ for all } B \in \mathcal{M}\}$. As in the preceding paragraph, $\mathcal{N}_3$ is a monotone class contained in $\mathcal{M}$. By the last sentence of the preceding paragraph, $\mathcal{N}_3$ contains $\mathcal{A}_0$. Hence $\mathcal{N}_3 = \mathcal{M}$.

We thus have that $\mathcal{M}$ is a monotone class closed under the operations of taking complements and taking intersections. This shows $\mathcal{M}$ is a σ-algebra, and so $\mathcal{A} \subset \mathcal{M}$. $\qquad\square$

2.3 Exercises

Exercise 2.1 Find an example of a set X and a monotone class $\mathcal{M}$ consisting of subsets of X such that $\emptyset \in \mathcal{M}$, $X \in \mathcal{M}$, but $\mathcal{M}$ is not a σ-algebra.

Exercise 2.2 Find an example of a set X and two σ-algebras $\mathcal{A}_1$ and $\mathcal{A}_2$, each consisting of subsets of X, such that $\mathcal{A}_1 \cup \mathcal{A}_2$ is not a σ-algebra.

Exercise 2.3 Suppose $\mathcal{A}_1 \subset \mathcal{A}_2 \subset \cdots$ are σ-algebras consisting of subsets of a set X. Is $\cup_{i=1}^{\infty} \mathcal{A}_i$ necessarily a σ-algebra? If not, give a counterexample.

Exercise 2.4 Suppose $\mathcal{M}_1 \subset \mathcal{M}_2 \subset \cdots$ are monotone classes. Let $\mathcal{M} = \cup_{n=1}^{\infty} \mathcal{M}_n$. Suppose $A_j \uparrow A$ and each $A_j \in \mathcal{M}$. Is A necessarily in $\mathcal{M}$? If not, give a counterexample.

Exercise 2.5 Let $(Y, \mathcal{A})$ be a measurable space and let f map X onto Y, but do not assume that f is one-to-one. Define $\mathcal{B} = \{f^{-1}(A) : A \in \mathcal{A}\}$. Prove that $\mathcal{B}$ is a σ-algebra of subsets of X.

Exercise 2.6 Suppose $\mathcal{A}$ is a σ-algebra with the property that whenever $A \in \mathcal{A}$, there exist $B, C \in \mathcal{A}$ with $B \cap C = \emptyset$, $B \cup C = A$, and neither B nor C is empty. Prove that $\mathcal{A}$ is uncountable.

Exercise 2.7 Suppose $\mathcal{F}$ is a collection of real-valued functions on X such that the constant functions are in $\mathcal{F}$ and $f + g$, fg, and cf

are in $\mathcal{F}$ whenever $f, g \in \mathcal{F}$ and $c \in \mathbb{R}$. Suppose $f \in \mathcal{F}$ whenever $f_n \to f$ and each $f_n \in \mathcal{F}$. Define the function

$$\chi_A(x) = \begin{cases} 1, & x \in A; \\ 0, & x \notin A. \end{cases}$$

Prove that $\mathcal{A} = \{A \subset X : \chi_A \in \mathcal{F}\}$ is a σ-algebra.

Exercise 2.8 Does there exist a σ-algebra which has countably many elements, but not finitely many?

Chapter 3

Measures

In this chapter we give the definition of a measure, some examples, and some of the simplest properties of measures. Constructing measures is often quite difficult and we defer the construction of the most important one, Lebesgue measure, until Chapter 4

3.1 Definitions and examples

Definition 3.1 Let X be a set and $\mathcal{A}$ a σ-algebra consisting of subsets of X. A *measure* on $(X, \mathcal{A})$ is a function $\mu : \mathcal{A} \to [0, \infty]$ such that
(1) $\mu(\emptyset) = 0$;
(2) if $A_i \in \mathcal{A}$, $i = 1, 2, \ldots$, are pairwise disjoint, then

$$\mu(\cup_{i=1}^{\infty} A_i) = \sum_{i=1}^{\infty} \mu(A_i).$$

Saying the A_i are pairwise disjoint means that $A_i \cap A_j = \emptyset$ if $i \neq j$.

Definition 3.1(2) is known as *countable additivity*. We say a set function is *finitely additive* if $\mu(\cup_{i=1}^{n} A_i) = \sum_{i=1}^{n} \mu(A_i)$ whenever $A_1, \ldots, A_n$ are in $\mathcal{A}$ and are pairwise disjoint.

The triple $(X, \mathcal{A}, \mu)$ is called a *measure space*.

Example 3.2 Let X be any set, $\mathcal{A}$ the collection of all subsets of X, and $\mu(A)$ the number of elements in A. μ is called *counting measure.*

Example 3.3 Let $X = \mathbb{R}$, $\mathcal{A}$ the collection of all subsets of $\mathbb{R}$, $x_1, x_2, \ldots \in \mathbb{R}$, and $a_1, a_2, \ldots \geq 0$. Set

$$\mu(A) = \sum_{\{i : x_i \in A\}} a_i.$$

Example 3.4 Let $\delta_x(A) = 1$ if $x \in A$ and 0 otherwise. This measure is called *point mass* at x.

Proposition 3.5 *The following hold:*
(1) If $A, B \in \mathcal{A}$ with $A \subset B$, then $\mu(A) \leq \mu(B)$.
(2) If $A_i \in \mathcal{A}$ and $A = \cup_{i=1}^{\infty} A_i$, then $\mu(A) \leq \sum_{i=1}^{\infty} \mu(A_i)$.
(3) Suppose $A_i \in \mathcal{A}$ and $A_i \uparrow A$. Then $\mu(A) = \lim_{n \to \infty} \mu(A_n)$.
(4) Suppose $A_i \in \mathcal{A}$ and $A_i \downarrow A$. If $\mu(A_1) < \infty$, then we have $\mu(A) = \lim_{n \to \infty} \mu(A_n)$.

Proof. (1) Let $A_1 = A$, $A_2 = B - A$, and $A_3 = A_4 = \cdots = \emptyset$. Now use part (2) of the definition of measure to write

$$\mu(B) = \mu(A) + \mu(B - A) + 0 + 0 + \cdots \geq \mu(A).$$

(2) Let $B_1 = A_1$, $B_2 = A_2 - A_1$, $B_3 = A_3 - (A_1 \cup A_2)$, $B_4 = A_4 - (A_1 \cup A_2 \cup A_3)$, and in general $B_i = A_i - (\cup_{j=1}^{i-1} A_j)$. The B_i are pairwise disjoint, $B_i \subset A_i$ for each i, and $\cup_{i=1}^{\infty} B_i = \cup_{i=1}^{\infty} A_i$. Hence

$$\mu(A) = \mu(\cup_{i=1}^{\infty} B_i) = \sum_{i=1}^{\infty} \mu(B_i) \leq \sum_{i=1}^{\infty} \mu(A_i).$$

(3) Define the B_i as in (2). Recall that if a_i are non-negative real numbers, then $\sum_{i=1}^{\infty} a_i$ is defined to be $\lim_{n \to \infty} \sum_{i=1}^{n} a_i$. Since $\cup_{i=1}^{n} B_i = \cup_{i=1}^{n} A_i$, then

$$\mu(A) = \mu(\cup_{i=1}^{\infty} A_i) = \mu(\cup_{i=1}^{\infty} B_i) = \sum_{i=1}^{\infty} \mu(B_i)$$

$$= \lim_{n \to \infty} \sum_{i=1}^{n} \mu(B_i) = \lim_{n \to \infty} \mu(\cup_{i=1}^{n} B_i) = \lim_{n \to \infty} \mu(\cup_{i=1}^{n} A_i).$$

(4) Apply (3) to the sets $A_1 - A_i$, $i = 1, 2, \ldots$. The sets $A_1 - A_i$ increase to $A_1 - A$, and so

$$\mu(A_1) - \mu(A) = \mu(A_1 - A) = \lim_{n \to \infty} \mu(A_1 - A_n)$$
$$= \lim_{n \to \infty} [\mu(A_1) - \mu(A_n)].$$

Now subtract $\mu(A_1)$ from both sides and then multiply both sides by -1. □

Example 3.6 To see that $\mu(A_1) < \infty$ is necessary in Proposition 3.5, let X be the positive integers, μ counting measure, and $A_i = \{i, i+1, \ldots\}$. Then the A_i decrease, $\mu(A_i) = \infty$ for all i, but $\mu(\cap_i A_i) = \mu(\emptyset) = 0$.

Definition 3.7 A measure μ is a *finite measure* if $\mu(X) < \infty$. A measure μ is σ-*finite* if there exist sets $E_i \in \mathcal{A}$ for $i = 1, 2, \ldots$ such that $\mu(E_i) < \infty$ for each i and $X = \cup_{i=1}^{\infty} E_i$. If μ is a finite measure, then $(X, \mathcal{A}, \mu)$ is called a finite measure space, and similarly, if μ is a σ-finite measure, then $(X, \mathcal{A}, \mu)$ is called a σ-finite measure space.

Suppose X is σ-finite so that $X = \cup_{i=1}^{\infty} E_i$ with $\mu(E_i) < \infty$ and $E_i \in \mathcal{A}$ for each i. If we let $F_n = \cup_{i=1}^{n} E_i$, then $\mu(F_n) < \infty$ for each n and $F_n \uparrow X$. Therefore there is no loss of generality in supposing the sets E_i in Definition 3.7 are increasing.

Let $(X, \mathcal{A}, \mu)$ be a measure space. A subset $A \subset X$ is a *null set* if there exists a set $B \in \mathcal{A}$ with $A \subset B$ and $\mu(B) = 0$. We do not require A to be in $\mathcal{A}$. If $\mathcal{A}$ contains all the null sets, then $(X, \mathcal{A}, \mu)$ is said to be a *complete measure space*. The *completion* of $\mathcal{A}$ is the smallest σ-algebra $\overline{\mathcal{A}}$ containing $\mathcal{A}$ such that $(X, \overline{\mathcal{A}}, \mu)$ is complete. Sometimes one just says that $\mathcal{A}$ is complete or that μ is complete when $(X, \mathcal{A}, \mu)$ is complete.

A *probability* or *probability measure* is a measure μ such that $\mu(X) = 1$. In this case we usually write $(\Omega, \mathcal{F}, \mathbb{P})$ instead of $(X, \mathcal{A}, \mu)$, and $\mathcal{F}$ is called a σ-*field*, which is the same thing as a σ-algebra.

3.2   Exercises

Exercise 3.1 Suppose $(X, \mathcal{A})$ is a measurable space and μ is a non-negative set function that is finitely additive and such that $\mu(\emptyset) = 0$. Suppose that whenever A_i is an increasing sequence of sets in $\mathcal{A}$, then $\mu(\cup_i A_i) = \lim_{i \to \infty} \mu(A_i)$. Show that μ is a measure.

Exercise 3.2 Suppose $(X, \mathcal{A})$ is a measurable space and μ is a non-negative set function that is finitely additive and such that $\mu(\emptyset) = 0$ and $\mu(X) < \infty$. Suppose that whenever A_i is a sequence of sets in $\mathcal{A}$ that decrease to $\emptyset$, then $\lim_{i \to \infty} \mu(A_i) = 0$. Show that μ is a measure.

Exercise 3.3 Let X be an uncountable set and let $\mathcal{A}$ be the collection of subsets A of X such that either A or A^c is countable. Define $\mu(A) = 0$ if A is countable and $\mu(A) = 1$ if A is uncountable. Prove that μ is a measure.

Exercise 3.4 Suppose $(X, \mathcal{A}, \mu)$ is a measure space and $A, B \in \mathcal{A}$. Prove that

$$\mu(A) + \mu(B) = \mu(A \cup B) + \mu(A \cap B).$$

Exercise 3.5 Prove that if $\mu_1, \mu_2, \dots$ are measures on a measurable space and $a_1, a_2, \dots \in [0, \infty)$, then $\sum_{n=1}^{\infty} a_n \mu_n$ is also a measure.

Exercise 3.6 Prove that if $(X, \mathcal{A}, \mu)$ is a measure space, $B \in \mathcal{A}$, and we define $\nu(A) = \mu(A \cap B)$ for $A \in \mathcal{A}$, then ν is a measure.

Exercise 3.7 Suppose $\mu_1, \mu_2, \dots$ are measures on a measurable space $(X, \mathcal{A})$ and $\mu_n(A) \uparrow$ for each $A \in \mathcal{A}$. Define

$$\mu(A) = \lim_{n \to \infty} \mu_n(A).$$

Is μ necessarily a measure? If not, give a counterexample. What if $\mu_n(A) \downarrow$ for each $A \in \mathcal{A}$ and $\mu_1(X) < \infty$?

Exercise 3.8 Let $(X, \mathcal{A}, \mu)$ be a measure space, let $\mathcal{N}$ be the collection of null sets with respect to $\mathcal{A}$, and let $\mathcal{B} = \sigma(\mathcal{A} \cup \mathcal{N})$. Show that $(X, \mathcal{B}, \mu)$ is complete and is the completion of $(X, \mathcal{A}, \mu)$.

Exercise 3.9 Suppose X is the set of real numbers, $\mathcal{B}$ is the Borel σ-algebra, and m and n are two measures on $(X, \mathcal{B})$ such that $m((a, b)) = n((a, b)) < \infty$ whenever $-\infty < a < b < \infty$. Prove that $m(A) = n(A)$ whenever $A \in \mathcal{B}$.

Exercise 3.10 Suppose $(X, \mathcal{A})$ is a measurable space and $\mathcal{C}$ is an arbitrary subset of $\mathcal{A}$. Suppose m and n are two σ-finite measures on $(X, \mathcal{A})$ such that $m(A) = n(A)$ for all $A \in \mathcal{C}$. Is it true that $m(A) = n(A)$ for all $A \in \sigma(\mathcal{C})$? What if m and n are finite measures?

Chapter 4

Construction of measures

Our goal in this chapter is to give a method for constructing measures. This is a complicated procedure, and involves the concept of outer measure, which we introduce in Section 4.1.

Our most important example will be one-dimensional Lebesgue measure, which we consider in Section 4.2. Further results and some examples related to Lebesgue measure are given in Section 4.3.

One cannot define the Lebesgue measure of every subset of the reals. This is shown in Section 4.4.

The methods used to construct measures via outer measures have other applications besides the construction of Lebesgue measure. The Carathéodory extension theorem is a tool developed in Section 4.5 that can be used in constructing measures.

Let us present some of the ideas used in the construction of Lebesgue measure on the line. We want the measure m of an open interval to be the length of the interval. Since every open subset of the reals is the countable union of disjoint open intervals (see Proposition 1.5), if $G = \cup_{i=1}^{\infty}(a_i, b_i)$, where the intervals (a_i, b_i) are pairwise disjoint, we must have

$$m(G) = \sum_{i=1}^{\infty}(b_i - a_i).$$

We then set

$$m(E) = \inf\{m(G) : G \text{ open}, E \subset G\}$$

for arbitrary subsets $E \subset \mathbb{R}$. The difficulty is that m is not a measure on the σ-algebra consisting of all subsets of the reals; this is proved in Section 4.4. We resolve this by considering a strictly smaller σ-algebra. This is the essential idea behind the construction of Lebesgue measure, but it is technically simpler to work with intervals of the form $(a, b]$ rather than open intervals.

4.1   Outer measures

We begin with the notion of outer measure.

Definition 4.1 Let X be a set. An *outer measure* is a function μ^* defined on the collection of all subsets of X satisfying
(1) $\mu^*(\emptyset) = 0$;
(2) if $A \subset B$, then $\mu^*(A) \leq \mu^*(B)$;
(3) $\mu^*(\cup_{i=1}^\infty A_i) \leq \sum_{i=1}^\infty \mu^*(A_i)$ whenever $A_1, A_2, \ldots$ are subsets of X.

A set N is a *null set* with respect to μ^* if $\mu^*(N) = 0$.

A common way to generate outer measures is as follows.

Proposition 4.2 *Suppose $\mathcal{C}$ is a collection of subsets of X such that $\emptyset$ and X are both in $\mathcal{C}$. Suppose $\ell : \mathcal{C} \to [0, \infty]$ with $\ell(\emptyset) = 0$. Define*

$$\mu^*(E) = \inf\Big\{ \sum_{i=1}^\infty \ell(A_i) : A_i \in \mathcal{C} \text{ for each } i \text{ and } E \subset \cup_{i=1}^\infty A_i \Big\}.$$

(4.1)

Then μ^* is an outer measure.

Proof. (1) and (2) of the definition of outer measure are obvious. To prove (3), let $A_1, A_2, \ldots$ be subsets of X and let $\varepsilon > 0$. For each i there exist $C_{i1}, C_{i2}, \ldots \in \mathcal{C}$ such that $A_i \subset \cup_{j=1}^\infty C_{ij}$ and

$\sum_j \ell(C_{ij}) \leq \mu^*(A_i) + \varepsilon/2^i$. Then $\cup_{i=1}^\infty A_i \subset \cup_i \cup_j C_{ij}$ and

$$\mu^*(\cup_{i=1}^\infty A_i) \leq \sum_{i,j} \ell(C_{ij}) = \sum_i \left(\sum_j \ell(C_{ij}) \right)$$
$$\leq \sum_{i=1}^\infty \mu^*(A_i) + \sum_{i=1}^\infty \varepsilon/2^i$$
$$= \sum_{i=1}^\infty \mu^*(A_i) + \varepsilon.$$

Since ε is arbitrary, $\mu^*(\cup_{i=1}^\infty A_i) \leq \sum_{i=1}^\infty \mu^*(A_i)$. $\qquad\square$

Example 4.3 Let $X = \mathbb{R}$ and let $\mathcal{C}$ be the collection of intervals of the form $(a, b]$, that is, intervals that are open on the left and closed on the right. Let $\ell(I) = b - a$ if $I = (a, b]$. Define μ^* by (4.1). Proposition 4.2 shows that μ^* is an outer measure, but we will see in Section 4.4 that μ^* is not a measure on the collection of all subsets of $\mathbb{R}$. We will also see, however, that if we restrict μ^* to a σ-algebra $\mathcal{L}$ which is strictly smaller than the collection of all subsets of $\mathbb{R}$, then μ^* will be a measure on $\mathcal{L}$. That measure is what is known as *Lebesgue measure*. The σ-algebra $\mathcal{L}$ is called the *Lebesgue σ-algebra*.

Example 4.4 Let $X = \mathbb{R}$ and let $\mathcal{C}$ be the collection of intervals of the form $(a, b]$ as in the previous example. Let $\alpha : \mathbb{R} \to \mathbb{R}$ be an increasing right continuous function on $\mathbb{R}$. Thus $\alpha(x) = \lim_{y \to x+} \alpha(y)$ for each x and $\alpha(x) \leq \alpha(y)$ if $x < y$. Let $\ell(I) = \alpha(b) - \alpha(a)$ if $I = (a, b]$. Again define μ^* by (4.1). Again Proposition 4.2 shows that μ^* is an outer measure. Restricting μ^* to a smaller σ-algebra gives us what is known as *Lebesgue-Stieltjes measure* corresponding to α. The special case where $\alpha(x) = x$ for all x is Lebesgue measure.

In general we need to restrict μ^* to a strictly smaller σ-algebra than the collection of all subsets of $\mathbb{R}$, but not always. For example, if $\alpha(x) = 0$ for $x < 0$ and 1 for $x \geq 1$, then the corresponding Lebesgue-Stieltjes measure is point mass at 0 (defined in Example 3.4), and the corresponding σ-algebra is the collection of all subsets of $\mathbb{R}$.

Definition 4.5 Let μ^* be an outer measure. A set $A \subset X$ is μ^*-*measurable* if

$$\mu^*(E) = \mu^*(E \cap A) + \mu^*(E \cap A^c) \tag{4.2}$$

for all $E \subset X$.

Theorem 4.6 *If μ^* is an outer measure on X, then the collection $\mathcal{A}$ of μ^*-measurable sets is a σ-algebra. If μ is the restriction of μ^* to $\mathcal{A}$, then μ is a measure. Moreover, $\mathcal{A}$ contains all the null sets.*

This is sometimes known as the *Carathéodory theorem*, but do not confuse this with the Carathéodory extension theorem in Section 4.5.

Proof. By Definition 4.1,

$$\mu^*(E) \le \mu^*(E \cap A) + \mu^*(E \cap A^c)$$

for all $E \subset X$. Thus to check (4.2) it is enough to show

$$\mu^*(E) \ge \mu^*(E \cap A) + \mu^*(E \cap A^c).$$

This will be trivial in the case $\mu^*(E) = \infty$.

Step 1. First we show $\mathcal{A}$ is an algebra. If $A \in \mathcal{A}$, then $A^c \in \mathcal{A}$ by symmetry and the definition of $\mathcal{A}$. Suppose $A, B \in \mathcal{A}$ and $E \subset X$. Then

$$\begin{aligned}
\mu^*(E) &= \mu^*(E \cap A) + \mu^*(E \cap A^c) \\
&= [\mu^*(E \cap A \cap B) + \mu^*(E \cap A \cap B^c)] \\
&\quad + [\mu^*(E \cap A^c \cap B) + \mu^*(E \cap A^c \cap B^c)].
\end{aligned}$$

The second equality follows from the definition of $\mathcal{A}$ with E first replaced by $E \cap A$ and then by $E \cap A^c$. The first three summands on the right of the second equals sign have a sum greater than or equal to $\mu^*(E \cap (A \cup B))$ because $A \cup B \subset (A \cap B) \cup (A \cap B^c) \cup (A^c \cap B)$. Since $A^c \cap B^c = (A \cup B)^c$, then

$$\mu^*(E) \ge \mu^*(E \cap (A \cup B)) + \mu^*(E \cap (A \cup B)^c),$$

which shows $A \cup B \in \mathcal{A}$. Therefore $\mathcal{A}$ is an algebra.

Step 2. Next we show $\mathcal{A}$ is a σ-algebra. Let A_i be pairwise disjoint sets in $\mathcal{A}$, let $B_n = \cup_{i=1}^n A_i$, and $B = \cup_{i=1}^\infty A_i$. If $E \subset X$,

$$\mu^*(E \cap B_n) = \mu^*(E \cap B_n \cap A_n) + \mu^*(E \cap B_n \cap A_n^c)$$
$$= \mu^*(E \cap A_n) + \mu^*(E \cap B_{n-1}).$$

Similarly, $\mu^*(E \cap B_{n-1}) = \mu^*(E \cap A_{n-1}) + \mu^*(E \cap B_{n-2})$, and continuing, we obtain

$$\mu^*(E \cap B_n) \geq \sum_{i=1}^n \mu^*(E \cap A_i).$$

Since $B_n \in \mathcal{A}$, then

$$\mu^*(E) = \mu^*(E \cap B_n) + \mu^*(E \cap B_n^c) \geq \sum_{i=1}^n \mu^*(E \cap A_i) + \mu^*(E \cap B^c).$$

Let $n \to \infty$. Recalling that μ^* is an outer measure,

$$\mu^*(E) \geq \sum_{i=1}^\infty \mu^*(E \cap A_i) + \mu^*(E \cap B^c) \qquad (4.3)$$
$$\geq \mu^*(\cup_{i=1}^\infty (E \cap A_i)) + \mu^*(E \cap B^c)$$
$$= \mu^*(E \cap B) + \mu^*(E \cap B^c)$$
$$\geq \mu^*(E).$$

This shows $B \in \mathcal{A}$.

Now if $C_1, C_2, \ldots$ are sets in $\mathcal{A}$, let $A_1 = C_1$, $A_2 = C_2 - A_1$, $A_3 = C_3 - (A_1 \cup A_2)$, and in general $A_i = C_i - (\cup_{j=1}^{i-1} A_j)$. Since each $C_i \in \mathcal{A}$ and $\mathcal{A}$ is an algebra, then $A_i = C_i \cap C_{i-1}^c \in \mathcal{A}$. The A_i are pairwise disjoint, so from the previous paragraph,

$$\cup_{i=1}^\infty C_i = \cup_{i=1}^\infty A_i \in \mathcal{A}.$$

Also, $\cap_{i=1}^\infty C_i = (\cup_{i=1}^\infty C_i^c)^c \in \mathcal{A}$, and therefore $\mathcal{A}$ is a σ-algebra.

Step 3. We now show μ^* restricted to $\mathcal{A}$ is a measure. The only way (4.3) can hold is if all the inequalities there are actually equalities, and in particular,

$$\mu^*(E) = \sum_{i=1}^\infty \mu^*(E \cap A_i) + \mu^*(E \cap B^c).$$

Taking $E = B$, we obtain

$$\mu^*(B) = \sum_{i=1}^{\infty} \mu^*(A_i).$$

Recalling that $B = \cup_{i=1}^{\infty} A_i$, this shows that μ^* is countably additive on $\mathcal{A}$.

Step 4. Finally, if $\mu^*(A) = 0$ and $E \subset X$, then

$$\mu^*(E \cap A) + \mu^*(E \cap A^c) = \mu^*(E \cap A^c) \leq \mu^*(E),$$

which shows $\mathcal{A}$ contains all the null sets.   □

4.2   Lebesgue-Stieltjes measures

Let $X = \mathbb{R}$ and let $\mathcal{C}$ be the collection of intervals of the form $(a, b]$, that is, intervals that are open on the left and closed on the right. Let $\alpha(x)$ be an increasing right continuous function. This means that $\alpha(x) \leq \alpha(y)$ if $x < y$ and $\lim_{z \to x+} \alpha(z) = \alpha(x)$ for all x. We do not require α to be strictly increasing. Define

$$\ell((a, b]) = \alpha(b) - \alpha(a).$$

Define

$$m^*(E) = \inf \left\{ \sum_{i=1}^{\infty} \ell(A_i) : A_i \in \mathcal{C} \text{ for each } i \text{ and } E \subset \cup_{i=1}^{\infty} A_i \right\}.$$

(In this book we usually use m instead of μ when we are talking about Lebesgue-Stieltjes measures.) We use Proposition 4.2 to tell us that m^* is an outer measure. We then use Theorem 4.6 to show that m^* is a measure on the collection of m^*-measurable sets. Note that if K and L are adjacent intervals, that is, if $K = (a, b]$ and $L = (b, c]$, then $K \cup L = (a, c]$ and

$$\ell(K) + \ell(L) = [\alpha(b) - \alpha(a)] + [\alpha(c) - \alpha(b)] \qquad (4.4)$$
$$= \alpha(c) - \alpha(a) = \ell(K \cup L)$$

by the definition of ℓ.

The next step in the construction of Lebesgue-Stieltjes measure corresponding to α is the following.

Proposition 4.7 *Every set in the Borel σ-algebra on $\mathbb{R}$ is m^*-measurable.*

Proof. Since the collection of m^*-measurable sets is a σ-algebra, it suffices to show that every interval J of the form $(c, d]$ is m^*-measurable. Let E be any set with $m^*(E) < \infty$; we need to show

$$m^*(E) \geq m^*(E \cap J) + m^*(E \cap J^c). \qquad (4.5)$$

Choose $I_1, I_2, \ldots$, each of the form $(a_i, b_i]$, such that $E \subset \cup_i I_i$ and

$$m^*(E) \geq \sum_i [\alpha(b_i) - \alpha(a_i)] - \varepsilon.$$

Since $E \subset \cup_i I_i$, we have

$$m^*(E \cap J) \leq \sum_i m^*(I_i \cap J)$$

and

$$m^*(E \cap J^c) \leq \sum_i m^*(I_i \cap J^c).$$

Adding we have

$$m^*(E \cap J) + m^*(E \cap J^c) \leq \sum_i [m^*(I_i \cap J) + m^*(I_i \cap J^c)].$$

Now $I_i \cap J$ is an interval that is open on the left and closed on the right, and $I_i \cap J^c$ is the union of zero, one, or two such intervals, depending on the relative locations of I_i and J. Using (4.4) either zero, one, or two times, we see that

$$m^*(I_i \cap J) + m^*(I_i \cap J^c) = m^*(I_i).$$

Thus

$$m^*(E \cap J) + m^*(E \cap J^c) \leq \sum_i m^*(I_i) \leq m^*(E) + \varepsilon.$$

Since ε is arbitrary, this proves (4.5). $\qquad \square$

After all this work, it would be upsetting if the measure of a half-open interval $(e, f]$ were not what it is supposed to be. Fortunately, everything is fine, due to Proposition 4.9. First we need the following lemma.

Lemma 4.8 *Let $J_k = (a_k, b_k)$, $k = 1, \ldots, n$, be a finite collection of finite open intervals covering a finite closed interval $[C, D]$. Then*

$$\sum_{k=1}^{n} [\alpha(b_k) - \alpha(a_k)] \geq \alpha(D) - \alpha(C). \tag{4.6}$$

Proof. Since $\{J_k\}$ is a cover of $[C, D]$, there exists at least one interval, say, J_{k_1}, such that $C \in J_{k_1}$. If J_{k_1} covers $[C, D]$, we stop. Otherwise, $b_{k_1} \leq D$, and there must be at least one interval, say, J_{k_2}, such that $b_{k_1} \in J_{k_2}$. If $[C, D] \subset J_{k_1} \cup J_{k_2}$, we stop. If not, then $b_{k_1} < b_{k_2} \leq D$, and there must be at least one interval, say, J_{k_3} that contains b_{k_2}. At each stage we choose J_{k_j} so that $b_{k_{j-1}} \in J_{k_j}$. We continue until we have covered $[C, D]$ with intervals $J_{k_1}, \ldots, J_{k_m}$. Since $\{J_k\}$ is a finite cover, we will stop for some $m \leq n$.

By our construction we have

$$a_{k_1} \leq C < b_{k_1}, \qquad a_{k_m} < D < b_{k_m},$$

and for $2 \leq j \leq m$,

$$a_{k_j} < b_{k_{j-1}} < b_{k_j}.$$

Then

$$\begin{aligned}
\alpha(D) - \alpha(C) &\leq \alpha(b_{k_m}) - \alpha(a_{k_1}) \\
&= [\alpha(b_{k_m}) - \alpha(b_{k_{m-1}})] + [\alpha(b_{k_{m-1}}) - \alpha(b_{k_{m-2}})] + \cdots \\
&\quad + [\alpha(b_{k_2}) - \alpha(b_{k_1})] + [\alpha(b_{k_1}) - \alpha(a_{k_1})] \\
&\leq [\alpha(b_{k_m}) - \alpha(a_{k_m})] + [\alpha(b_{k_{m-1}}) - \alpha(a_{k_{m-1}})] + \cdots \\
&\quad + [\alpha(b_{k_2}) - \alpha(a_{k_2})] + [\alpha(b_{k_1}) - \alpha(a_{k_1})].
\end{aligned}$$

Since $\{J_{k_1}, \ldots, J_{k_m}\} \subset \{J_1, \ldots, J_n\}$, this proves (4.6). $\qquad \square$

Proposition 4.9 *If e and f are finite and $I = (e, f]$, then $m^*(I) = \ell(I)$.*

Proof. First we show $m^*(I) \leq \ell(I)$. This is easy. Let $A_1 = I$ and $A_2 = A_3 = \cdots = \emptyset$. Then $I \subset \cup_{i=1}^{\infty} A_i$, hence

$$m^*(I) \leq \sum_{i=1}^{\infty} \ell(A_i) = \ell(A_1) = \ell(I).$$

For the other direction, suppose $I \subset \cup_{i=1}^{\infty} A_i$, where $A_i = (c_i, d_i]$. Let $\varepsilon > 0$ and choose $C \in (e, f)$ such that $\alpha(C) - \alpha(e) < \varepsilon/2$. This is possible by the right continuity of α. Let $D = f$. For each i, choose $d'_i > d_i$ such that $\alpha(d'_i) - \alpha(d_i) < \varepsilon/2^{i+1}$ and let $B_i = (c_i, d'_i)$.

Then $[C, D]$ is compact and $\{B_i\}$ is an open cover for $[C, D]$. Use compactness to choose a finite subcover $\{J_1, \ldots, J_n\}$ of $\{B_i\}$. We now apply Lemma 4.8. We conclude that

$$\ell(I) \leq \alpha(D) - \alpha(C) + \varepsilon/2 \leq \sum_{k=1}^{n} (\alpha(d'_k) - \alpha(c_k)) + \varepsilon/2 \leq \sum_{i=1}^{\infty} \ell(A_i) + \varepsilon.$$

Taking the infimum over all countable collections $\{A_i\}$ that cover I, we obtain

$$\ell(I) \leq m^*(I) + \varepsilon.$$

Since ε is arbitrary, $\ell(I) \leq m^*(I)$. □

We now drop the asterisks from m^* and call m *Lebesgue-Stieltjes measure*. In the special case where $\alpha(x) = x$, m is *Lebesgue measure*. In the special case of Lebesgue measure, the collection of m^*-measurable sets is called the *Lebesgue σ-algebra*. A set is *Lebesgue measurable* if it is in the Lebesgue σ-algebra.

Given a measure μ on $\mathbb{R}$ such that $\mu(K) < \infty$ whenever K is compact, define $\alpha(x) = \mu((0, x])$ if $x \geq 0$ and $\alpha(x) = -\mu((x, 0])$ if $x < 0$. Then α is increasing, right continuous, and Exercise 4.1 asks you to show that μ is Lebesgue-Stieltjes measure corresponding to α.

4.3 Examples and related results

Example 4.10 Let m be Lebesgue measure. If $x \in \mathbb{R}$, then $\{x\}$ is a closed set and hence is Borel measurable. Moreover

$$m(\{x\}) = \lim_{n \to \infty} m((x - (1/n), x]) = \lim_{n \to \infty} [x - (x - (1/n))] = 0.$$

We then conclude

$$m([a, b]) = m((a, b]) + m(\{a\}) = b - a + 0 = b - a$$

and

$$m((a, b)) = m((a, b]) - m(\{b\}) = b - a - 0 = b - a.$$

Since σ-algebras are closed under the operation of countable unions, then countable sets are Borel measurable. Adding 0 to itself countably many times is still 0, so the Lebesgue measure of a countable set is 0.

However there are uncountable sets which have Lebesgue measure 0. See the next example.

Example 4.11 Recall from undergraduate analysis that the *Cantor set* is constructed as follows. Let F_0 be the interval $[0, 1]$. We let F_1 be what remains if we remove the middle third, that is,

$$F_1 = F_0 - (\tfrac{1}{3}, \tfrac{2}{3}).$$

F_1 consists of two intervals of length $\frac{1}{3}$ each. We remove the middle third of each of these two intervals and let

$$F_2 = F_1 - [(\tfrac{1}{9}, \tfrac{2}{9}) \cup (\tfrac{7}{9}, \tfrac{8}{9})].$$

We continue removing middle thirds, and the Cantor set F is $\cap_n F_n$. Recall that the Cantor set is closed, uncountable, and every point is a limit point. Moreover, it contains no intervals.

The measure of F_1 is $2(\frac{1}{3})$, the measure of F_2 is $4(\frac{1}{9})$, and the measure of F_n is $(\frac{2}{3})^n$. Since the Cantor set C is the intersection of all these sets, the Lebesgue measure of C is 0.

Suppose we define f_0 to be $\frac{1}{2}$ on the interval $(\frac{1}{3}, \frac{2}{3})$, to be $\frac{1}{4}$ on the interval $(\frac{1}{9}, \frac{2}{9})$, to be $\frac{3}{4}$ on the interval $(\frac{7}{9}, \frac{8}{9})$, and so on. Define $f(x) = \inf\{f_0(y) : y \geq x\}$ for $x < 1$. Define $f(1) = 1$. Notice $f = f_0$ on the complement of the Cantor set. f is increasing, so it has only jump discontinuities; see Proposition 1.6. But if it has a jump continuity, there is a rational of the form $k/2^n$ with $k \leq 2^n$ that is not in the range of f. On the other hand, by the construction, each of the values $\{k/2^n : n \geq 0, k \leq 2^n\}$ is taken by f_0 for some point in the complement of C, and so is taken by f. The only way this can happen is if f is continuous. This function f is called the *Cantor-Lebesgue function* or sometimes simply the *Cantor function*. We will use it in examples later on. For now, we note that it is a function that increases only on the Cantor set, which is a set of Lebesgue measure 0, yet f is continuous.

Example 4.12 Let $q_1, q_2, \ldots$ be an enumeration of the rationals, let $\varepsilon > 0$, and let I_i be the interval $(q_i - \varepsilon/2^i, q_i + \varepsilon/2^i)$. Then the measure of I_i is $\varepsilon/2^{i-1}$, so the measure of $\cup_i I_i$ is at most 2ε. (It is not equal to that because there is a lot of overlap.) Therefore the measure of $A = [0, 1] - \cup_i I_i$ is larger than $1 - 2\varepsilon$. But A contains no rational numbers.

Example 4.13 Let us follow the construction of the Cantor set, with this difference. Instead of removing the middle third at the first stage, remove the middle fourth, i.e., remove $(\frac{3}{8}, \frac{5}{8})$. On each of the two intervals that remain, remove the middle sixteenths. On each of the four intervals that remain, remove the middle interval of length $\frac{1}{64}$, and so on. The total that we removed is

$$\tfrac{1}{4} + 2(\tfrac{1}{16}) + 4(\tfrac{1}{64}) + \cdots = \tfrac{1}{2}.$$

The set that remains contains no intervals, is closed, every point is a limit point, is uncountable, and has measure $1/2$. Such a set is called a *generalized Cantor set*. Of course, other choices than $\frac{1}{4}$, $\frac{1}{16}$, etc. are possible.

Let $A \subset [0, 1]$ be a Borel measurable set. We will show that A is "almost equal" to the countable intersection of open sets and "almost equal" to the countable union of closed sets. (A similar argument to what follows is possible for subsets of $\mathbb{R}$ that have infinite measure; see Exercise 4.2.)

Proposition 4.14 *Suppose $A \subset [0, 1]$ is a Borel measurable set. Let m be Lebesgue measure.*

(1) Given $\varepsilon > 0$, there exists an open set G so that $m(G - A) < \varepsilon$ and $A \subset G$.

(2) Given $\varepsilon > 0$, there exists a closed set F so that $m(A - F) < \varepsilon$ and $F \subset A$.

(3) There exists a set H which contains A that is the countable intersection of a decreasing sequence of open sets and $m(H - A) = 0$.

(4) There exists a set F which is contained in A that is the countable union of an increasing sequence of closed sets which is contained in A and $m(A - F) = 0$.

Proof. (1) There exists a set of the form $E = \cup_{j=1}^{\infty}(a_j, b_j]$ such that $A \subset E$ and $m(E - A) < \varepsilon/2$. Let $G = \cup_{j=1}^{\infty}(a_j, b_j + \varepsilon 2^{-j-1})$. Then G is open and contains A and

$$m(G - E) < \sum_{j=1}^{\infty} \varepsilon 2^{-j-1} = \varepsilon/2.$$

Therefore

$$m(G - A) \leq m(G - E) + m(E - A) < \varepsilon.$$

(2) Find an open set G such that $A' \subset G$ and $m(G - A') < \varepsilon$, where $A' = [0, 1] - A$. Let $F = [0, 1] - G$. Then F is closed, $F \subset A$, and $m(A - F) \leq m(G - A') < \varepsilon$.

(3) By (1), for each i, there is an open set G_i that contains A and such that $m(G_i - A) < 2^{-i}$. Then $H_i = \cap_{j=1}^{i} G_j$ will contain A, is open, and since it is contained in G_i, then $m(H_i - A) < 2^{-i}$. Let $H = \cap_{i=1}^{\infty} H_i$. H need not be open, but it is the intersection of countably many open sets. The set H is a Borel set, contains A, and $m(H - A) \leq m(H_i - A) < 2^{-i}$ for each i, hence $m(H - A) = 0$.

(4) If $A' = [0, 1] - A$, apply (3) to A' to find a set H containing A' that is the countable intersection of a decreasing sequence of open sets and such that $m(H - A') = 0$. Let $J = [0, 1] - H$. It is left to the reader to verify that J has the desired properties. □

The countable intersections of open sets are sometimes called G_δ sets; the G is for *geoffnet*, the German word for "open" and the δ for *Durchschnitt*, the German word for "intersection." The countable unions of closed sets are called F_σ sets, the F coming from *fermé*, the French word for "closed," and the σ coming from *Summe*, the German word for "union."

Therefore, when trying to understand Lebesgue measure, we can look at G_δ or F_σ sets, which are not so bad, and at null sets, which can be quite bad but don't have positive measure.

4.4   Nonmeasurable sets

Theorem 4.15 *Let m^* be defined by (4.1), where $\mathcal{C}$ is the collection of intervals that are open on the left and closed on the right* and $\ell((a, b]) = b - a$. m^* *is not a measure on the collection of all subsets of* $\mathbb{R}$.

Proof. Suppose m^* is a measure. Define $x \sim y$ if $x - y$ is rational. It is easy to see that this is an equivalence relationship on $[0, 1]$. For each equivalence class, pick an element out of that class (we need to use the axiom of choice to do this). Call the collection of such points A. Given a set B, define $B + x = \{y + x : y \in B\}$. Note that $\ell((a + q, b + q]) = b - a = \ell((a, b])$ for each a, b, and q, and so by the definition of m^*, we have $m^*(A + q) = m^*(A)$ for each set A and each q. Moreover, the sets $A + q$ are disjoint for different rationals q.

Now

$$[0, 1] \subset \cup_{q \in [-1, 1] \cap \mathbb{Q}} (A + q),$$

where the union is only over rational q, so

$$1 \leq \sum_{q \in [-1, 1], q \in \mathbb{Q}} m^*(A + q),$$

and therefore $m^*(A) > 0$. But

$$\cup_{q \in [-1, 1] \cap \mathbb{Q}} (A + q) \subset [-1, 2],$$

where again the union is only over rational q, so if m^* is a measure, then

$$3 \geq \sum_{q \in [0, 1], q \in \mathbb{Q}} m^*(A + q),$$

which implies $m^*(A) = 0$, a contradiction.   □

4.5 The Carathéodory extension theorem

We prove the Carathéodory extension theorem in this section. This theorem abstracts some of the techniques used above to give a tool for constructing measures in a variety of contexts.

Let $\mathcal{A}_0$ be an algebra but not necessarily a σ-algebra. Saying ℓ is a measure on $\mathcal{A}_0$ means the following: (1) of Definition 3.1

holds and if $A_1, A_2, \ldots$ are pairwise disjoint elements of $\mathcal{A}_0$ and also $\cup_{i=1}^{\infty} A_i \in \mathcal{A}_0$, then $\ell(\cup_{i=1}^{\infty} A_i) = \sum_{i=1}^{\infty} \ell(A_i)$. Sometimes one calls a measure on an algebra a *premeasure*. Recall $\sigma(\mathcal{A}_0)$ is the σ-algebra generated by $\mathcal{A}_0$.

Theorem 4.16 *Suppose $\mathcal{A}_0$ is an algebra and $\ell : \mathcal{A}_0 \to [0, \infty]$ is a measure on $\mathcal{A}_0$. Define*

$$\mu^*(E) = \inf\left\{\sum_{i=1}^{\infty} \ell(A_i) : each\ A_i \in \mathcal{A}_0, E \subset \cup_{i=1}^{\infty} A_i\right\}$$

for $E \subset X$. Then
(1) μ^* is an outer measure;
(2) $\mu^*(A) = \ell(A)$ if $A \in \mathcal{A}_0$;
(3) every set in $\mathcal{A}_0$ is μ^*-measurable;
(4) if ℓ is σ-finite, then there is a unique extension to $\sigma(\mathcal{A}_0)$.

Proof. (1) is Proposition 4.2. We turn to (2). Suppose $E \in \mathcal{A}_0$. We know $\mu^*(E) \leq \ell(E)$ since we can take $A_1 = E$ and $A_2, A_3, \ldots$ empty in the definition of μ^*. If $E \subset \cup_{i=1}^{\infty} A_i$ with $A_i \in \mathcal{A}_0$, let $B_n = E \cap (A_n - (\cup_{i=1}^{n-1} A_i))$. Then the B_n are pairwise disjoint, they are each in $\mathcal{A}_0$, and their union is E. Therefore

$$\ell(E) = \sum_{i=1}^{\infty} \ell(B_i) \leq \sum_{i=1}^{\infty} \ell(A_i).$$

Taking the infimum over all such sequences $A_1, A_2, \ldots$ shows that $\ell(E) \leq \mu^*(E)$.

Next we look at (3). Suppose $A \in \mathcal{A}_0$. Let $\varepsilon > 0$ and let $E \subset X$. Pick $B_1, B_2, \ldots \in \mathcal{A}_0$ such that $E \subset \cup_{i=1}^{\infty} B_i$ and $\sum_i \ell(B_i) \leq \mu^*(E) + \varepsilon$. Then

$$\mu^*(E) + \varepsilon \geq \sum_{i=1}^{\infty} \ell(B_i) = \sum_{i=1}^{\infty} \ell(B_i \cap A) + \sum_{i=1}^{\infty} \ell(B_i \cap A^c)$$
$$\geq \mu^*(E \cap A) + \mu^*(E \cap A^c).$$

Since ε is arbitrary, $\mu^*(E) \geq \mu^*(E \cap A) + \mu^*(E \cap A^c)$. Thus A is μ^*-measurable.

Finally, we look at (4). Suppose we have two extensions to $\sigma(\mathcal{A}_0)$, the smallest σ-algebra containing $\mathcal{A}_0$. One is μ^* and let the other extension be called ν. We will show that if E is in $\sigma(\mathcal{A}_0)$, then $\mu^*(E) = \nu(E)$.

Let us first assume that μ^* is a finite measure. The μ^*-measurable sets form a σ-algebra containing $\mathcal{A}_0$. Because $E \in \mathcal{A}_0$, E must be μ^*-measurable and

$$\mu^*(E) = \inf\left\{ \sum_{i=1}^{\infty} \ell(A_i) : E \subset \cup_{i=1}^{\infty} A_i, \text{ each } A_i \in \mathcal{A}_0 \right\}.$$

But $\ell = \nu$ on $\mathcal{A}_0$, so $\sum_i \ell(A_i) = \sum_i \nu(A_i)$. Therefore if $E \subset \cup_{i=1}^{\infty} A_i$ with each $A_i \in \mathcal{A}_0$, then

$$\nu(E) \leq \sum_i \nu(A_i) = \sum_i \ell(A_i),$$

which implies

$$\nu(E) \leq \mu^*(E). \tag{4.7}$$

Since we do not know that ν is constructed via an outer measure, we must use a different argument to get the reverse inequality. Let $\varepsilon > 0$ and choose $A_i \in \mathcal{A}_0$ such that $\mu^*(E) + \varepsilon \geq \sum_i \ell(A_i)$ and $E \subset \cup_i A_i$. Let $A = \cup_{i=1}^{\infty} A_i$ and $B_k = \cup_{i=1}^{k} A_i$. Observe

$$\mu^*(E) + \varepsilon \geq \sum_i \ell(A_i) = \sum_i \mu^*(A_i) \geq \mu^*(\cup_i A_i) = \mu^*(A),$$

hence $\mu^*(A - E) \leq \varepsilon$. We have

$$\mu^*(A) = \lim_{k \to \infty} \mu^*(B_k) = \lim_{k \to \infty} \nu(B_k) = \nu(A).$$

Then

$$\mu^*(E) \leq \mu^*(A) = \nu(A) = \nu(E) + \nu(A - E)$$
$$\leq \nu(E) + \mu^*(A - E) \leq \nu(E) + \varepsilon,$$

using (4.7) in the next to last inequality. Since ε is arbitrary, this completes the proof when ℓ is finite.

It remains to consider the case when ℓ is σ-finite. Write $X = \cup_i K_i$, where $K_i \uparrow X$ and $\ell(K_i) < \infty$ for each i. By the preceding paragraph we have uniqueness for the measure ℓ_i defined by $\ell_i(A) = \ell(A \cap K_i)$. If μ and ν are two extensions of ℓ and $A \in \sigma(\mathcal{A}_0)$, then

$$\mu(A) = \lim_{i \to \infty} \mu(A \cap K_i) = \lim_{i \to \infty} \ell_i(A) = \lim_{i \to \infty} \nu(A \cap K_i) = \nu(A),$$

which proves $\mu = \nu$. $\qquad \square$

4.6   Exercises

Exercise 4.1 Let μ be a measure on the Borel σ-algebra of $\mathbb{R}$ such that $\mu(K) < \infty$ whenever K is compact, define $\alpha(x) = \mu((0, x])$ if $x \geq 0$ and $\alpha(x) = -\mu((x, 0])$ if $x < 0$. Show that μ is the Lebesgue-Stieltjes measure corresponding to α.

Exercise 4.2 Let m be Lebesgue measure and A a Lebesgue measurable subset of $\mathbb{R}$ with $m(A) < \infty$. Let $\varepsilon > 0$. Show there exist G open and F closed such that $F \subset A \subset G$ and $m(G - F) < \varepsilon$.

Exercise 4.3 If $(X, \mathcal{A}, \mu)$ is a measure space, define

$$\mu^*(A) = \inf\{\mu(B) : A \subset B, B \in \mathcal{A}\}$$

for all subsets A of X. Show that μ^* is an outer measure. Show that each set in $\mathcal{A}$ is μ^*-measurable and μ^* agrees with the measure μ on $\mathcal{A}$.

Exercise 4.4 Let m be Lebesgue-Stieltjes measure corresponding to a right continuous increasing function α. Show that for each x,

$$m(\{x\}) = \alpha(x) - \lim_{y \to x-} \alpha(y).$$

Exercise 4.5 Suppose m is Lebesgue measure. Define $x + A = \{x + y : y \in A\}$ and $cA = \{cy : y \in A\}$ for $x \in \mathbb{R}$ and c a real number. Show that if A is a Lebesgue measurable set, then $m(x + A) = m(A)$ and $m(cA) = |c|m(A)$.

Exercise 4.6 Let m be Lebesgue measure. Suppose for each n, A_n is a Lebesgue measurable subset of $[0, 1]$. Let B consist of those points x that are in infinitely many of the A_n.
(1) Show B is Lebesgue measurable.
(2) If $m(A_n) > \delta > 0$ for each n, show $m(B) \geq \delta$.
(3) If $\sum_{n=1}^{\infty} m(A_n) < \infty$, prove that $m(B) = 0$.
(4) Give an example where $\sum_{n=1}^{\infty} m(A_n) = \infty$, but $m(B) = 0$.

Exercise 4.7 Suppose $\varepsilon \in (0, 1)$ and m is Lebesgue measure. Find a measurable set $E \subset [0, 1]$ such that the closure of E is $[0, 1]$ and $m(E) = \varepsilon$.

Exercise 4.8 If X is a metric space, $\mathcal{B}$ is the Borel σ-algebra, and μ is a measure on $(X, \mathcal{B})$, then the *support* of μ is the smallest closed set F such that $\mu(F^c) = 0$. Show that if F is a closed subset of $[0, 1]$, then there exists a finite measure on $[0, 1]$ whose support is F.

Exercise 4.9 Let m be Lebesgue measure. Find an example of Lebesgue measurable subsets $A_1, A_2, \ldots$ of $[0, 1]$ such that $m(A_n) > 0$ for each n, $m(A_n \triangle A_m) > 0$ if $n \neq m$, and $m(A_n \cap A_m) = m(A_n)m(A_m)$ if $n \neq m$.

Exercise 4.10 Let $\varepsilon \in (0, 1)$, let m be Lebesgue measure, and suppose A is a Borel measurable subset of $\mathbb{R}$. Prove that if

$$m(A \cap I) \leq (1 - \varepsilon)m(I)$$

for every interval I, then $m(A) = 0$.

Exercise 4.11 Suppose m is Lebesgue measure and A is a Borel measurable subset of $\mathbb{R}$ with $m(A) > 0$. Prove that if

$$B = \{x - y : x, y \in A\},$$

then B contains a non-empty open interval centered at the origin.

Exercise 4.12 Let m be Lebesgue measure. Construct a Borel subset A of $\mathbb{R}$ such that $0 < m(A \cap I) < m(I)$ for every open interval I.

Exercise 4.13 Let N be the non-measurable set defined in Section 4.4. Prove that if $A \subset N$ and A is Lebesgue measurable, then $m(A) = 0$.

Exercise 4.14 Let m be Lebesgue measure. Prove that if A is a Lebesgue measurable subset of $\mathbb{R}$ and $m(A) > 0$, then there is a subset of A that is non-measurable.

Exercise 4.15 Let X be a set and $\mathcal{A}$ a collection of subsets of X that form an algebra of sets. Suppose ℓ is a measure on $\mathcal{A}$ such that $\ell(X) < \infty$. Define μ^* using ℓ as in (4.1). Prove that a set A is μ^*-measurable if and only if

$$\mu^*(A) = \ell(X) - \mu^*(A^c).$$

Exercise 4.16 Suppose μ^* is an outer measure. Show that if $A_n \uparrow A$, then $\mu^*(A_n) \uparrow \mu^*(A)$. Given an example to show that even if μ^* is finite, $A_n \downarrow A$ does not necessarily imply $\mu^*(A_n) \downarrow \mu^*(A)$.

Exercise 4.17 Suppose A is a Lebesgue measurable subset of $\mathbb{R}$ and

$$B = \cup_{x \in A}[x - 1, x + 1].$$

Prove that B is Lebesgue measurable.

Chapter 5

Measurable functions

We are now ready to move from sets to functions.

5.1 Measurability

Suppose we have a measurable space $(X, \mathcal{A})$.

Definition 5.1 A function $f : X \to \mathbb{R}$ is *measurable* or $\mathcal{A}$ *measurable* if $\{x : f(x) > a\} \in \mathcal{A}$ for all $a \in \mathbb{R}$. A complex-valued function is measurable if both its real and imaginary parts are measurable.

Example 5.2 Suppose f is real-valued and identically constant. Then the set $\{x : f(x) > a\}$ is either empty or all of X, so f is measurable.

Example 5.3 Suppose $f(x) = 1$ if $x \in A$ and 0 otherwise. Then the set $\{x : f(x) > a\}$ is either $\emptyset$, A, or X. Hence f is measurable if and only if A is in $\mathcal{A}$.

Example 5.4 Suppose X is the real line with the Borel σ-algebra and $f(x) = x$. Then $\{x : f(x) > a\} = (a, \infty)$, and so f is measurable.

Proposition 5.5 *Suppose f is real-valued. The following conditions are equivalent.*

(1) $\{x : f(x) > a\} \in \mathcal{A}$ for all $a \in \mathbb{R}$;
(2) $\{x : f(x) \leq a\} \in \mathcal{A}$ for all $a \in \mathbb{R}$;
(3) $\{x : f(x) < a\} \in \mathcal{A}$ for all $a \in \mathbb{R}$;
(4) $\{x : f(x) \geq a\} \in \mathcal{A}$ for all $a \in \mathbb{R}$.

Proof. The equivalence of (1) and (2) and of (3) and (4) follow from taking complements, e.g., $\{x : f(x) \leq a\} = \{x : f(x) > a\}^c$. If f is measurable, then

$$\{x : f(x) \geq a\} = \cap_{n=1}^{\infty}\{x : f(x) > a - 1/n\}$$

shows that (4) holds if (1) does. If (4) holds, then (1) holds by using the equality

$$\{x : f(x) > a\} = \cup_{n=1}^{\infty}\{x : f(x) \geq a + 1/n\}.$$

This completes the proof.   □

Proposition 5.6 *If X is a metric space, $\mathcal{A}$ contains all the open sets, and $f : X \to \mathbb{R}$ is continuous, then f is measurable.*

Proof. Note that $\{x : f(x) > a\} = f^{-1}((a, \infty))$ is open, and hence in $\mathcal{A}$.   □

Proposition 5.7 *Let $c \in \mathbb{R}$. If f and g are measurable real-valued functions, then so are $f + g$, $-f$, cf, fg, $\max(f, g)$, and $\min(f, g)$.*

Proof. If $f(x) + g(x) < a$, then $f(x) < a - g(x)$, and there exists a rational r such that $f(x) < r < a - g(x)$. Hence

$$\{x : f(x) + g(x) < a\} = \cup_{r \in \mathbb{Q}}(\{x : f(x) < r\} \cap \{x : g(x) < a - r\}).$$

This proves $f + g$ is measurable.

Since $\{x : -f(x) > a\} = \{x : f(x) < -a\}$, then $-f$ is measurable using Proposition 5.5.

If $c > 0$, then $\{x : cf(x) > a\} = \{x : f(x) > a/c\}$ shows cf is measurable. When $c = 0$, cf is measurable by Example 5.2. When

$c < 0$, write $cf = -(|c|f)$, which is measurable by what we have already proved.

f^2 is measurable since for $a < 0$, $\{x : f(x) > a\} = X$, while for $a \geq 0$,

$$\{x : f(x)^2 > a) = \{x : f(x) > \sqrt{a}\} \cup \{x : f(x) < -\sqrt{a}\}.$$

The measurability of fg follows since

$$fg = \tfrac{1}{2}[(f + g)^2 - f^2 - g^2].$$

The equality

$$\{x : \max(f(x), g(x)) > a\} = \{x : f(x) > a\} \cup \{x : g(x) > a\}$$

shows $\max(f, g)$ is measurable, and the result for $\min(f, g)$ follows from $\min(f, g) = -\max(-f, -g)$. $\qquad\square$

Proposition 5.8 *If f_i is a measurable real-valued function for each i, then so are $\sup_i f_i$, $\inf_i f_i$, $\limsup_{i \to \infty} f_i$, and $\liminf_{i \to \infty} f_i$.*

Proof. The result will follow for $\limsup$ and $\liminf$ once we have the result for the sup and inf by using the definitions since $\limsup_i f_i = \inf_j \sup_{i \geq j} f_j$ and similarly for the $\liminf$. We have $\{x : \sup_i f_i > a\} = \cap_{i=1}^{\infty}\{x : f_i(x) > a\}$, so $\sup_i f_i$ is measurable, and the proof for $\inf f_i$ is similar. $\qquad\square$

Definition 5.9 We say $f = g$ *almost everywhere*, written $f = g$ a.e., if $\{x : f(x) \neq g(x)\}$ has measure zero. Similarly, we say $f_i \to f$ a.e. if the set of x where $f_i(x)$ does not converge to $f(x)$ has measure zero.

If X is a metric space, $\mathcal{B}$ is the Borel σ-algebra, and $f : X \to \mathbb{R}$ is measurable with respect to $\mathcal{B}$, we say f is Borel measurable. If $f : \mathbb{R} \to \mathbb{R}$ is measurable with respect to the Lebesgue σ-algebra, we say f is Lebesgue measurable.

We saw in Proposition 5.6 that all continuous functions are Borel measurable. The same is true for increasing functions on the real line.

Proposition 5.10 *If* $f : \mathbb{R} \to \mathbb{R}$ *is monotone, then* f *is Borel measurable.*

Proof. Let us suppose f is increasing, for otherwise we look at $-f$. Given $a \in \mathbb{R}$, let $x_0 = \sup\{y : f(y) \leq a\}$. If $f(x_0) \leq a$, then $\{x : f(x) > a\} = (x_0, \infty)$. If $f(x_0) > a$, then $\{x : f(x) > a\} = [x_0, \infty)$. In either case $\{x : f(x) > a\}$ is a Borel set. □

Proposition 5.11 *Let* $(X, \mathcal{A})$ *be a measurable space and let* $f : X \to \mathbb{R}$ *be an* $\mathcal{A}$ *measurable function. If A is in the Borel σ-algebra on* $\mathbb{R}$, *then* $f^{-1}(A) \in \mathcal{A}$.

Proof. Let $\mathcal{B}$ be the Borel σ-algebra on $\mathbb{R}$ and $\mathcal{C} = \{A \in \mathcal{B} : f^{-1}(A) \in \mathcal{A}\}$. If $A_1, A_2, \ldots \in \mathcal{C}$, then since

$$f^{-1}(\cup_i A_i) = \cup_i f^{-1}(A_i) \in \mathcal{A},$$

we have that $\mathcal{C}$ is closed under countable unions. Similarly $\mathcal{C}$ is closed under countable intersections and complements, so $\mathcal{C}$ is a σ-algebra. Since f is measurable, $\mathcal{C}$ contains (a, ∞) for every real a, hence $\mathcal{C}$ contains the σ-algebra generated by these intervals, that is, $\mathcal{C}$ contains $\mathcal{B}$. □

The above proposition says that if f is measurable, then the inverse image of a Borel set is measurable.

Example 5.12 Let us construct a set that is Lebesgue measurable, but not Borel measurable. Recall the Lebesgue measurable sets were constructed in Chapter 4 and include the completion of the Borel σ-algebra.

Let f be the Cantor-Lebesgue function of Example 4.11 and define

$$F(x) = \inf\{y : f(y) \geq x\}.$$

Although F is not continuous, observe that F is strictly increasing (hence one-to-one) and maps $[0, 1]$ into C, the Cantor set. Since F is increasing, F^{-1} maps Borel measurable sets to Borel measurable sets.

Let m be Lebesgue measure and let A be the non-measurable set we constructed in Proposition 4.15. Let $B = F(A)$. Since $F(A) \subset C$ and $m(C) = 0$, then $F(A)$ is a null set, hence is Lebesgue measurable. On the other hand, $F(A)$ is not Borel measurable, because if it were, then $A = F^{-1}(F(A))$ would be Borel measurable, a contradiction.

5.2 Approximation of functions

Definition 5.13 Let $(X, \mathcal{A})$ be a measurable space. If $E \in \mathcal{A}$, define the *characteristic function* of E by

$$\chi_E(x) = \begin{cases} 1, & x \in E; \\ 0, & x \notin E. \end{cases}$$

A *simple function* s is a function of the form

$$s(x) = \sum_{i=1}^{n} a_i \chi_{E_i}(x)$$

for real numbers a_i and measurable sets E_i.

Proposition 5.14 *Suppose f is a non-negative and measurable function. Then there exists a sequence of non-negative measurable simple functions s_n increasing to f.*

Proof. Let

$$E_{ni} = \{x : (i-1)/2^n \le f(x) < i/2^n\}$$

and

$$F_n = \{x : f(x) \ge n\}$$

for $n = 1, 2, \ldots$ and $i = 1, 2, \ldots, n2^n$. Then define

$$s_n = \sum_{i=1}^{n2^n} \frac{i-1}{2^n} \chi_{E_{ni}} + n\chi_{F_n}.$$

In words, $s_n(x) = n$ if $f(x) \ge n$. If $f(x)$ is between $(i-1)/2^n$ and $i/2^n$ for $i/2^n \le n$, we let $s_n(x) = (i-1)/2^n$.

It is easy to see that s_n has the desired properties. □

5.3   Lusin's theorem

The following theorem is known as Lusin's theorem. It is very pretty but usually other methods are better for solving problems. Example 5.16 will illustrate why this is a less useful theorem than at first glance.

We use m for Lebesgue measure. Recall that the support of a function f is the closure of the set $\{x : f(x) \neq 0\}$.

Theorem 5.15 *Suppose $f : [0,1] \to \mathbb{R}$ is Borel measurable, m is Lebesgue measure, and $\varepsilon > 0$. There exists a closed set $F \subset [0,1]$ such that $m([0,1] - F) < \varepsilon$ and the restriction of f to F is a continuous function on F.*

This theorem can be loosely interpreted as saying every measurable function is "almost continuous."

Proof.   First let us suppose that $f = \chi_A$, where A is a Borel measurable subset of $[0,1]$. By Proposition 4.14 we can find E closed and G open such that $E \subset A \subset G$ and $m(G - A) < \varepsilon/2$ and $m(A - E) < \varepsilon/2$. Let $\delta = \inf\{|x - y| : x \in E, y \in G^c\}$. Since $E \subset A \subset [0,1]$, E is compact and $\delta > 0$. Letting

$$g(x) = \left(1 - \frac{d(x,E)}{\delta}\right)^+,$$

where $y^+ = \max(y,0)$ and $d(x,E) = \inf\{|x - y| : y \in E\}$, we see that g is continuous, takes values in $[0,1]$, is equal to 1 on E, and equal to 0 on G^c. Take $F = (E \cup G^c) \cap [0,1]$. Then $m([0,1] - F) \leq m(G - E) < \varepsilon$, and $f = g$ on F.

Next suppose $f = \sum_{i=1}^{M} a_i \chi_{A_i}$ is simple, where each A_i is a measurable subset of $[0,1]$ and each $a_i \geq 0$. Choose F_i closed such that $m([0,1] - F_i) < \varepsilon/M$ and χ_{A_i} restricted to F_i is continuous. If we let $F = \cap_{i=1}^{M} F_i$, then F is closed, $m([0,1] - F) < \varepsilon$, and f restricted to F is continuous.

Now suppose f is non-negative, bounded by K, and has support in $[0,1]$. Let

$$A_{in} = \{x : (i-1)/2^n \leq f(x) < i/2^n\}.$$

Then

$$f_n(x) = \sum_{i=1}^{K2^n+1} \frac{i}{2^n} \chi_{A_{in}}(x)$$

are simple functions increasing to f. Note that

$$h_n(x) = f_{n+1}(x) - f_n(x)$$

is also a simple function and is bounded by 2^{-n}. Choose F_0 closed such that $m([0,1] - F_0) < \varepsilon/2$ and f_0 restricted to F_0 is continuous. For $n \geq 1$, choose F_n closed such that $m([0,1] - F_n) < \varepsilon/2^{n+1}$ and h_n restricted to F_n is continuous. Let $F = \cap_{n=0}^{\infty} F_n$. Then F, being the intersection of closed sets, will be closed, and

$$m([0,1] - F) \leq \sum_{n=0}^{\infty} m([0,1] - F_n) < \varepsilon.$$

On the set F, we have that $f_0(x) + \sum_{n=1}^{\infty} h_n(x)$ converges uniformly to $f(x)$ because each h_n is bounded by 2^{-n}. The uniform limit of continuous functions is continuous, hence f is continuous on F.

If $f \geq 0$, let $B_K = \{x : f(x) \leq K\}$. Since f is everywhere finite, $B_K \uparrow [0,1]$ as $K \to \infty$, hence $m(B_K) > 1 - \varepsilon/3$ if K is sufficiently large. Choose $D \subset B_K$ such that D is closed and $m(B_K - D) < \varepsilon/3$. Now choose $E \subset [0,1]$ closed such that $f \cdot \chi_D$ restricted to E is continuous and $m([0,1] - E) < \varepsilon/3$. Then $F = D \cap E$ is closed, $m([0,1] - F) < \varepsilon$, and f restricted to F is continuous.

Finally, for arbitrary measurable f write $f = f^+ - f^-$ and find F^+ and F^- closed such that $m([0,1] - F^+) < \varepsilon/2$, $m([0,1] - F^-) < \varepsilon/2$, and f^+ restricted to F^+ is continuous and similarly for f^-. Then $F = F^+ \cap F^-$ is the desired set. □

Example 5.16 Suppose $f = \chi_B$, where B consists of the irrationals in $[0,1]$. f is Borel measurable because $[0,1] - B$ is countable, hence the union of countably many points, and thus the union of countably many closed sets. Every point of $[0,1]$ is a point of discontinuity of f because for any $x \in [0,1]$, there are both rationals and irrationals in every neighborhood of x, hence f takes the values 0 and 1 in every neighborhood of x.

Recall Example 4.12. f restricted to the set A there is identically one, hence f restricted to A is a continuous function. A is closed because it is equal to the interval $[0,1]$ minus the union of open intervals.

This does not contradict Lusin's theorem. No claim is made that the function f is continuous at most points of $[0,1]$. What is asserted is that there is a closed set F with large measure so that f restricted to F is continuous when viewed as a function from F to $\mathbb{R}$.

5.4 Exercises

Exercise 5.1 Suppose $(X, \mathcal{A})$ is a measurable space, f is a real-valued function, and $\{x : f(x) > r\} \in \mathcal{A}$ for each rational number r. Prove that f is measurable.

Exercise 5.2 Let $f : (0, 1) \to \mathbb{R}$ be such that for every $x \in (0, 1)$ there exist $r > 0$ and a Borel measurable function g, both depending on x, such that f and g agree on $(x - r, x + r) \cap (0, 1)$. Prove that f is Borel measurable.

Exercise 5.3 Suppose f_n are measurable functions. Prove that

$$A = \{x : \lim_{n \to \infty} f_n(x) \text{ exists}\}$$

is a measurable set.

Exercise 5.4 If $f : \mathbb{R} \to \mathbb{R}$ is Lebesgue measurable, prove that there exists a Borel measurable function g such that $f = g$ a.e.

Exercise 5.5 Give an example of a collection of measurable non-negative functions $\{f_\alpha\}_{\alpha \in A}$ such that if g is defined by $g(x) = \sup_{\alpha \in A} f_\alpha(x)$, then g is finite for all x but g is non-measurable. (A is allowed to be uncountable.)

Exercise 5.6 Suppose $f : X \to \mathbb{R}$ is Lebesgue measurable and $g : \mathbb{R} \to \mathbb{R}$ is continuous. Prove that $g \circ f$ is Lebesgue measurable. Is this true if g is Borel measurable instead of continuous? Is this true if g is Lebesgue measurable instead of continuous?

Exercise 5.7 Suppose $f : \mathbb{R} \to \mathbb{R}$ is Borel measurable. Define $\mathcal{A}$ to be the smallest σ-algebra containing the sets $\{x : f(x) > a\}$ for every $a \in \mathbb{R}$. Suppose $g : \mathbb{R} \to \mathbb{R}$ is measurable with respect to $\mathcal{A}$, which means that $\{x : g(x) > a\} \in \mathcal{A}$ for every $a \in \mathbb{R}$. Prove that there exists a Borel measurable function $h : \mathbb{R} \to \mathbb{R}$ such that $g = h \circ f$.

Exercise 5.8 One can show that there exist discontinuous real-valued functions f such that

$$f(x + y) = f(x) + f(y) \qquad (5.1)$$

for all $x, y \in \mathbb{R}$. (The construction uses Zorn's lemma, which is equivalent to the axiom of choice.) Prove that if f satisfies (5.1) and in addition f is Lebesgue measurable, then f is continuous.

Chapter 6

The Lebesgue integral

In this chapter we define the Lebesgue integral. We only give the definition here; we consider the properties of the Lebesgue integral in later chapters.

6.1 Definitions

Definition 6.1 Let $(X, \mathcal{A}, \mu)$ be a measure space. If

$$s = \sum_{i=1}^{n} a_i \chi_{E_i}$$

is a non-negative measurable simple function, define the Lebesgue integral of s to be

$$\int s \, d\mu = \sum_{i=1}^{n} a_i \mu(E_i). \tag{6.1}$$

Here, if $a_i = 0$ and $\mu(E_i) = \infty$, we use the convention that $a_i \mu(E_i) = 0$. If $f \geq 0$ is a measurable function, define

$$\int f \, d\mu = \sup\left\{ \int s \, d\mu : 0 \leq s \leq f, s \text{ simple} \right\}. \tag{6.2}$$

Let f be measurable and let $f^+ = \max(f, 0)$ and $f^- = \max(-f, 0)$. Provided $\int f^+ \, d\mu$ and $\int f^- \, d\mu$ are not both infinite, define

$$\int f \, d\mu = \int f^+ \, d\mu - \int f^- \, d\mu. \tag{6.3}$$

Finally, if $f = u + iv$ is complex-valued and $\int(|u| + |v|)\,d\mu$ is finite, define

$$\int f\,d\mu = \int u\,d\mu + i \int v\,d\mu. \tag{6.4}$$

A few remarks are in order. A function s might be written as a simple function in more than one way. For example $s = \chi_{A \cup B} = \chi_A + \chi_B$ if A and B are disjoint. It is not hard to check that the definition of $\int s\,d\mu$ is unaffected by how s is written. If $s = \sum_{i=1}^m a_i \chi_{A_i} = \sum_{j=1}^n b_j \chi_{B_j}$, then we need to show

$$\sum_{i=1}^m a_i \mu(A_i) = \sum_{j=1}^n b_j \mu(B_j). \tag{6.5}$$

We leave the proof of this to the reader as Exercise 6.1.

Secondly, if s is a simple function, one has to think a moment to verify that the definition of $\int s\,d\mu$ by means of (6.1) agrees with its definition by means of (6.2).

Definition 6.2 If f is measurable and $\int |f|\,d\mu < \infty$, we say f is *integrable*.

The proof of the next proposition follows from the definitions.

Proposition 6.3 *(1) If f is a real-valued measurable function with $a \leq f(x) \leq b$ for all x and $\mu(X) < \infty$, then $a\mu(X) \leq \int f\,d\mu \leq b\mu(X)$;*

(2) If f and g are measurable, real-valued, and integrable and $f(x) \leq g(x)$ for all x, then $\int f\,d\mu \leq \int g\,d\mu$.

(3) If f is integrable, then $\int cf\,d\mu = c \int f\,d\mu$ for all complex c.

(4) If $\mu(A) = 0$ and f is integrable, then $\int f\chi_A\,d\mu = 0$.

The integral $\int f\chi_A\,d\mu$ is often written $\int_A f\,d\mu$. Other notation for the integral is to omit the μ and write $\int f$ if it is clear which measure is being used, to write $\int f(x)\,\mu(dx)$, or to write $\int f(x)\,d\mu(x)$.

When we are integrating a function f with respect to Lebesgue measure m, it is usual to write $\int f(x)\,dx$ for $\int f(x)\,m(dx)$ and to define

$$\int_a^b f(x)\,dx = \int_{[a,b]} f(x)\,m(dx).$$

Proposition 6.4 *If f is integrable,*

$$\left| \int f \right| \le \int |f|.$$

Proof. For the real case, this is easy. $f \le |f|$, so $\int f \le \int |f|$. Also $-f \le |f|$, so $-\int f \le \int |f|$. Now combine these two facts.

For the complex case, $\int f$ is a complex number. If it is 0, the inequality is trivial. If it is not, then $\int f = re^{i\theta}$ for some r and θ. Then

$$\left| \int f \right| = r = e^{-i\theta} \int f = \int e^{-i\theta} f.$$

From the definition of $\int f$ when f is complex, it follows that $\text{Re}\,(\int f) = \int \text{Re}\,(f)$. Since $|\int f|$ is real, we have

$$\left| \int f \right| = \text{Re}\left(\int e^{-i\theta} f \right) = \int \text{Re}\,(e^{-i\theta} f) \le \int |f|$$

as desired. $\qquad\qquad\square$

We do not yet know that $\int (f+g) = \int f + \int g$. We will see this in Theorem 7.4.

6.2 Exercises

Exercise 6.1 Verify (6.5).

Exercise 6.2 Suppose f is non-negative and measurable and μ is σ-finite. Show there exist simple functions s_n increasing to f at each point such that $\mu(\{x : s_n(x) \ne 0\}) < \infty$ for each n.

Exercise 6.3 Let f be a non-negative measurable function. Prove that

$$\lim_{n\to\infty} \int (f \wedge n) \to \int f.$$

Exercise 6.4 Let $(X, \mathcal{A}, \mu)$ be a measure space and suppose μ is σ-finite. Suppose f is integrable. Prove that given ε there exists δ such that

$$\int_A |f(x)| \, \mu(dx) < \varepsilon$$

whenever $\mu(A) < \delta$.

Exercise 6.5 Suppose $\mu(X) < \infty$ and f_n is a sequence of bounded real-valued measurable functions that converge to f uniformly. Prove that

$$\int f_n \, d\mu \to \int f \, d\mu.$$

This is sometimes called the *bounded convergence theorem*.

Exercise 6.6 If f_n is a sequence of non-negative integrable functions such that $f_n(x)$ decreases to $f(x)$ for every x, prove that $\int f_n \, d\mu \to \int f \, d\mu$.

Exercise 6.7 Let $(X, \mathcal{A}, \mu)$ be a measure space and suppose f is a non-negative, measurable function that is finite at each point of X, but not necessarily integrable. Prove that there exists a continuous increasing function $g : [0, \infty) \to [0, \infty)$ such that $\lim_{x \to \infty} g(x) = \infty$ and $g \circ f$ is integrable.

Chapter 7

Limit theorems

The main reason the Lebesgue integral is so much easier to work with than the Riemann integral is that it behaves nicely when taking limits. In this chapter we prove the monotone convergence theorem, Fatou's lemma, and the dominated convergence theorem. We also prove that the Lebesgue integral is linear.

7.1 Monotone convergence theorem

One of the most important results concerning Lebesgue integration is the *monotone convergence theorem.*

Theorem 7.1 *Suppose f_n is a sequence of non-negative measurable functions with $f_1(x) \leq f_2(x) \leq \cdots$ for all x and with*

$$\lim_{n \to \infty} f_n(x) = f(x)$$

for all x. Then $\int f_n \, d\mu \to \int f \, d\mu$.

Proof. By Proposition 6.3(2), $\int f_n$ is an increasing sequence of real numbers. Let L be the limit. Since $f_n \leq f$ for all n, then $L \leq \int f$. We must show $L \geq \int f$.

Let $s = \sum_{i=1}^{m} a_i \chi_{E_i}$ be any non-negative simple function less than or equal to f and let $c \in (0, 1)$. Let $A_n = \{x : f_n(x) \geq cs(x)\}$.

51

Since $f_n(x)$ increases to $f(x)$ for each x and $c < 1$, then $A_n \uparrow X$. For each n,

$$\int f_n \geq \int_{A_n} f_n \geq c \int_{A_n} s_n$$

$$= c \int_{A_n} \sum_{i=1}^{m} a_i \chi_{E_i}$$

$$= c \sum_{i=1}^{m} a_i \mu(E_i \cap A_n).$$

If we let $n \to \infty$, by Proposition 3.5(3) the right hand side converges to

$$c \sum_{i=1}^{m} a_i \mu(E_i) = c \int s.$$

Therefore $L \geq c \int s$. Since c is arbitrary in the interval $(0, 1)$, then $L \geq \int s$. Taking the supremum over all simple $s \leq f$, we obtain $L \geq \int f$.   □

Example 7.2 Let $X = [0, \infty)$ and $f_n(x) = -1/n$ for all x. Then $\int f_n = -\infty$, but $f_n \uparrow f$ where $f = 0$ and $\int f = 0$. The reason the monotone convergence theorem does not apply here is that the f_n are not non-negative.

Example 7.3 Suppose $f_n = n\chi_{(0,1/n)}$. Then $f_n \geq 0$, $f_n \to 0$ for each x, but $\int f_n = 1$ does not converge to $\int 0 = 0$. The reason the monotone convergence theorem does not apply here is that the f_n do not increase to f for each x.

7.2   Linearity of the integral

Once we have the monotone convergence theorem, we can prove that the Lebesgue integral is linear.

Theorem 7.4 *If f and g are non-negative and measurable or if f and g are integrable, then*

$$\int (f + g) \, d\mu = \int f \, d\mu + \int g \, d\mu.$$

Proof. First suppose f and g are non-negative and simple, say, $f = \sum_{i=1}^{m} a_i \chi_{A_i}$ and $g = \sum_{j=1}^{n} b_j \chi_{B_j}$. Without loss of generality we may assume that $A_1, \ldots, A_m$ are pairwise disjoint and that $B_1, \ldots, B_n$ are pairwise disjoint. Since $a_i = 0$ and $b_j = 0$ are permissible, we may also assume $\cup_{i=1}^{m} A_i = X = \cup_{j=1}^{n} B_j$. Then

$$f + g = \sum_{i=1}^{m} \sum_{j=1}^{n} (a_i + b_j) \chi_{A_i \cap B_j},$$

and we have

$$\int (f + g) = \sum_{i=1}^{m} \sum_{j=1}^{n} (a_i + b_j) \mu(A_i \cap B_j)$$

$$= \sum_{i=1}^{m} \sum_{j=1}^{n} a_i \mu(A_i \cap B_j) + \sum_{i=1}^{m} \sum_{j=1}^{n} b_j \mu(A_i \cap B_j)$$

$$= \sum_{i=1}^{m} a_i \mu(A_i) + \sum_{j=1}^{n} b_j \mu(B_j)$$

$$= \int f + \int g.$$

Thus the theorem holds in this case.

Next suppose f and g are non-negative. Take s_n non-negative, simple, and increasing to f and t_n non-negative, simple, and increasing to g. Then $s_n + t_n$ are simple functions increasing to $f + g$, so the result follows from the monotone convergence theorem and

$$\int (f+g) = \lim_{n \to \infty} \int (s_n + t_n) = \lim_{n \to \infty} \int s_n + \lim_{n \to \infty} \int t_n = \int f + \int g.$$

Suppose now that f and g are real-valued and integrable but take both positive and negative values. Since

$$\int |f + g| \leq \int (|f| + |g|) = \int |f| + \int |g| < \infty,$$

then $f + g$ is integrable. Write

$$(f + g)^+ - (f + g)^- = f + g = f^+ - f^- + g^+ - g^-,$$

so that

$$(f + g)^+ + f^- + g^- = f^+ + g^+ + (f + g)^-.$$

Using the result for non-negative functions,

$$\int (f+g)^+ + \int f^- + \int g^- = \int f^+ + \int g^+ + \int (f+g)^-.$$

Rearranging,

$$\int (f+g) = \int (f+g)^+ - \int (f+g)^-$$

$$= \int f^+ - \int f^- + \int g^+ - \int g^-$$

$$= \int f + \int g.$$

If f and g are complex-valued, apply the above to the real and imaginary parts.   □

Proposition 7.5 *Suppose f_n are non-negative measurable functions. Then*

$$\int \sum_{n=1}^{\infty} f_n = \sum_{n=1}^{\infty} \int f_n.$$

Proof. Let $F_N = \sum_{n=1}^{N} f_n$. Since $0 \le F_n(x) \uparrow \sum_{n=1}^{\infty} f_n(x)$, we can write

$$\int \sum_{n=1}^{\infty} f_n = \int \lim_{N \to \infty} \sum_{n=1}^{N} f_n$$

$$= \int \lim_{N \to \infty} F_N = \lim_{N \to \infty} \int F_N \qquad (7.1)$$

$$= \lim_{N \to \infty} \sum_{n=1}^{N} \int f_n = \sum_{n=1}^{\infty} \int f_n,$$

using the monotone convergence theorem and the linearity of the integral.   □

7.3   Fatou's lemma

The next theorem is known as *Fatou's lemma*.

Theorem 7.6 *Suppose the f_n are non-negative and measurable. Then*

$$\int \liminf_{n \to \infty} f_n \le \liminf_{n \to \infty} \int f_n.$$

Proof. Let $g_n = \inf_{i \ge n} f_i$. Then the g_n are non-negative and g_n increases to $\liminf_n f_n$. Clearly $g_n \le f_i$ for each $i \ge n$, so $\int g_n \le \int f_i$. Therefore

$$\int g_n \le \inf_{i \ge n} \int f_i. \tag{7.2}$$

If we take the limit in (7.2) as $n \to \infty$, on the left hand side we obtain $\int \liminf_n f_n$ by the monotone convergence theorem, while on the right hand side we obtain $\liminf_n \int f_n$. □

A typical use of Fatou's lemma is the following. Suppose we have $f_n \to f$ and $\sup_n \int |f_n| \le K < \infty$. Then $|f_n| \to |f|$, and by Fatou's lemma, $\int |f| \le K$.

7.4 Dominated convergence theorem

Another very important theorem is the *dominated convergence theorem*.

Theorem 7.7 *Suppose that f_n are measurable real-valued functions and $f_n(x) \to f(x)$ for each x. Suppose there exists a non-negative integrable function g such that $|f_n(x)| \le g(x)$ for all x. Then*

$$\lim_{n \to \infty} \int f_n \, d\mu \to \int f \, d\mu.$$

Proof. Since $f_n + g \ge 0$, by Fatou's lemma,

$$\int f + \int g = \int (f + g) \le \liminf_{n \to \infty} \int (f_n + g) = \liminf_{n \to \infty} \int f_n + \int g.$$

Since g is integrable,

$$\int f \le \liminf_{n \to \infty} \int f_n. \tag{7.3}$$

Similarly, $g - f_n \geq 0$, so

$$\int g - \int f = \int (g - f) \leq \liminf_{n \to \infty} \int (g - f_n) = \int g + \liminf_{n \to \infty} \int (-f_n),$$

and hence

$$- \int f \leq \liminf_{n \to \infty} \int (-f_n) = -\limsup_{n \to \infty} \int f_n.$$

Therefore

$$\int f \geq \limsup_{n \to \infty} \int f_n,$$

which with (7.3) proves the theorem.   □

Exercise 7.1 asks you to prove a version of the dominated convergence theorem for complex-valued functions.

Example 7.3 is an example where the limit of the integrals is not the integral of the limit because there is no dominating function g.

If in the monotone convergence theorem or dominated convergence theorem we have only $f_n(x) \to f(x)$ almost everywhere, the conclusion still holds. For example, if the f_n are measurable, non-negative, and $f_n \uparrow f$ a.e., let $A = \{x : f_n(x) \to f(x)\}$. Then $f_n \chi_A(x) \uparrow f \chi_A(x)$ for each x. Since A^c has measure 0, we see from Proposition 6.3(4) and the monotone convergence theorem that

$$\lim_n \int f_n = \lim_n \int f_n \chi_A = \int f \chi_A = \int f.$$

7.5   Exercises

Exercise 7.1 State and prove a version of the dominated convergence theorem for complex-valued functions.

Exercise 7.2 The following generalized dominated convergence theorem is often useful. Suppose f_n, g_n, f, and g are integrable, $f_n \to f$ a.e., $g_n \to g$ a.e., $|f_n| \leq g_n$ for each n, and $\int g_n \to \int g$. Prove that $\int f_n \to \int f$.

Exercise 7.3 Give an example of a sequence of non-negative functions f_n tending to 0 pointwise such that $\int f_n \to 0$, but there is no integrable function g such that $f_n \leq g$ for all n.

Exercise 7.4 Suppose $(X, \mathcal{A}, \mu)$ is a measure space, each f_n is integrable and non-negative, $f_n \to f$ a.e., and $\int f_n \to \int f$. Prove that for each $A \in \mathcal{A}$

$$\int_A f_n \, d\mu \to \int_A f \, d\mu.$$

Exercise 7.5 Suppose f_n and f are integrable, $f_n \to f$ a.e., and $\int |f_n| \to \int |f|$. Prove that

$$\int |f_n - f| \to 0.$$

Exercise 7.6 Suppose $f : \mathbb{R} \to \mathbb{R}$ is integrable, $a \in \mathbb{R}$, and we define

$$F(x) = \int_a^x f(y) \, dy.$$

Show that F is a continuous function.

Exercise 7.7 Let f_n be a sequence of non-negative Lebesgue measurable functions on $\mathbb{R}$. Is it necessarily true that

$$\limsup_{n \to \infty} \int f_n \, dx \le \int \limsup_{n \to \infty} f_n \, dx?$$

If not, give a counterexample.

Exercise 7.8 Find the limit

$$\lim_{n \to \infty} \int_0^n \left(1 + \frac{x}{n}\right)^{-n} \log(2 + \cos(x/n)) \, dx$$

and justify your reasoning.

Exercise 7.9 Find the limit

$$\lim_{n \to \infty} \int_0^n \left(1 - \frac{x}{n}\right)^n \log(2 + \cos(x/n)) \, dx$$

and justify your reasoning.

Exercise 7.10 Prove that the limit exists and find its value:

$$\lim_{n \to \infty} \int_0^1 \frac{1 + nx^2}{(1 + x^2)^n} \log(2 + \cos(x/n)) \, dx.$$

Exercise 7.11 Prove the limit exists and determine its value:

$$\lim_{n \to \infty} \int_0^\infty n e^{-nx} \sin(1/x) \, dx.$$

Exercise 7.12 Let $g : \mathbb{R} \to \mathbb{R}$ be integrable and let $f : \mathbb{R} \to \mathbb{R}$ be bounded, measurable, and continuous at 1. Prove that

$$\lim_{n \to \infty} \int_{-n}^n f\left(1 + \frac{x}{n^2}\right) g(x) \, dx$$

exists and determine its value.

Exercise 7.13 Suppose $\mu(X) < \infty$, f_n converges to f uniformly, and each f_n is integrable. Prove that f is integrable and $\int f_n \to \int f$. Is the condition $\mu(X) < \infty$ necessary?

Exercise 7.14 Prove that

$$\sum_{k=1}^\infty \frac{1}{(p+k)^2} = -\int_0^1 \frac{x^p}{1-x} \log x \, dx$$

for $p > 0$.

Exercise 7.15 Let $\{f_n\}$ be a sequence of real-valued functions on $[0, 1]$ that is uniformly bounded.
(1) Show that if A is a Borel subset of $[0, 1]$, then there exists a subsequence n_j such that $\int_A f_{n_j}(x) \, dx$ converges.
(2) Show that if $\{A_i\}$ is a countable collection of Borel subsets of $[0, 1]$, then there exists a subsequence n_j such that $\int_{A_i} f_{n_j}(x) \, dx$ converges for each i.
(3) Show that there exists a subsequence n_j such that $\int_A f_{n_j}(x) \, dx$ converges for each Borel subset A of $[0, 1]$.

Exercise 7.16 Let $(X, \mathcal{A}, \mu)$ be a measure space. A family of measurable functions $\{f_n\}$ is *uniformly integrable* if given ε there exists M such that

$$\int_{\{x : |f_n(x)| > M\}} |f_n(x)| \, d\mu < \varepsilon$$

for each n. The sequence is *uniformly absolutely continuous* if given ε there exists δ such that

$$\left| \int_A f_n \, d\mu \right| < \varepsilon$$

for each n if $\mu(A) < \delta$.

Suppose μ is a finite measure. Prove that $\{f_n\}$ is uniformly integrable if and only if $\sup_n \int |f_n| \, d\mu < \infty$ and $\{f_n\}$ is uniformly absolutely continuous.

Exercise 7.17 The following is known as the *Vitali convergence theorem*. Suppose μ is a finite measure, $f_n \to f$ a.e., and $\{f_n\}$ is uniformly integrable. Prove that $\int |f_n - f| \to 0$.

Exercise 7.18 Suppose μ is a finite measure, $f_n \to f$ a.e., each f_n is integrable, f is integrable, and $\int |f_n - f| \to 0$. Prove that $\{f_n\}$ is uniformly integrable.

Exercise 7.19 Suppose μ is a finite measure and for some $\varepsilon > 0$

$$\sup_n \int |f_n|^{1+\varepsilon} \, d\mu < \infty.$$

Prove that $\{f_n\}$ is uniformly integrable.

Exercise 7.20 Suppose f_n is a uniformly integrable sequence of functions defined on $[0, 1]$. Prove that there is a subsequence n_j such that $\int_0^1 f_{n_j} g \, dx$ converges whenever g is a real-valued bounded measurable function.

Exercise 7.21 Suppose μ_n is a sequence of measures on $(X, \mathcal{A})$ such that $\mu_n(X) = 1$ for all n and $\mu_n(A)$ converges as $n \to \infty$ for each $A \in \mathcal{A}$. Call the limit $\mu(A)$.
(1) Prove that μ is a measure.
(2) Prove that $\int f \, d\mu_n \to \int f \, d\mu$ whenever f is bounded and measurable.
(3) Prove that

$$\int f \, d\mu \leq \liminf_{n \to \infty} \int f \, d\mu_n$$

whenever f is non-negative and measurable.

Exercise 7.22 Let $(X, \mathcal{A}, \mu)$ be a measure space and let f be non-negative and integrable. Define ν on $\mathcal{A}$ by

$$\nu(A) = \int_A f \, d\mu.$$

(1) Prove that ν is a measure.

(2) Prove that if g is integrable with respect to ν, then fg is integrable with respect to μ and

$$\int g \, d\nu = \int fg \, d\mu.$$

Exercise 7.23 Suppose μ and ν are positive measures on the Borel σ-algebra on $[0, 1]$ such that $\int f \, d\mu = \int f \, d\nu$ whenever f is real-valued and continuous on $[0, 1]$. Prove that $\mu = \nu$.

Exercise 7.24 Let $\mathcal{B}$ be the Borel σ-algebra on $[0, 1]$. Let μ_n be a sequence of finite measures on $([0, 1], \mathcal{B})$ and let μ be another finite measure on $([0, 1], \mathcal{B})$. Suppose $\mu_n([0, 1]) \to \mu([0, 1])$. Prove that the following are equivalent:

(1) $\int f \, d\mu_n \to \int f \, d\mu$ whenever f is a continuous real-valued function on $[0, 1]$;

(2) $\limsup_{n \to \infty} \mu_n(F) \leq \mu(F)$ for all closed subsets F of $[0, 1]$;

(3) $\liminf_{n \to \infty} \mu_n(G) \geq \mu(G)$ for all open subsets G of $[0, 1]$;

(4) $\lim_{n \to \infty} \mu_n(A) = \mu(A)$ whenever A is a Borel subset of $[0, 1]$ such that $\mu(\partial A) = 0$, where $\partial A = \overline{A} - A^o$ is the boundary of A;

(5) $\lim_{n \to \infty} \mu_n([0, x]) = \mu([0, x])$ for every x such that $\mu(\{x\}) = 0$.

Exercise 7.25 Let $\mathcal{B}$ be the Borel σ-algebra on $[0, 1]$. Suppose μ_n are finite measures on $([0, 1], \mathcal{B})$ such that $\int f \, d\mu_n \to \int_0^1 f \, dx$ whenever f is a real-valued continuous function on $[0, 1]$. Suppose that g is a bounded measurable function such that the set of discontinuities of g has measure 0. Prove that

$$\int g \, d\mu_n \to \int_0^1 g \, dx.$$

Exercise 7.26 Let $\mathcal{B}$ be the Borel σ-algebra on $[0, 1]$. Let μ_n be a sequence of finite measures on $([0, 1], \mathcal{B})$ with $\sup_n \mu_n([0, 1]) < \infty$. Define $\alpha_n(x) = \mu_n([0, x])$.

(1) If r is a rational in $[0, 1]$, prove that there exists a subsequence $\{n_j\}$ such that $\alpha_{n_j}(r)$ converges.

(2) Prove that there exists a subsequence $\{n_j\}$ such that $\alpha_n(r)$ converges for every rational in $[0, 1]$.

(3) Let $\overline{\alpha}(r) = \lim_{n \to \infty} \alpha_n(r)$ for r rational and define

$$\alpha(x) = \lim_{r \to x+, r \in \mathbb{Q}} \overline{\alpha}(r).$$

This means, since clearly $\overline{\alpha}(r) \leq \overline{\alpha}(s)$ if $r < s$, that

$$\alpha(x) = \inf\{\overline{\alpha}(r) : r > x, r \in \mathbb{Q}\}.$$

Let μ be the Lebesgue-Stieltjes measure associated with α. Prove that

$$\int f \, d\mu_n \to \int f \, d\mu$$

whenever f is a continuous real-valued function on $[0, 1]$.

Chapter 8

Properties of Lebesgue integrals

We present some propositions which imply that a function is zero a.e. and we give an approximation result.

8.1 Criteria for a function to be zero a.e.

The following two propositions are very useful.

Proposition 8.1 *Suppose f is real-valued and measurable and for every measurable set A we have $\int_A f \, d\mu = 0$. Then $f = 0$ almost everywhere.*

Proof. Let $A = \{x : f(x) > \varepsilon\}$. Then

$$0 = \int_A f \geq \int_A \varepsilon = \varepsilon \mu(A)$$

since $f\chi_A \geq \varepsilon\chi_A$. Hence $\mu(A) = 0$. We use this argument for $\varepsilon = 1/n$ and $n = 1, 2, \ldots$ to conclude

$$\mu\{x : f(x) > 0\} = \mu(\cup_{n=1}^{\infty}\{x : f(x) > 1/n\})$$

$$\leq \sum_{n=1}^{\infty} \mu(\{x : f(x) > 1/n\}) = 0.$$

Similarly $\mu\{x : f(x) < 0\} = 0$. $\square$

Proposition 8.2 *Suppose f is measurable and non-negative and $\int f \, d\mu = 0$. Then $f = 0$ almost everywhere.*

Proof. If f is not equal to 0 almost everywhere, there exists an n such that $\mu(A_n) > 0$ where $A_n = \{x : f(x) > 1/n\}$. But since f is non-negative,

$$0 = \int f \geq \int_{A_n} f \geq \frac{1}{n}\mu(A_n),$$

a contradiction. $\square$

As a corollary to Proposition 8.1 we have the following.

Corollary 8.3 *Let m be Lebesgue measure and $a \in \mathbb{R}$. Suppose $f : \mathbb{R} \to \mathbb{R}$ is integrable and $\int_a^x f(y)\,dy = 0$ for all x. Then $f = 0$ a.e.*

Proof. For any interval $[c, d]$,

$$\int_c^d f = \int_a^d f - \int_a^c f = 0.$$

By linearity, if G is the finite union of disjoint intervals, then $\int_G f = 0$. By the dominated convergence theorem and Proposition 1.5, $\int_G f = 0$ for any open set G. Again by the dominated convergence theorem, if G_n are open sets decreasing to H, then $\int_H f = \lim_n \int_{G_n} f = 0$.

If E is any Borel measurable set, Proposition 4.14 tells us that there exists a sequence G_n of open sets that decrease to a set H where H differs from E by a null set. Then

$$\int_E f = \int f\chi_E = \int f\chi_H = \int_H f = 0.$$

This with Proposition 8.1 implies f is zero a.e. $\square$

8.2 An approximation result

We give a result on approximating a function on $\mathbb{R}$ by continuous functions.

Theorem 8.4 *Suppose f is a Borel measurable real-valued integrable function on $\mathbb{R}$. Let $\varepsilon > 0$. Then there exists a continuous function g with compact support such that*

$$\int |f - g| < \varepsilon.$$

Proof. If we write $f = f^+ - f^-$, it is enough to find continuous functions g_1 and g_2 with compact support such that $\int |f^+ - g_1| < \varepsilon/2$ and $\int |f^- - g_2| < \varepsilon/2$ and to let $g = g_1 - g_2$. Hence we may assume $f \geq 0$.

By the monotone convergence theorem, $\int f \cdot \chi_{[-n,n]}$ increases to $\int f$, so by taking n large enough, the difference of the integrals will be less than $\varepsilon/2$. If we find g continuous with compact support such that $\int |f \cdot \chi_{[-n,n]} - g| < \varepsilon/2$, then $\int |f - g| < \varepsilon$. Therefore we may in addition assume that f is 0 outside some bounded interval.

Suppose $f = \chi_A$, where A is a bounded Borel measurable set. We can choose G open and F closed such that $F \subset A \subset G$ and $m(G - F) < \varepsilon$ by Proposition 4.14. Without loss of generality, we may assume G is also a bounded set. Since F is compact, there is a minimum distance between F and G^c, say, δ. Let

$$g(x) = \left(1 - \frac{\operatorname{dist}(x, F)}{\delta}\right)^+.$$

Then g is continuous, $0 \leq g \leq 1$, g is 1 on F, g is 0 on G^c, and g has compact support. We have

$$|g - \chi_A| \leq \chi_G - \chi_F,$$

so

$$\int |g - \chi_A| \leq \int (\chi_G - \chi_F) = m(G - F) < \varepsilon.$$

Thus our result holds for characteristic functions of bounded sets.

If $f = \sum_{i=1}^{p} a_i \chi_{A_i}$, where each A_i is contained in a bounded interval and each $a_i > 0$, and we find g_i continuous with compact support such that $\int |\chi_{A_i} - g_i| < \varepsilon/a_i p$, then $g = \sum_{i=1}^{p} a_i g_i$ will be the desired function. Thus our theorem holds for non-negative simple functions with compact support.

If f is non-negative and has compact support, we can find simple functions s_m supported in a bounded interval increasing to f whose integrals increase to $\int f$. Let s_m be a simple function such that $s_m \leq f$ and $\int s_m \geq \int f - \varepsilon/2$. We choose continuous g with compact support such that $\int |s_m - g| < \varepsilon/2$ using the preceding paragraphs, and then $\int |f - g| < \varepsilon$. $\qquad\square$

The method of proof, where one proves a result for characteristic functions, then simple functions, then non-negative functions, and then finally integrable functions, is very common.

8.3 Exercises

Exercise 8.1 This exercise gives a change of variables formula in two simple cases. Show that if f is an integrable function on the reals and a is a non-zero real number, then

$$\int_{\mathbb{R}} f(x + a)\, dx = \int_{\mathbb{R}} f(x)\, dx$$

and

$$\int_{\mathbb{R}} f(ax)\, dx = a^{-1} \int_{\mathbb{R}} f(x)\, dx.$$

Exercise 8.2 Let $(X, \mathcal{A}, \mu)$ be a σ-finite measure space. Suppose f is non-negative and integrable. Prove that if $\varepsilon > 0$, there exists $A \in \mathcal{A}$ such that $\mu(A) < \infty$ and

$$\varepsilon + \int_A f\, d\mu > \int f\, d\mu.$$

Exercise 8.3 Suppose A is a Borel measurable subset of $[0, 1]$, m is Lebesgue measure, and $\varepsilon \in (0, 1)$. Prove that there exists a continuous function $f : [0, 1] \to \mathbb{R}$ such that $0 \leq f \leq 1$ and

$$m(\{x : f(x) \neq \chi_A(x)\}) < \varepsilon.$$

Exercise 8.4 Suppose f is a non-negative integrable function on a measure space $(X, \mathcal{A}, \mu)$. Prove that

$$\lim_{t \to \infty} t \, \mu(\{x : f(x) \geq t\}) = 0.$$

Exercise 8.5 Find a non-negative function f on $[0, 1]$ such that

$$\lim_{t \to \infty} t \, m(\{x : f(x) \geq t\}) = 0$$

but f is not integrable, where m is Lebesgue measure.

Exercise 8.6 Suppose μ is a finite measure. Prove that a measurable non-negative function f is integrable if and only if

$$\sum_{n=1}^{\infty} \mu(\{x : f(x) \geq n\}) < \infty.$$

Exercise 8.7 Let μ be a measure, not necessarily σ-finite, and suppose f is real-valued and integrable with respect to μ. Prove that $A = \{x : f(x) \neq 0\}$ has σ-finite measure, that is, there exists $F_n \uparrow A$ such that $\mu(F_n) < \infty$ for each n.

Exercise 8.8 Recall that a function $f : \mathbb{R} \to \mathbb{R}$ is *convex* if

$$f(\lambda x + (1 - \lambda)y) \leq \lambda f(x) + (1 - \lambda)f(y)$$

whenever $x < y \in \mathbb{R}$ and $\lambda \in [0, 1]$.
(1) Prove that if f is convex and $x \in \mathbb{R}$, there exists a real number c such that $f(y) \geq f(x) + c(y - x)$ for all $y \in \mathbb{R}$. Graphically, this says that the graph of f lies above the line with slope c that passes through the point $(x, f(x))$.
(2) Let $(X, \mathcal{A}, \mu)$ be a measure space, suppose $\mu(X) = 1$, and let $f : \mathbb{R} \to \mathbb{R}$ be convex. Let $g : X \to \mathbb{R}$ be integrable. Prove *Jensen's inequality*:

$$f\left(\int g \, d\mu\right) \leq \int_X f \circ g \, d\mu.$$

Exercise 8.9 Suppose f is a real-valued function on $\mathbb{R}$ such that

$$f\left(\int_0^1 g(x) \, dx\right) \leq \int_0^1 f(g(x)) \, dx$$

whenever g is bounded and measurable. Prove that f is convex.

Exercise 8.10 Suppose $g : [0, 1] \to \mathbb{R}$ is bounded and measurable and

$$\int_0^1 f(x)g(x)\, dx = 0$$

whenever f is continuous and $\int_0^1 f(x)\, dx = 0$. Prove that g is equal to a constant a.e.

Chapter 9

Riemann integrals

We compare the Lebesgue integral and the Riemann integral. We show that the Riemann integral of a function exists if and only if the set of discontinuities of the function have Lebesgue measure zero, and in that case the Riemann integral and Lebesgue integral agree.

9.1 Comparison with the Lebesgue integral

We only consider bounded measurable functions from $[a, b]$ into $\mathbb{R}$. If we are looking at the Lebesgue integral, we write $\int f$, while, temporarily, if we are looking at the Riemann integral, we write $R(f)$. Recall that the Riemann integral on $[a, b]$ is defined as follows: if $P = \{x_0, x_1, \ldots, x_n\}$ with $x_0 = a$ and $x_n = b$ is a partition of $[a, b]$, let

$$U(P, f) = \sum_{i=1}^{n} \left(\sup_{x_{i-1} \leq x \leq x_i} f(x) \right) (x_i - x_{i-1})$$

and

$$L(P, f) = \sum_{i=1}^{n} \left(\inf_{x_{i-1} \leq x \leq x_i} f(x) \right) (x_i - x_{i-1}).$$

Set

$$\overline{R}(f) = \inf\{U(P, f) : P \text{ is a partition}\}$$

and
$$\underline{R}(f) = \sup\{L(P, f) : P \text{ is a partition}\}.$$

The Riemann integral exists if $\overline{R}(f) = \underline{R}(f)$, and the common value is the Riemann integral, which we denote $R(f)$.

Theorem 9.1 *A bounded Borel measurable real-valued function f on $[a, b]$ is Riemann integrable if and only if the set of points at which f is discontinuous has Lebesgue measure 0, and in that case, the Riemann integral is equal in value to the Lebesgue integral.*

Proof. *Step 1.* First we show that if f is Riemann integrable, then f is continuous a.e. and $R(f) = \int f$. If P is a partition, define

$$T_P(x) = \sum_{i=1}^{n} \Big(\sup_{x_{i-1} \leq y \leq x_i} f(y) \Big) \chi_{[x_{i-1}, x_i)}(x),$$

and

$$S_P(x) = \sum_{i=1}^{n} \Big(\inf_{x_{i-1} \leq y \leq x_i} f(y) \Big) \chi_{[x_{i-1}, x_i)}(x).$$

We observe that $\int T_P = U(P, f)$ and $\int S_P = L(P, f)$.

If f is Riemann integrable, there exists a sequence of partitions Q_i such that $U(Q_i, f) \downarrow R(f)$ and a sequence Q_i' such that $L(Q_i', f) \uparrow R(f)$. It is not hard to check that adding points to a partition increases L and decreases U, so if we let $P_i = \cup_{j \leq i}(Q_j \cup Q_j')$, then P_i is an increasing sequence of partitions, $U(P_i, f) \downarrow R(f)$ and $L(P_i, f) \uparrow R(f)$. We see also that $T_{P_i}(x)$ decreases at each point, say, to $T(x)$, and $S_{P_i}(x)$ increases at each point, say, to $S(x)$. Also $T(x) \geq f(x) \geq S(x)$. By the dominated convergence theorem (recall that f is bounded)

$$\int (T - S) = \lim_{i \to \infty} \int (T_{P_i} - S_{P_i}) = \lim_{i \to \infty} (U(P_i, f) - L(P_i, f)) = 0.$$

We conclude $T = S = f$ a.e.

If x is not in the null set where $T(x) \neq S(x)$ nor in $\cup_i P_i$, which is countable and hence of Lebesgue measure 0, then $T_{P_i}(x) \downarrow f(x)$ and $S_{P_i}(x) \uparrow f(x)$. We claim that f is continuous at such x. To prove the claim, given ε, choose i large enough so that $T_{P_i}(x) - S_{P_i}(x) < \varepsilon$

and then choose δ small enough so that $(x - \delta, x + \delta)$ is contained in the subinterval of P_i that contains x. Finally, since

$$R(f) = \lim_{i \to \infty} U(P_i, f) = \lim_{i \to \infty} \int T_{P_i} = \int f$$

by the dominated convergence theorem, we see that the Riemann integral and Lebesgue integral agree.

Step 2. Now suppose that f is continuous a.e. Let $\varepsilon > 0$. Let P_i be the partition where we divide $[a, b]$ into 2^i equal parts. If x is not in the null set where f is discontinuous, nor in $\cup_{i=1}^{\infty} P_i$, then $T_{P_i}(x) \downarrow f(x)$ and $S_{P_i}(x) \uparrow f(x)$. By the dominated convergence theorem,

$$U(P_i, f) = \int T_{P_i} \to \int f$$

and

$$L(P_i, f) = \int S_{P_i} \to \int f.$$

This does it. $\qquad\qquad\qquad\qquad\qquad\qquad\qquad\qquad\qquad\qquad\quad\square$

Example 9.2 Let $[a, b] = [0, 1]$ and $f = \chi_A$, where A is the set of irrational numbers in $[0, 1]$. If $x \in [0, 1]$, every neighborhood of x contains both rational and irrational points, so f is continuous at no point of $[0, 1]$. Therefore f is not Riemann integrable.

Example 9.3 Define $f(x)$ on $[0, 1]$ to be 0 if x is irrational and to be $1/q$ if x is rational and equals p/q when in reduced form. f is discontinuous at every rational. If x is irrational and $\varepsilon > 0$, there are only finitely many rationals r for which $f(r) \geq \varepsilon$, so taking δ less than the distance from x to any of this finite collection of rationals shows that $|f(y) - f(x)| < \varepsilon$ if $|y - x| < \delta$. Hence f is continuous at x. Therefore the set of discontinuities is a countable set, hence of measure 0, hence f is Riemann integrable.

9.2   Exercises

Exercise 9.1 Find a measurable function $f : [0, 1] \to \mathbb{R}$ such that $\overline{R}(f) \neq \int_0^1 f(x) \, dx$ and $\underline{R}(f) \neq \int_0^1 f(x) \, dx$.

Exercise 9.2 Find a function $f : (0,1] \to \mathbb{R}$ that is continuous, is not Lebesgue integrable, but where the improper Riemann integral exists. Thus we want f such that $\int_0^1 |f(x)|\, m(dx) = \infty$ but $\lim_{a \to 0+} R(f\chi_{[a,1]})$ exists.

Exercise 9.3 Suppose $f : [0,1] \to \mathbb{R}$ is integrable, f is bounded on $(a,1]$ for each $a > 0$, and the improper Riemann integral

$$\lim_{a \to 0+} R(f\chi_{(a,1]})$$

exists. Show that the limit is equal to $\int_0^1 f(x)\, dx$.

Exercise 9.4 Divide $[a,b]$ into 2^n equal subintervals and pick a point x_i out of each subinterval. Let μ_n be the measure defined by

$$\mu_n(dx) = 2^{-n} \sum_{i=1}^{2^n} \delta_{x_i}(dx),$$

where δ_y is point mass at y. Note that if f is a bounded measurable real-valued function on $[a,b]$, then

$$\int_a^b f(x)\, \mu_n(dx) = \sum_{i=1}^{2^n} f(x_i) 2^{-n} \tag{9.1}$$

is a Riemann sum approximation to $R(f)$.

(1) Prove that $\mu_n([0,x]) \to m([0,x])$ for every $x \in [0,1]$. Conclude by Exercise 7.24 that $\int f\, d\mu_n \to \int_0^1 f\, dx$ whenever f is continuous.

(2) Use Exercise 7.25 to see that if f is a bounded and measurable function on $[a,b]$ whose set of discontinuities has measure 0, then the Riemann sum approximation of f given in (9.1) converges to the Lebesgue integral of f. This provides an alternative proof of Step 2 of Theorem 9.1.

Exercise 9.5 Let f be a bounded, real-valued, and measurable function. Prove that if

$$\overline{f} = \lim_{\delta \to 0} \sup_{|y-x|<\delta, a \leq y \leq b} f(y),$$

then $\overline{f} = T$ a.e., where we use the notation of Theorem 9.1. Conclude $\overline{f}$ is Lebesgue measurable.

Exercise 9.6 Define $\underline{f} = \lim_{\delta \to 0} \inf_{|y-x|<\delta, a\leq y\leq b} f(y)$ and let $\overline{f}$ be defined as in Exercise 9.5.

(1) Suppose that the set of discontinuities of a bounded real-valued measurable function f has positive Lebesgue measure. Prove that there exists $\varepsilon > 0$ such that if

$$A_\varepsilon = \{x \in [a, b] : \overline{f}(x) - \underline{f}(x) > \varepsilon\},$$

then $m(A_\varepsilon) > 0$.

(2) Prove that $U(P, f) - L(P, f) > \varepsilon m(A_\varepsilon)$ for every partition P on $[a, b]$, using the notation of Theorem 9.1. Conclude that f is not Riemann integrable. This provides another proof of Step 1 of Theorem 9.1.

Exercise 9.7 A real-valued function on a metric space is *lower semicontinuous* if $\{x : f(x) > a\}$ is open whenever $a \in \mathbb{R}$ and *upper semicontinuous* if $\{x : f(x) < a\}$ is open whenever $a \in \mathbb{R}$.

(1) Prove that if f_n is a sequence of real-valued continuous functions increasing to f, then f is lower semicontinuous.

(2) Find a bounded lower semicontinuous function $f : [0, 1] \to \mathbb{R}$ such that f is continuous everywhere except at $x = 1/2$.

(3) Find a bounded lower semicontinuous real-valued function f defined on $[0, 1]$ such that the set of discontinuities of f is equal to the set of rationals in $[0, 1]$.

(4) Find a bounded lower semicontinuous function $f : [0, 1] \to \mathbb{R}$ such that the set of discontinuities of f has positive measure.

(5) Does there exist a bounded lower semicontinuous function $f : [0, 1] \to \mathbb{R}$ such that f is discontinuous a.e.?

Exercise 9.8 Find a sequence f_n of continuous functions mapping $[0, 1]$ into $[0, 1]$ such that the f_n increase to a bounded function f which is not Riemann integrable. Such an example shows there is no monotone convergence theorem or dominated convergence theorem for Riemann integrals.

Exercise 9.9 Let $M > 0$ and let $\mathcal{B}$ be the σ-algebra on $[-M, M]^2$ generated by the collection of sets of the form $[a, b] \times [c, d]$ with $-M \leq a \leq b \leq M$ and $-M \leq c \leq d \leq M$. Suppose μ is a measure on $([-M, M]^2, \mathcal{B})$ such that

$$\mu([a, b] \times [c, d]) = (b - a)(d - c).$$

(We will construct such a measure μ in Chapter 11.) Prove that if f is continuous with support in $[-M, M]^2$, then the Lebesgue integral of f with respect to μ is equal to the double (Riemann) integral of f and the two multiple (Riemann) integrals of f.

Chapter 10

Types of convergence

There are various ways in which a sequence of functions f_n can converge, and we compare some of them.

10.1 Definitions and examples

Definition 10.1 If μ is a measure, we say a sequence of measurable functions f_n *converges almost everywhere* to f and write $f_n \to f$ a.e. if there is a set of measure 0 such that for x not in this set we have $f_n(x) \to f(x)$.

We say f_n *converges in measure* to f if for each $\varepsilon > 0$

$$\mu(\{x : |f_n(x) - f(x)| > \varepsilon\}) \to 0$$

as $n \to \infty$.

Let $1 \le p < \infty$. We say f_n *converges in L^p* to f if

$$\int |f_n - f|^p \, d\mu \to 0$$

as $n \to \infty$.

Proposition 10.2 *(1) Suppose μ is a finite measure. If $f_n \to f$ a.e., then f_n converges to f in measure.*

(2) If μ is a measure, not necessarily finite, and $f_n \to f$ in measure, there is a subsequence n_j such that $f_{n_j} \to f$ a.e.

75

Proof. Let $\varepsilon > 0$ and suppose $f_n \to f$ a.e. If

$$A_n = \{x : |f_n(x) - f(x)| > \varepsilon\},$$

then $\chi_{A_n} \to 0$ a.e., and by the dominated convergence theorem,

$$\mu(A_n) = \int \chi_{A_n}(x)\,\mu(dx) \to 0.$$

This proves (1).

To prove (2), suppose $f_n \to f$ in measure, let $n_1 = 1$, and choose $n_j > n_{j-1}$ by induction so that

$$\mu(\{x : |f_{n_j}(x) - f(x)| > 1/j\}) \le 2^{-j}.$$

Let $A_j = \{x : |f_{n_j}(x) - f(x)| > 1/j\}$. If we set

$$A = \cap_{k=1}^{\infty} \cup_{j=k}^{\infty} A_j,$$

then by Proposition 3.5

$$\mu(A) = \lim_{k\to\infty} \mu(\cup_{j=k}^{\infty} A_j) \le \lim_{k\to\infty} \sum_{j=k}^{\infty} \mu(A_j) \le \lim_{k\to\infty} 2^{-k+1} = 0.$$

Therefore A has measure 0. If $x \notin A$, then $x \notin \cup_{j=k}^{\infty} A_j$ for some k, and so $|f_{n_j}(x) - f(x)| \le 1/j$ for $j \ge k$. This implies $f_{n_j} \to f$ on A^c. $\qquad\square$

Example 10.3 Part (1) of the above proposition is not true if $\mu(X) = \infty$. To see this, let $X = \mathbb{R}$ and let $f_n = \chi_{(n,n+1)}$. We have $f_n \to 0$ a.e., but f_n does not converge in measure.

The next proposition compares convergence in L^p to convergence in measure. Before we prove this, we prove an easy preliminary result known as *Chebyshev's inequality*.

Lemma 10.4 *If $1 \le p < \infty$, then*

$$\mu(\{x : |f(x)| \ge a\}) \le \frac{\int |f|^p \, d\mu}{a^p}.$$

Proof. Let $A = \{x : |f(x)| \geq a\}$. Since $\chi_A \leq |f|^p \chi_A / a^p$, we have

$$\mu(A) \leq \int_A \frac{|f|^p}{a^p} \, d\mu \leq \frac{1}{a^p} \int |f|^p \, d\mu.$$

This is what we wanted. $\qquad\square$

Proposition 10.5 *If f_n converges to f in L^p, then it converges in measure.*

Proof. If $\varepsilon > 0$, by Chebyshev's inequality

$$\mu(\{x : |f_n(x) - f(x)| > \varepsilon\}) = \mu(\{x : |f_n(x) - f(x)|^p > \varepsilon^p\})$$
$$\leq \frac{\int |f_n - f|^p}{\varepsilon^p} \to 0$$

as required. $\qquad\square$

Example 10.6 Let $f_n = n^2 \chi_{(0,1/n)}$ on $[0, 1]$ and let μ be Lebesgue measure. This gives an example where f_n converges to 0 a.e. and in measure, but does not converge in L^p for any $p \geq 1$.

Example 10.7 We give an example where $f_n \to f$ in measure and in L^p, but not almost everywhere. Let $S = \{e^{i\theta} : 0 \leq \theta < 2\pi\}$ be the unit circle in the complex plane and define

$$\mu(A) = m(\{\theta \in [0, 2\pi) : e^{i\theta} \in A\})$$

to be arclength measure on S, where m is Lebesgue measure on $[0, 2\pi)$.

Let $X = S$ and let $f_n(x) = \chi_{F_n}(x)$, where

$$F_n = \left\{ e^{i\theta} : \sum_{j=1}^{n} \frac{1}{j} \leq \theta \leq \sum_{j=1}^{n+1} \frac{1}{j} \right\}.$$

Let $f(e^{i\theta}) = 0$ for all θ.

Then $\mu(F_n) \leq 1/(n+1) \to 0$, so $f_n \to f$ in measure. Also, since f_n is either 1 or 0,

$$\int |f_n - f|^p \, d\mu = \int \chi_{F_n} \, d\mu = \mu(F_n) \to 0.$$

But because $\sum_{j=1}^{\infty} 1/j = \infty$, each point of S is in infinitely many F_n, and each point of S is in $S - F_n$ for infinitely many n, so f_n does not converge to f at any point.

The F_n are arcs whose length tends to 0, but such that $\cup_{n \geq m} F_n$ contains S for each m.

The following is known as *Egorov's theorem.*

Theorem 10.8 *Suppose μ is a finite measure, $\varepsilon > 0$, and $f_n \to f$ a.e. Then there exists a measurable set A such that $\mu(A) < \varepsilon$ and $f_n \to f$ uniformly on A^c.*

This type of convergence is sometimes known as *almost uniform convergence*. Egorov's theorem is not as useful for solving problems as one might expect, and students have a tendency to try to use it when other methods work much better.

Proof. Let

$$A_{nk} = \cup_{m=n}^{\infty} \{x : |f_m(x) - f(x)| > 1/k\}.$$

For fixed k, A_{nk} decreases as n increases. The intersection $\cap_n A_{nk}$ has measure 0 because for almost every x, $|f_m(x) - f(x)| \leq 1/k$ if m is sufficiently large. Therefore $\mu(A_{nk}) \to 0$ as $n \to \infty$. We can thus find an integer n_k such that $\mu(A_{n_k k}) < \varepsilon 2^{-k}$. Let

$$A = \cup_{k=1}^{\infty} A_{n_k k}.$$

Hence $\mu(A) < \varepsilon$. If $x \notin A$, then $x \notin A_{n_k k}$, and so $|f_n(x) - f(x)| \leq 1/k$ if $n \geq n_k$. Thus $f_n \to f$ uniformly on A^c.   □

10.2   Exercises

Exercise 10.1 Suppose that f_n is a sequence that is Cauchy in measure. This means that given ε and $a > 0$, there exists N such that if $m, n \geq N$, then

$$\mu(\{x : |f_n(x) - f_m(x)| > a\}) < \varepsilon.$$

Prove that f_n converges in measure.

Exercise 10.2 Suppose $\mu(X) < \infty$. Define

$$d(f, g) = \int \frac{|f - g|}{1 + |f - g|} \, d\mu.$$

Prove that d is a metric on the space of measurable functions, except for the fact that $d(f, g) = 0$ only implies that $f = g$ a.e., not necessarily everywhere. Prove that $f_n \to f$ in measure if and only if $d(f_n, f) \to 0$.

Exercise 10.3 Prove that if $f_n \to f$ in measure and each f_n is non-negative, then

$$\int f \leq \liminf_{n \to \infty} \int f_n.$$

Exercise 10.4 Prove that if A_n is measurable, $\mu(A_n) < \infty$ for each n, and χ_{A_n} converges to f in measure, then there exists a measurable set A such that $f = \chi_A$ a.e.

Exercise 10.5 Suppose for each ε there exists a measurable set F such that $\mu(F^c) < \varepsilon$ and f_n converges to f uniformly on F. Prove that f_n converges to f a.e.

Exercise 10.6 Suppose that f_n and f are measurable functions such that for each $\varepsilon > 0$ we have

$$\sum_{n=1}^{\infty} \mu(\{x : |f_n(x) - f(x)| > \varepsilon\}) < \infty.$$

Prove that $f_n \to f$ a.e.

Exercise 10.7 Let f_n be a sequence of measurable functions and define

$$g_n(x) = \sup_{m \geq n} |f_m(x) - f_n(x)|.$$

Prove that if g_n converges in measure to 0, then f_n converges a.e.

Exercise 10.8 If $(X, \mathcal{A}, \mu)$ is a measure space and f_n is a sequence of real-valued measurable functions such that $\int f_n g \, d\mu$ converges to 0 for every integrable g, is it necessarily true that f_n converges to 0 in measure? If not, give a counterexample.

Exercise 10.9 Suppose $(X, \mathcal{A}, \mu)$ is a measure space and X is a countable set. Prove that if f_n is a sequence of measurable functions converging to f in measure, then f_n also converges to f a.e.

Chapter 11

Product measures

We have defined Lebesgue measure on the line. Now we give a method for constructing measures on the plane, in n-dimensional Euclidean spaces, and many other product spaces. The main theorem, the Fubini theorem, which allows one to interchange the order of integration, is one of the most important theorems in real analysis.

11.1   Product σ-algebras

Suppose $(X, \mathcal{A}, \mu)$ and $(Y, \mathcal{B}, \nu)$ are two measure spaces and suppose also that μ and ν are σ-finite measures. A *measurable rectangle* is a set of the form $A \times B$, where $A \in \mathcal{A}$ and $B \in \mathcal{B}$.

Let $\mathcal{C}_0$ be the collection of finite unions of disjoint measurable rectangles. Thus every element of $\mathcal{C}_0$ is of the form $\cup_{i=1}^n (A_i \times B_i)$, where $A_i \in \mathcal{A}$, $B_i \in \mathcal{B}$, and if $i \neq j$, then $(A_i \times B_i) \cap (A_j \times B_j) = \emptyset$. Since $(A \times B)^c = (A \times B^c) \cup (A^c \times Y)$ and the intersection of two measurable rectangles is a measurable rectangle, it is easy to check that $\mathcal{C}_0$ is an algebra of sets. We define the *product σ-algebra*

$$\mathcal{A} \times \mathcal{B} = \sigma(\mathcal{C}_0).$$

If $E \in \mathcal{A} \times \mathcal{B}$, we define the *$x$-section* of E by

$$s_x(E) = \{y \in Y : (x, y) \in E\}$$

and similarly define the *y-section*:

$$t_y(E) = \{x : (x, y) \in E\}.$$

Given a function $f : X \times Y \to \mathbb{R}$ that is $\mathcal{A} \times \mathcal{B}$ measurable, for each x and y we define $S_x f : Y \to \mathbb{R}$ and $T_y f : X \to \mathbb{R}$ by

$$S_x f(y) = f(x, y), \qquad T_y f(x) = f(x, y).$$

Lemma 11.1 *(1) If $E \in \mathcal{A} \times \mathcal{B}$, then $s_x(E) \in \mathcal{B}$ for each x and $t_y(E) \in \mathcal{A}$ for each y.*
(2) If f is $\mathcal{A} \times \mathcal{B}$ measurable, then $S_x f$ is $\mathcal{B}$ measurable for each x and $T_y f$ is $\mathcal{A}$ measurable for each y.

Proof. (1) Let $\mathcal{C}$ be the collection of sets in $\mathcal{A} \times \mathcal{B}$ for which $s_x(E) \in \mathcal{B}$ for each x. We will show that $\mathcal{C}$ is a σ-algebra containing the measurable rectangles, and hence is all of $\mathcal{A} \times \mathcal{B}$.

If $E = A \times B$, then $s_x(E)$ is equal to B if $x \in A$ and equal to $\emptyset$ if $x \notin A$. Hence $s_x(E) \in \mathcal{B}$ for each x when E is a measurable rectangle.

If $E \in \mathcal{C}$, then $y \in s_x(E^c)$ if and only if $(x, y) \in E^c$, which happens if and only if $y \notin s_x(E)$. Therefore $s_x(E^c) = (s_x(E))^c$, and $\mathcal{C}$ is closed under the operation of taking complements. Similarly, it is easy to see that $s_x(\cup_{i=1}^\infty E_i) = \cup_{i=1}^\infty s_x(E_i)$, and so $\mathcal{C}$ is closed under the operation of countable unions.

Therefore $\mathcal{C}$ is a σ-algebra containing the measurable rectangles, and hence is equal to $\mathcal{A} \times \mathcal{B}$. The argument for $t_y(E)$ is the same.

(2) Fix x. If $f = \chi_E$ for $E \in \mathcal{A} \times \mathcal{B}$, note that $S_x f(y) = \chi_{s_x(E)}(y)$, which is $\mathcal{B}$ measurable. By linearity, $S_x f$ is $\mathcal{B}$ measurable when f is a simple function. If f is non-negative, take $\mathcal{A} \times \mathcal{B}$ measurable simple functions r_n increasing to f, and since $S_x r_n \uparrow S_x f$, then $S_x f$ is $\mathcal{B}$ measurable. Writing $f = f^+ - f^-$ and using linearity again shows that $S_x f$ is $\mathcal{B}$ measurable. The argument for $T_y f$ is the same. $\square$

Let $E \in \mathcal{A} \times \mathcal{B}$ and let

$$h(x) = \nu(s_x(E)), \qquad k(y) = \mu(t_y(E)).$$

Proposition 11.2 *(1)* h *is* $\mathcal{A}$ *measurable and* k *is* $\mathcal{B}$ *measurable.*
(2) We have

$$\int h(x)\,\mu(dx) = \int k(y)\,\nu(dy). \tag{11.1}$$

Since $\chi_{s_x(E)}(y) = S_x\chi_E(y)$ for all x and y, (11.1) could be written as

$$\int \left[\int S_x\chi_E(y)\,\nu(dy)\right]\mu(dx) = \int \left[\int T_y\chi_E(x)\,\mu(dx)\right]\nu(dy).$$

We will usually write this as

$$\int\int \chi_E(x,y)\,\nu(dy)\,\mu(dx) = \int\int \chi_E(x,y)\,\mu(dx)\,\nu(dy).$$

Proof. First suppose μ and ν are finite measures. Let $\mathcal{C}$ be the collection of sets in $\mathcal{A}\times\mathcal{B}$ for which (1) and (2) hold. We will prove that $\mathcal{C}$ contains $\mathcal{C}_0$ and is a monotone class. This will prove that $\mathcal{C}$ is the smallest σ-algebra containing $\mathcal{C}_0$ and hence is equal to $\mathcal{A}\times\mathcal{B}$.

If $E = A \times B$, with $A \in \mathcal{A}$ and $B \in \mathcal{B}$, then $h(x) = \chi_A(x)\nu(B)$, which is $\mathcal{A}$ measurable, and $\int h(x)\,\mu(dx) = \mu(A)\nu(B)$. Similarly, $k(y) = \mu(A)\chi_B(y)$ is $\mathcal{B}$ measurable and $\int k(y)\,\nu(dy) = \mu(A)\nu(B)$. Therefore (1) and (2) hold for measurable rectangles.

If $E = \cup_{i=1}^n E_i$, where each E_i is a measurable rectangle and the E_i are disjoint, then $s_x(E) = \cup_{i=1}^n s_x(E_i)$, and since the $s_x(E_i)$ are disjoint, then

$$h(x) = \nu(s_x(E)) = \nu(\cup_{i=1}^n s_x(E_i)) = \sum_{i=1}^n \nu(s_x(E_i)).$$

This shows that h is $\mathcal{A}$ measurable, since it is the sum of $\mathcal{A}$ measurable functions. Similarly $k(y)$ is $\mathcal{B}$ measurable. If we let $h_i(x) = \nu(s_x(E_i))$ and define $k_i(y)$ similarly, then

$$\int h_i(x)\,\mu(dx) = \int k_i(y)\,\nu(dy)$$

by the preceding paragraph, and then (2) holds for E by linearity. Therefore $\mathcal{C}$ contains $\mathcal{C}_0$.

Suppose $E_n \uparrow E$ and each $E_n \in \mathcal{C}$. If we let $h_n(x) = \nu(s_x(E_n))$ and let $k_n(y) = \mu(t_n(E_n))$, then $h_n \uparrow h$ and $k_n \uparrow k$. Therefore h is

$\mathcal{A}$ measurable and k is $\mathcal{B}$ measurable. We have (11.1) holding when h and k are replaced by h_n and k_n, resp. We let $n \to \infty$ and use the monotone convergence theorem to see that (11.1) holds with h and k.

If $E_n \downarrow E$ with each $E_n \in \mathcal{C}$, almost the same argument shows that h and k are measurable with respect to $\mathcal{A}$ and $\mathcal{B}$, and that (11.1) holds. The only difference is that we use the dominated convergence theorem in place of the monotone convergence theorem. This is where we need μ and ν to be finite measures.

We have shown $\mathcal{C}$ is a monotone class containing $\mathcal{C}_0$. By the monotone class theorem (Theorem 2.10), $\mathcal{C}$ is equal to $\sigma(\mathcal{C}_0)$, which is $\mathcal{A} \times \mathcal{B}$.

Finally suppose μ and ν are σ-finite. Then there exist $F_i \uparrow X$ and $G_i \uparrow Y$ such that each F_i is $\mathcal{A}$ measurable and has finite μ measure and each G_i is $\mathcal{B}$ measurable and has finite ν measure. Let $\mu_i(A) = \mu(A \cap F_i)$ for each $A \in \mathcal{A}$ and $\nu_i(A) = \nu(A \cap G_i)$ for each $B \in \mathcal{B}$. Let $h_i(x) = \nu_i(s_x(E)) = \nu(s_x(E) \cap G_i)$ and similarly define $k_i(y)$. By what we have proved above, h_i is $\mathcal{A}$ measurable, k_i is $\mathcal{B}$ measurable, and (11.1) holds if we replace h and k by h_i and k_i, resp. Now $h_i \uparrow h$ and $k_i \uparrow k$, which proves the measurability of h and k. Applying the monotone convergence theorem proves that (11.1) holds with h and k. $\qquad\square$

We now define $\mu \times \nu$ by

$$\mu \times \nu(E) = \int h(x) \, \mu(dx) = \int k(y) \, \nu(dy). \qquad (11.2)$$

Clearly $\mu \times \nu(\emptyset) = 0$. If $E_1, \ldots, E_n$ are disjoint and in $\mathcal{A} \times \mathcal{B}$ and $E = \cup_{i=1}^{n} E_i$, then we saw in the proof of Proposition 11.2 that $\nu(s_x(E)) = \sum_{i=1}^{n} \nu(s_x(E_i))$. We conclude that

$$\mu \times \nu(E) = \int \nu(s_x(E)) \, \mu(dx) = \sum_{i=1}^{n} \int \nu(s_x(E_i)) \, \mu(dx)$$

$$= \sum_{i=1}^{n} \mu \times \nu(E_i),$$

or $\mu \times \nu$ is finitely additive. If $E_n \uparrow E$ with each $E_n \in \mathcal{A} \times \mathcal{B}$ and we let $h_n(x) = \nu(s_x(E_n))$, then $h_n \uparrow h$, and by the monotone convergence theorem, $\mu \times \nu(E_n) \uparrow \mu \times \nu(E)$. Therefore $\mu \times \nu$ is a measure.

Note that if $E = A \times B$ is a measurable rectangle, then $h(x) = \chi_A(x)\nu(B)$ and so

$$\mu \times \nu(A \times B) = \mu(A)\nu(B),$$

which is what it should be.

11.2 The Fubini theorem

The main result of this chapter is the *Fubini theorem*, which allows one to interchange the order of integration. This is sometimes called the *Fubini-Tonelli theorem*.

Theorem 11.3 *Suppose* $f : X \times Y \to \mathbb{R}$ *is measurable with respect to* $\mathcal{A} \times \mathcal{B}$. *If either*

(a) f is non-negative, or
(b) $\int |f(x,y)| \, d(\mu \times \nu)(x,y) < \infty$,
then
(1) for each x, the function $y \mapsto f(x,y)$ is measurable with respect to $\mathcal{B}$;
(2) for each y, the function $x \mapsto f(x,y)$ is measurable with respect to $\mathcal{A}$;
(3) the function $g(x) = \int f(x,y)\,\nu(dy)$ is measurable with respect to $\mathcal{A}$;
(4) the function $h(y) = \int f(x,y)\,\mu(dx)$ is measurable with respect to $\mathcal{B}$;
(5) we have

$$\int f(x,y)\,d(\mu \times \nu)(x,y) = \int \left[\int f(x,y)\,d\mu(x) \right] d\nu(y) \quad (11.3)$$

$$= \int \left[\int f(x,y)\,d\nu(y) \right] \mu(dx).$$

The last integral in (11.3) should be interpreted as

$$\int \left[\int S_x f(y)\,\nu(dy) \right] \mu(dx)$$

and similarly for the second integral in (11.3). Since no confusion results, most often the brackets are omitted in (11.3).

Proof. If f is the characteristic function of a set in $\mathcal{A} \times \mathcal{B}$, then (1)–(5) are merely a restatement of Lemma 11.1 and Proposition

11.2. By linearity, (1)–(5) hold if f is a simple function. Since the increasing limit of measurable functions is measurable, then writing a non-negative function as the increasing limit of simple functions and using the monotone convergence theorem shows that (1)–(5) hold when f is non-negative. In the case where $\int |f| \, d(\mu \times \nu) < \infty$, writing $f = f^+ - f^-$ and using linearity proves (1)–(5) for this case, too. □

Observe that if we know

$$\int \int |f(x,y)| \, \mu(dx) \, \nu(dy) < \infty,$$

then since $|f(x,y)|$ is non-negative the Fubini theorem tells us that

$$\int |f(x,y)| \, d(\mu \times \nu) = \int \int |f(x,y)| \, \mu(dx) \, \nu(dy) < \infty$$

We can then apply the Fubini theorem again to conclude

$$\int f(x,y) \, d(\mu \times \nu) = \int \int f(x,y) \, d\mu \, d\nu = \int \int f(x,y) \, d\nu \, d\mu.$$

Thus in the hypotheses of the Fubini theorem, we could as well assume $\int \int |f(x,y)| \, d\mu \, d\nu < \infty$ or $\int \int |f(x,y)| \, d\nu \, d\mu < \infty$.

When f is measurable with respect to $\mathcal{A} \times \mathcal{B}$, we sometimes say that f is *jointly measurable*.

Even when $(X, \mathcal{A}, \mu)$ and $(Y, \mathcal{B}, \nu)$ are complete, it will not be the case in general that $(X \times Y, \mathcal{A} \times \mathcal{B}, \mu \times \nu)$ is complete. For example, let $(X, \mathcal{A}, \mu) = (Y, \mathcal{B}, \nu)$ be Lebesgue measure on $[0,1]$ with the Lebesgue σ-algebra. Let A be a non-measurable set in $[0,1]$ and let $E = A \times \{1/2\}$. Then E is not a measurable set with respect to $\mathcal{A} \times \mathcal{B}$, or else $A = t_{1/2}(E)$ would be in $\mathcal{A}$ by Lemma 11.1. On the other hand, $E \subset [0,1] \times \{1/2\}$, which has zero measure with respect to $\mu \times \nu$, so E is a null set.

One can take the completion of $(X \times Y, \mathcal{A} \times \mathcal{B}, \mu \times \nu)$ without great difficulty. See [8] for details.

There is no difficulty extending the Fubini theorem to the product of n measures. If we have $\mu_1, \ldots, \mu_n$ all equal to m, Lebesgue measure on $\mathbb{R}$ with the Lebesgue σ-algebra $\mathcal{L}$, then the completion of $(\mathbb{R}^n, \mathcal{L} \times \cdots \times \mathcal{L}, m \times \cdots \times m)$ is called n-*dimensional Lebesgue measure*.

For a general change of variables theorem, see [4].

11.3 Examples

We give two examples to show that the hypotheses of the Fubini theorem are necessary.

Example 11.4 Let $X = Y = [0,1]$ with μ and ν both being Lebesgue measure. Let g_i be continuous functions with support in $(1/(i+1), 1/i)$ such that $\int_0^1 g_i(x)\,dx = 1$, $i = 1, 2, \ldots$. Let

$$f(x,y) = \sum_{i=1}^{\infty} [g_i(x) - g_{i+1}(x)] g_i(y).$$

For each point (x, y) at most two terms in the sum are non-zero, so the sum is actually a finite one. If we first integrate with respect to y, we get

$$\int_0^1 f(x,y)\,dy = \sum_{i=1}^{\infty} [g_i(x) - g_{i+1}(x)].$$

This is a telescoping series, and sums to $g_1(x)$. Therefore

$$\int_0^1 \int_0^1 f(x,y)\,dy\,dx = \int_0^1 g_1(x)\,dx = 1.$$

On the other hand, integrating first with respect to x gives 0, so

$$\int_0^1 \int_0^1 f(x,y)\,dx\,dy = 0.$$

This doesn't contradict the Fubini theorem because

$$\int_0^1 \int_0^1 |f(x,y)|\,dx\,dy = \infty.$$

Example 11.5 For this example, you have to take on faith a bit of set theory. There exists a set X together with a partial order "$\leq$" such that X is uncountable but for any $y \in X$, the set $\{x \in X : x \leq y\}$ is countable. An example is to let X be the set of countable ordinals. The σ-algebra is the collection of subsets A of X such that either A or A^c is countable. Define μ on X by $\mu(A) = 0$ if A is countable and 1 if A is uncountable. Define f on $X \times X$ by $f(x,y) = 1$ if $x \leq y$ and zero otherwise. Then $\int \int f(x,y)\,dy\,dx = 1$ but $\int \int f(x,y)\,dx\,dy = 0$. The reason there is no contradiction is that f is not measurable with respect to the product σ-algebra.

11.4   Exercises

Exercise 11.1 State and prove a version of the Fubini theorem for complex-valued functions.

Exercise 11.2 Let $(X, \mathcal{A})$ and $(Y, \mathcal{B})$ be two measurable spaces and let $f \geq 0$ be measurable with respect to $\mathcal{A} \times \mathcal{B}$. Let $g(x) = \sup_{y \in Y} f(x, y)$ and suppose $g(x) < \infty$ for each x. Is g necessarily measurable with respect to $\mathcal{A}$? If not, find a counterexample.

Exercise 11.3 Prove the equality

$$\int_{-\infty}^{\infty} |f(x)| \, dx = \int_0^{\infty} m(\{x : |f(x)| \geq t\}) \, dt,$$

where m is Lebesgue measure.

Exercise 11.4 Let A be a Lebesgue measurable subset of $[0, 1]^2$ with $m_2(A) = 1$, where m_2 is two-dimensional Lebesgue measure. Show that for almost every $x \in [0, 1]$ (with respect to one dimensional Lebesgue measure) the set $s_x(A)$ has one-dimensional Lebesgue measure one.

Exercise 11.5 Let $f : [0, 1]^2 \to \mathbb{R}$ be such that for every $x \in [0, 1]$ the function $y \to f(x, y)$ is Lebesgue measurable on $[0, 1]$ and for every $y \in [0, 1]$ the function $x \to f(x, y)$ is continuous on $[0, 1]$. Prove that f is measurable with respect to the the completion of the product σ-algebra $\mathcal{L} \times \mathcal{L}$ on $[0, 1]^2$. Here $\mathcal{L}$ is the Lebesgue σ-algebra on $[0, 1]$.

Exercise 11.6 Suppose f is real-valued and integrable with respect to two-dimensional Lebesgue measure on $[0, 1]^2$ and

$$\int_0^a \int_0^b f(x, y) \, dy \, dx = 0$$

for all $a \in [0, 1]$ and $b \in [0, 1]$. Prove that $f = 0$ a.e.

Exercise 11.7 Prove that

$$\int_0^1 \int_0^1 \frac{x^2 - y^2}{(x^2 + y^2)^{3/2}} \log(4 + \sin x) \, dy \, dx$$

$$= \int_0^1 \int_0^1 \frac{x^2 - y^2}{(x^2 + y^2)^{3/2}} \log(4 + \sin x) \, dx \, dy.$$

Exercise 11.8 Let $X = Y = [0,1]$ and let $\mathcal{B}$ be the Borel σ-algebra. Let m be Lebesgue measure and μ counting measure on $[0,1]$.
(1) If $D = \{(x,y) : x = y\}$, show that D is measurable with respect to $\mathcal{B} \times \mathcal{B}$.
(2) Show that

$$\int_X \int_Y \chi_D(x,y)\,\mu(dy)\,m(dx) \neq \int_Y \int_X \chi_D(x,y)\,m(dx)\,\mu(dy).$$

Why does this not contradict the Fubini theorem?

Exercise 11.9 Let $X = Y = \mathbb{R}$ and let $\mathcal{B}$ be the Borel σ-algebra. Define

$$f(x,y) = \begin{cases} 1, & x \geq 0 \text{ and } x \leq y < x+1; \\ -1, & x \geq 0 \text{ and } x+1 \leq y < x+2; \\ 0, & \text{otherwise.} \end{cases}$$

Show that

$$\int \int f(x,y)\,dy\,dx \neq \int \int f(x,y)\,dx\,dy.$$

Why does this not contradict the Fubini theorem?

Exercise 11.10 Find a real-valued function f that is integrable on $[0,1]^2$ such that

$$\int_0^a \int_0^1 f(x,y)\,dy\,dx = 0, \qquad \int_0^1 \int_0^b f(x,y)\,dy\,dx = 0$$

for every $a,b \in [0,1]$, but f is not zero almost everywhere with respect to 2-dimensional Lebesgue measure.

Exercise 11.11 Let μ be a finite measure on $\mathbb{R}$ and let $f(x) = \mu((-\infty,x])$. Show

$$\int [f(x+c) - f(x)]\,dx = c\mu(\mathbb{R}).$$

Exercise 11.12 Use

$$\frac{1}{x} = \int_0^\infty e^{-xy}\,dy$$

and the Fubini theorem to calculate

$$\int_0^b \int_0^\infty e^{-xy} \sin x \, dy \, dx$$

two different ways. Then prove that

$$\lim_{b \to \infty} \int_0^b \frac{\sin x}{x} \, dx = \frac{\pi}{2}.$$

Recall that

$$\int e^{au} \sin u \, du = \frac{e^{au}(a \sin u - \cos u)}{1 + a^2} + C.$$

Exercise 11.13 Let $X = \{1, 2, \ldots\}$ and let μ be counting measure on X. Define $f : X \times X \to \mathbb{R}$ by

$$f(x, y) = \begin{cases} 1, & x = y; \\ -1, & x = y + 1; \\ 0, & \text{otherwise.} \end{cases}$$

Show that

$$\int_X \int_X f(x, y) \, \mu(dx) \, \mu(dy) \neq \int_X \int_X f(x, y) \, \mu(dy) \, \mu(dx).$$

Why is this not a contradiction to the Fubini theorem?

Exercise 11.14 Let $\{a_n\}$ and $\{r_n\}$ be two sequences of real numbers such that $\sum_{n=1}^\infty |a_n| < \infty$. Prove that

$$\sum_{n=1}^\infty \frac{a_n}{\sqrt{|x - r_n|}}$$

converges absolutely for almost every $x \in \mathbb{R}$.

Exercise 11.15 Let $(X, \mathcal{A}, \mu)$ and $(Y, \mathcal{B}, \nu)$ be measure spaces. Prove that if λ is a measure on $\mathcal{A} \times \mathcal{B}$ such that

$$\lambda(A \times B) = \mu(A)\nu(B)$$

whenever $A \in \mathcal{A}$ and $B \in \mathcal{B}$, then $\lambda = \mu \times \nu$ on $\mathcal{A} \times \mathcal{B}$.

Exercise 11.16 Let S be the unit circle $\{e^{i\theta} : 0 \le \theta < 2\pi\}$ and define a measure μ on S by $\mu(A) = m(\{\theta : e^{i\theta} \in A\})$, where m is Lebesgue measure on $[0, 2\pi)$. Let m_2 be two-dimensional Lebesgue measure. Show that if A is a Borel subset of S and $R > 0$, then

$$m_2(\{re^{i\theta} : 0 < r < R, e^{i\theta} \in A\}) = \mu(A)R^2/2.$$

Exercise 11.17 Use Exercise 11.16 to prove that if f is a continuous real-valued function with support in the ball $B(0, R) = \{(x, y) : x^2 + y^2 < R^2\}$, then

$$\int\int_{B(0,R)} f(x, y)\, dy\, dx = \int_0^{2\pi} \int_0^R f(r\cos\theta, r\sin\theta)\, r\, dr\, d\theta.$$

Exercise 11.18 Prove that

$$\int_0^\infty e^{-x^2/2}\, dx = \sqrt{\pi/2}$$

by filling in the missing steps and making rigorous the following. If $I = \int_0^\infty e^{-x^2/2}\, dx$, then

$$I^2 = \int_0^\infty \int_0^\infty e^{-(x^2+y^2)/2}\, dy\, dx = \int_0^{\pi/2} \int_0^\infty e^{-r^2/2} r\, dr\, d\theta = \pi/2.$$

Exercise 11.19 If $M = (M_{ij})_{i,j=1}^n$ is a $n \times n$ matrix and $x = (x_1, \ldots, x_n) \in \mathbb{R}^n$, define Mx to be the element of $\mathbb{R}^n$ whose i^{th} coordinate is $\sum_{j=1}^n M_{ij}x_j$. (This is just the usual matrix multiplication of a $n \times n$ matrix and a $n \times 1$ matrix.) If A is a Borel subset of $\mathbb{R}^n$, let $M(A) = \{Mx : x \in A\}$.
(1) If $c \in \mathbb{R}$ and

$$M_{ij} = \begin{cases} c, & i = j = 1; \\ 1, & i = j \ne 1; \\ 0, & i \ne j; \end{cases}$$

show

$$m_n(M(A)) = |c|m_n(A) = |\det M|m_n(A),$$

where we use m_n for n-dimensional Lebesgue measure. (Multiplication by M multiplies the first coordinate by c.)
(2) If $1 \le k \le n$ and

$$M_{ij} = \begin{cases} 1, & i = 1 \text{ and } j = k; \\ 1, & j = 1 \text{ and } i = k; \\ 1, & i = j \text{ and neither equals } k; \\ 0, & \text{otherwise}; \end{cases}$$

show
$$m_n(M(A)) = m_n(A) = |\det M| m_n(A).$$

(Multiplication by M interchanges the first and k^{th} coordinates.)
(3) If $c \in \mathbb{R}$ and
$$M_{ij} = \begin{cases} 1, & i = j; \\ c, & i = 1, j = 2; \\ 0, & \text{otherwise}, \end{cases}$$

show
$$m_n(M(A)) = m_n(A) = |\det M| m_n(A).$$

(Multiplication by M replaces x_1 by $x_1 + cx_2$.)
(4) Since every $n \times n$ matrix can be written as the product of matrices each of which has the form given in (1), (2), or (3), conclude that if M is any $n \times n$ matrix, then

$$m_n(M(A)) = |\det M| m_n(A).$$

(5) If M is an orthogonal matrix, so that M times its transpose is the identity, show $m_n(M(A)) = m_n(A)$. (Multiplication by an orthogonal matrix is a rotation of $\mathbb{R}^n$.)

Chapter 12

Signed measures

Signed measures have the countable additivity property of measures, but are allowed to take negative as well as positive values. We will see shortly that an example of a signed measure is $\nu(A) = \int_A f \, d\mu$, where f is integrable and takes both positive and negative values.

12.1 Positive and negative sets

Definition 12.1 Let $\mathcal{A}$ be a σ-algebra. A *signed measure* is a function $\mu : \mathcal{A} \to (-\infty, \infty]$ such that $\mu(\emptyset) = 0$ and $\mu(\cup_{i=1}^{\infty} A_i) = \sum_{i=1}^{\infty} \mu(A_i)$ whenever the A_i are pairwise disjoint and all the A_i are in $\mathcal{A}$.

When we want to emphasize that a measure is defined as in Definition 3.1 and only takes non-negative values, we refer to it as a *positive measure*.

Definition 12.2 Let μ be a signed measure. A set $A \in \mathcal{A}$ is called a *positive set* for μ if $\mu(B) \geq 0$ whenever $B \subset A$ and $B \in \mathcal{A}$. We say $A \in \mathcal{A}$ is a *negative set* if $\mu(B) \leq 0$ whenever $B \subset A$ and $B \in \mathcal{A}$. A *null set* A is one where $\mu(B) = 0$ whenever $B \subset A$ and $B \in \mathcal{A}$.

Note that if μ is a signed measure, then

$$\mu(\cup_{i=1}^{\infty} A_i) = \lim_{n \to \infty} \mu(\cup_{i=1}^{n} A_i).$$

The proof is the same as in the case of positive measures.

Example 12.3 Suppose m is Lebesgue measure and

$$\mu(A) = \int_A f \, dm$$

for some integrable f. If we let $P = \{x : f(x) \geq 0\}$, then P is easily seen to be a positive set, and if $N = \{x : f(x) < 0\}$, then N is a negative set. The Hahn decomposition which we give below is a decomposition of our space (in this case $\mathbb{R}$) into the positive and negative sets P and N. This decomposition is unique, except that $C = \{x : f(x) = 0\}$ could be included in N instead of P, or apportioned partially to P and partially to N. Note, however, that C is a null set. The Jordan decomposition below is a decomposition of μ into μ^+ and μ^-, where $\mu^+(A) = \int_A f^+ \, dm$ and $\mu^-(A) = \int_A f^- \, dm$.

Proposition 12.4 Let μ be a signed measure which takes values in $(-\infty, \infty]$. Let E be measurable with $\mu(E) < 0$. Then there exists a measurable subset F of E that is a negative set with $\mu(F) < 0$.

Proof. If E is a negative set, we are done. If not, there exists a measurable subset with positive measure. Let n_1 be the smallest positive integer such that there exists $E_1 \subset E$ with $\mu(E_1) \geq 1/n_1$. We then define pairwise disjoint measurable sets $E_2, E_3, \ldots$ by induction as follows. Let $k \geq 2$ and suppose $E_1, \ldots, E_{k-1}$ are pairwise disjoint measurable sets with $\mu(E_i) > 0$ for $i = 1, \ldots, k-1$. If $F_k = E - (E_1 \cup \cdots \cup E_{k-1})$ is a negative set, then

$$\mu(F_k) = \mu(E) - \sum_{i=1}^{k-1} \mu(E_i) \leq \mu(E) < 0$$

and F_k is the desired set F. If F_k is not a negative set, let n_k be the smallest positive integer such that there exists $E_k \subset F_k$ with E_k measurable and $\mu(E_k) \geq 1/n_k$.

We stop the construction if there exists k such that F_k is a negative set with $\mu(F_k) < 0$. If not, we continue and let $F = \cap_k F_k = E - (\cup_k E_k)$. Since $0 > \mu(E) > -\infty$ and $\mu(E_k) \geq 0$, then

$$\mu(E) = \mu(F) + \sum_{k=1}^{\infty} \mu(E_k).$$

Then $\mu(F) \leq \mu(E) < 0$, so the sum converges.

It remains to show that F is a negative set. Suppose $G \subset F$ is measurable with $\mu(G) > 0$. Then $\mu(G) \geq 1/N$ for some N. But this contradicts the construction, since for some k, $n_k > N$, and we would have chosen the set G instead of the set E_k at stage k. Therefore F must be a negative set. $\qquad\square$

12.2   Hahn decomposition theorem

Recall that we write $A \triangle B$ for $(A - B) \cup (B - A)$. The following is known as the *Hahn decomposition*.

Theorem 12.5 *(1) Let μ be a signed measure taking values in $(-\infty, \infty]$. There exist disjoint measurable sets E and F in $\mathcal{A}$ whose union is X and such that E is a negative set and F is a positive set.*

(2) If E' and F' are another such pair, then $E \triangle E' = F \triangle F'$ is a null set with respect to μ.

(3) If μ is not a positive measure, then $\mu(E) < 0$. If $-\mu$ is not a positive measure, then $\mu(F) > 0$.

Proof.   (1) Let $L = \inf\{\mu(A) : A \text{ is a negative set}\}$. Choose negative sets A_n such that $\mu(A_n) \to L$. Let $E = \cup_{n=1}^{\infty} A_n$. Let $B_n = A_n - (B_1 \cup \cdots \cup B_{n-1})$ for each n. Since A_n is a negative set, so is each B_n. Also, the B_n are disjoint and $\cup_n B_n = \cup_n A_n = E$. If $C \subset E$, then

$$\mu(C) = \lim_{n \to \infty} \mu(C \cap (\cup_{i=1}^{n} B_i)) = \lim_{n \to \infty} \sum_{i=1}^{n} \mu(C \cap B_i) \leq 0.$$

Thus E is a negative set.

Since E is a negative set,

$$\mu(E) = \mu(A_n) + \mu(E - A_n) \leq \mu(A_n).$$

Letting $n \to \infty$, we obtain $\mu(E) = L$.

Let $F = E^c$. If F were not a positive set, there would exist $B \subset F$ with $\mu(B) < 0$. By Proposition 12.4 there exists a negative set C contained in B with $\mu(C) < 0$. But then $E \cup C$ would be a negative set with $\mu(E \cup C) < \mu(E) = L$, a contradiction.

(2) To prove uniqueness, if E', F' are another such pair of sets and $A \subset E - E' \subset E$, then $\mu(A) \leq 0$. But $A \subset E - E' = F' - F \subset F'$, so $\mu(A) \geq 0$. Therefore $\mu(A) = 0$. The same argument works if $A \subset E' - E$, and any subset of $E \triangle E'$ can be written as the union of A_1 and A_2, where $A_1 \subset E - E'$ and $A_2 \subset E' - E$.

(3) Suppose μ is not a positive measure but $\mu(E) = 0$. If $A \in \mathcal{A}$, then

$$\mu(A) = \mu(A \cap E) + \mu(A \cap F) \geq \mu(E) + \mu(A \cap F) \geq 0,$$

which says that μ must be a positive measure, a contradiction. A similar argument applies for $-\mu$ and F.   □

Let us say two measures μ and ν are *mutually singular* if there exist two disjoint sets E and F in $\mathcal{A}$ whose union is X with $\mu(E) = \nu(F) = 0$. This is often written $\mu \perp \nu$.

Example 12.6 If μ is Lebesgue measure restricted to $[0, 1/2]$, that is, $\mu(A) = m(A \cap [0, 1/2])$, and ν is Lebesgue measure restricted to $[1/2, 1]$, then μ and ν are mutually singular. We let $E = [0, 1/2]$ and $F = (1/2, 1]$. This example works because the Lebesgue measure of $\{1/2\}$ is 0.

Example 12.7 A more interesting example is the following. Let f be the Cantor-Lebesgue function where we define $f(x) = 1$ if $x \geq 1$ and $f(x) = 0$ if $x \leq 0$ and let ν be the Lebesgue-Stieltjes measure associated with f. Let μ be Lebesgue measure. Then $\mu \perp \nu$. To see this, we let $E = C$, where C is the Cantor set, and $F = C^c$. We already know that $m(E) = 0$ and we need to show $\nu(F) = 0$. To do that, we need to show $\nu(I) = 0$ for every open interval contained in F. This will follow if we show $\nu(J) = 0$ for every interval of the form $J = (a, b]$ contained in F. But f is constant on every such interval, so $f(b) = f(a)$, and therefore $\nu(J) = f(b) - f(a) = 0$.

12.3 Jordan decomposition theorem

The following is known as the *Jordan decomposition theorem.*

Theorem 12.8 *If μ is a signed measure on a measurable space $(X, \mathcal{A})$, there exist positive measures μ^+ and μ^- such that $\mu = \mu^+ - \mu^-$ and μ^+ and μ^- are mutually singular. This decomposition is unique.*

Proof. Let E and F be negative and positive sets, resp., for μ so that $X = E \cup F$ and $E \cap F = \emptyset$. Let $\mu^+(A) = \mu(A \cap F)$, $\mu^-(A) = -\mu(A \cap E)$. This gives the desired decomposition.

If $\mu = \nu^+ - \nu^-$ is another such decomposition with ν^+, ν^- mutually singular, let E' be a set such that $\nu^+(E') = 0$ and $\nu^-((E')^c) = 0$. Set $F' = (E')^c$. Hence $X = E' \cup F'$ and $E' \cap F' = \emptyset$. If $A \subset F'$, then $\nu^-(A) \leq \nu^-(F') = 0$, and so

$$\mu(A) = \nu^+(A) - \nu^-(A) = \nu^+(A) \geq 0,$$

and consequently F' is a positive set for μ. Similarly, E' is a negative set for μ. Thus E', F' gives another Hahn decomposition of X. By the uniqueness part of the Hahn decomposition theorem, $F \triangle F'$ is a null set with respect to μ. Since $\nu^+(E') = 0$ and $\nu^-(F') = 0$, if $A \in \mathcal{A}$, then

$$\nu^+(A) = \nu^+(A \cap F') = \nu^+(A \cap F') - \nu^-(A \cap F')$$
$$= \mu(A \cap F') = \mu(A \cap F) = \mu^+(A),$$

and similarly $\nu^- = \mu^-$. $\qquad\qquad\square$

The measure

$$|\mu| = \mu^+ + \mu^- \qquad\qquad (12.1)$$

is called the *total variation measure* of μ and $|\mu|(X)$ is called the *total variation* of μ.

12.4 Exercises

Exercise 12.1 Suppose μ is a signed measure. Prove that A is a null set with respect to μ if and only if $|\mu|(A) = 0$.

Exercise 12.2 Let μ be a signed measure. Define

$$\int f \, d\mu = \int f \, d\mu^+ - \int f \, d\mu^-.$$

Prove that

$$\left| \int f \, d\mu \right| \le \int |f| \, d|\mu|.$$

Exercise 12.3 Let μ be a signed measure on $(X, \mathcal{A})$. Prove that

$$|\mu|(A) = \sup \left\{ \left| \int_A f \, d\mu \right| : |f| \le 1 \right\}.$$

Exercise 12.4 Let μ be a positive measure and ν a signed measure. Prove that $\nu \ll \mu$ if and only if $\nu^+ \ll \mu$ and $\nu^- \ll \mu$.

Exercise 12.5 Let $(X, \mathcal{A})$ be a measurable space. Suppose $\lambda = \mu - \nu$, where μ and ν are finite positive measures. Prove that $\mu(A) \ge \lambda^+(A)$ and $\nu(A) \ge \lambda^-(A)$ for every $A \in \mathcal{A}$.

Exercise 12.6 Let $(X, \mathcal{A})$ be a measurable space. Prove that if μ and ν are finite signed measures, then $|\mu + \nu|(A) \le |\mu(A)| + |\nu(A)|$ for every $A \in \mathcal{A}$.

Exercise 12.7 Suppose that μ is a signed measure on $(X, \mathcal{A})$. Prove that if $A \in \mathcal{A}$, then

$$\mu^+(A) = \sup\{\mu(B) : B \in \mathcal{A}, B \subset A\}$$

and

$$\mu^-(A) = -\inf\{\mu(B) : B \in \mathcal{A}, B \subset A\}.$$

Exercise 12.8 Suppose that μ is a signed measure on $(X, \mathcal{A})$. Prove that if $A \in \mathcal{A}$, then

$$|\mu|(A) = \sup \left\{ \sum_{j=1}^{n} |\mu(B_j)| : \text{ each } B_j \in \mathcal{A}, \right.$$

$$\left. \text{the } B_j \text{ are disjoint}, \cup_{j=1}^n B_j = A \right\}.$$

Chapter 13

The Radon-Nikodym theorem

Suppose f is non-negative and integrable with respect to μ. If we define ν by

$$\nu(A) = \int_A f \, d\mu, \tag{13.1}$$

then ν is a measure. The only part that needs thought is the countable additivity. If A_n are disjoint measurable sets, we have

$$\nu(\cup_n A_n) = \int_{\cup_n A_n} f \, d\mu = \sum_{n=1}^{\infty} \int_{A_n} f \, d\mu = \sum_{n=1}^{\infty} \nu(A_n)$$

by using Proposition 7.5. Moreover, $\nu(A)$ is zero whenever $\mu(A)$ is.

In this chapter we consider the converse. If we are given two measures μ and ν, when does there exist f such that (13.1) holds? The Radon-Nikodym theorem answers this question.

13.1 Absolute continuity

Definition 13.1 A measure ν is said to be *absolutely continuous* with respect to a measure μ if $\nu(A) = 0$ whenever $\mu(A) = 0$. We write $\nu \ll \mu$.

Proposition 13.2 *Let ν be a finite measure. Then ν is absolutely continuous with respect to μ if and only if for all ε there exists δ such that $\mu(A) < \delta$ implies $\nu(A) < \varepsilon$.*

Proof. Suppose for each ε, there exists δ such that $\mu(A) < \delta$ implies $\nu(A) < \varepsilon$. If $\mu(A) = 0$, then $\nu(A) < \varepsilon$ for all ε, hence $\nu(A) = 0$, and thus $\nu \ll \mu$.

Suppose now that $\nu \ll \mu$. If there exists an ε for which no corresponding δ exists, then there exists E_k such that $\mu(E_k) < 2^{-k}$ but $\nu(E_k) \geq \varepsilon$. Let $F = \cap_{n=1}^{\infty} \cup_{k=n}^{\infty} E_k$. Then

$$\mu(F) = \lim_{n \to \infty} \mu(\cup_{k=n}^{\infty} E_k) \leq \lim_{n \to \infty} \sum_{k=n}^{\infty} 2^{-k} = 0,$$

but

$$\nu(F) = \lim_{n \to \infty} \nu(\cup_{k=n}^{\infty} E_k) \geq \varepsilon;$$

This contradicts the absolute continuity. $\qquad\square$

13.2   The main theorem

Lemma 13.3 *Let μ and ν be finite positive measures on a measurable space $(X, \mathcal{A})$. Either $\mu \perp \nu$ or else there exists $\varepsilon > 0$ and $G \in \mathcal{A}$ such that $\mu(G) > 0$ and G is a positive set for $\nu - \varepsilon\mu$.*

Proof. Consider the Hahn decomposition for $\nu - \frac{1}{n}\mu$. Thus there exists a negative set E_n and a positive set F_n for this measure, E_n and F_n are disjoint, and their union is X. Let $F = \cup_n F_n$ and $E = \cap_n E_n$. Note $E^c = \cup_n E_n^c = \cup_n F_n = F$.

For each n, $E \subset E_n$, so

$$\nu(E) \leq \nu(E_n) \leq \tfrac{1}{n}\mu(E_n) \leq \tfrac{1}{n}\mu(X).$$

Since ν is a positive measure, this implies $\nu(E) = 0$.

One possibility is that $\mu(E^c) = 0$, in which case $\mu \perp \nu$. The other possibility is that $\mu(E^c) > 0$. In this case, $\mu(F_n) > 0$ for some n. Let $\varepsilon = 1/n$ and $G = F_n$. Then from the definition of F_n, G is a positive set for $\nu - \varepsilon\mu$. $\qquad\square$

We now are ready for the *Radon-Nikodym theorem.*

Theorem 13.4 *Suppose μ is a σ-finite positive measure on a measurable space $(X, \mathcal{A})$ and ν is a finite positive measure on $(X, \mathcal{A})$ such that ν is absolutely continuous with respect to μ. Then there exists a μ-integrable non-negative function f which is measurable with respect to $\mathcal{A}$ such that*

$$\nu(A) = \int_A f \, d\mu$$

for all $A \in \mathcal{A}$. Moreover, if g is another such function, then $f = g$ almost everywhere with respect to μ.

The function f is called the *Radon-Nikodym derivative* of ν with respect to μ or sometimes the *density* of ν with respect to μ, and is written $f = d\nu/d\mu$. Sometimes one writes

$$d\nu = f \, d\mu.$$

The idea of the proof is to look at the set of f such that $\int_A f \, d\mu \leq \nu(A)$ for each $A \in \mathcal{A}$, and then to choose the one such that $\int_X f \, d\mu$ is largest.

Proof. *Step 1.* Let us first prove the uniqueness assertion. Suppose f and g are two functions such that

$$\int_A f \, d\mu = \nu(A) = \int_A g \, d\mu$$

for all $A \in \mathcal{A}$. For every set A we have

$$\int_A (f - g) \, d\mu = \nu(A) - \nu(A) = 0.$$

By Proposition 8.1 we have $f - g = 0$ a.e. with respect to μ.

Step 2. Let us assume μ is a finite measure for now. In this step we define the function f. Define

$$\mathcal{F} = \left\{ g \text{ measurable} : g \geq 0, \int_A g \, d\mu \leq \nu(A) \text{ for all } A \in \mathcal{A} \right\}.$$

$\mathcal{F}$ is not empty because $0 \in \mathcal{F}$. Let $L = \sup\{\int g \, d\mu : g \in \mathcal{F}\}$, and let g_n be a sequence in $\mathcal{F}$ such that $\int g_n \, d\mu \to L$. Let $h_n = \max(g_1, \ldots, g_n)$.

We claim that if g_1 and g_2 are in $\mathcal{F}$, then $h_2 = \max(g_1, g_2)$ is also in $\mathcal{F}$. To see this, let $B = \{x : g_1(x) \geq g_2(x)\}$, and write

$$
\begin{aligned}
\int_A h_2 \, d\mu &= \int_{A \cap B} h_2 \, d\mu + \int_{A \cap B^c} h_2 \, d\mu \\
&= \int_{A \cap B} g_1 \, d\mu + \int_{A \cap B^c} g_2 \, d\mu \\
&\leq \nu(A \cap B) + \nu(A \cap B^c) \\
&= \nu(A).
\end{aligned}
$$

Therefore $h_2 \in \mathcal{F}$.

By an induction argument, h_n is in $\mathcal{F}$.

The h_n increase, say to f. By the monotone convergence theorem,

$$
\int_A f \, d\mu \leq \nu(A) \tag{13.2}
$$

for all $A \in \mathcal{A}$ and

$$
\int f \, d\mu \geq \int h_n \, d\mu \geq \int g_n \, d\mu
$$

for each n, so $\int f \, d\mu = L$.

Step 3. Next we prove that f is the desired function. Define a measure λ by

$$
\lambda(A) = \nu(A) - \int_A f \, d\mu.
$$

λ is a positive measure since $f \in \mathcal{F}$.

Suppose λ is not mutually singular to μ. By Lemma 13.3, there exists $\varepsilon > 0$ and G such that G is measurable, $\mu(G) > 0$, and G is a positive set for $\lambda - \varepsilon\mu$. For any $A \in \mathcal{A}$,

$$
\nu(A) - \int_A f \, d\mu = \lambda(A) \geq \lambda(A \cap G) \geq \varepsilon\mu(A \cap G) = \int_A \varepsilon\chi_G \, d\mu,
$$

or

$$
\nu(A) \geq \int_A (f + \varepsilon\chi_G) \, d\mu.
$$

Hence $f + \varepsilon\chi_G \in \mathcal{F}$. But

$$
\int_X (f + \varepsilon\chi_G) \, d\mu = L + \varepsilon\mu(G) > L,
$$

a contradiction to the definition of L.

Therefore $\lambda \perp \mu$. Then there must exist $H \in \mathcal{A}$ such that $\mu(H) = 0$ and $\lambda(H^c) = 0$. Since $\nu \ll \mu$, then $\nu(H) = 0$, and hence

$$\lambda(H) = \nu(H) - \int_H f \, d\mu = 0.$$

This implies $\lambda = 0$, or $\nu(A) = \int_A f \, d\mu$ for all A.

Step 4. We now suppose μ is σ-finite. There exist $F_i \uparrow X$ such that $\mu(F_i) < \infty$ for each i. Let μ_i be the restriction of μ to F_i, that is, $\mu_i(A) = \mu(A \cap F_i)$. Define ν_i, the restriction of ν to F_i, similarly. If $\mu_i(A) = 0$, then $\mu(A \cap F_i) = 0$, hence $\nu(A \cap F_i) = 0$, and thus $\nu_i(A) = 0$. Therefore $\nu_i \ll \mu_i$. If f_i is the function such that $d\nu_i = f_i \, d\mu_i$, the argument of Step 1 shows that $f_i = f_j$ on F_i if $i \leq j$. Define f by $f(x) = f_i(x)$ if $x \in F_i$. Then for each $A \in \mathcal{A}$,

$$\nu(A \cap F_i) = \nu_i(A) = \int_A f_i \, d\mu_i = \int_{A \cap F_i} f \, d\mu.$$

Letting $i \to \infty$ shows that f is the desired function. $\qquad\square$

13.3 Lebesgue decomposition theorem

The proof of the *Lebesgue decomposition theorem* is almost the same as the proof of the Radon-Nikodym theorem.

Theorem 13.5 *Suppose μ is a σ-finite positive measure and ν is a finite positive measure. There exist positive measures λ, ρ such that $\nu = \lambda + \rho$, ρ is absolutely continuous with respect to μ, and λ and μ are mutually singular.*

Proof. As in the proof of Theorem 13.4 we reduce to the case where μ is a finite measure. Define $\mathcal{F}$ and L and construct f as in the proof of the Radon-Nikodym theorem. Let $\rho(A) = \int_A f \, d\mu$ and let $\lambda = \nu - \rho$. Our construction shows that

$$\int_A f \, d\mu \leq \nu(A),$$

so $\lambda(A) \geq 0$ for all A. We have $\rho + \lambda = \nu$. We need to show μ and λ are mutually singular.

If not, by Lemma 13.3, there exists $\varepsilon > 0$ and $F \in \mathcal{A}$ such that $\mu(F) > 0$ and F is a positive set for $\lambda - \varepsilon\mu$. We get a contradiction exactly as in the proof of the Radon-Nikodym theorem. We conclude that $\lambda \perp \mu$.   □

13.4   Exercises

Exercise 13.1 This exercise asks you to prove the Radon-Nikodym theorem for signed measures. Let $(X, \mathcal{A})$ be a measurable space. Suppose ν is a finite signed measure, μ is a finite positive measure, and $\nu(A) = 0$ whenever $\mu(A) = 0$ and $A \in \mathcal{A}$. Show there exists an integrable real-valued function f such that $\nu(A) = \int_A f \, d\mu$ for all $A \in \mathcal{A}$.

Exercise 13.2 State and prove a version of the Lebesgue decomposition theorem for signed measures.

Exercise 13.3 We define a *complex measure* μ on a measurable space $(X, \mathcal{A})$ to be a bounded map from $\mathcal{A}$ to $\mathbb{C}$ such that $\mu(\emptyset) = 0$ and $\mu(\cup_{i=1}^{\infty} A_i) = \sum_{i=1}^{\infty} \mu(A_i)$ whenever the A_i are in $\mathcal{A}$ and are pairwise disjoint. Note that this implies that $\mu(A)$ is finite for each measurable set A. Formulate and prove a Radon-Nikodym theorem for complex measures.

Exercise 13.4 Let μ be a complex measure on a measurable space $(X, \mathcal{A})$. The *total variation measure* is defined to be a positive measure $|\mu|$ such that $d|\mu| = |f| \, d\rho$, where ρ is a positive measure and f is a measurable function such that $d\mu = f \, d\rho$. The quantity $|\mu|(X)$ is called the *total variation* of μ. Prove that the definition of $|\mu|$ is independent of the choice of ρ, that is, if $d\mu = f_1 \, d\rho_1 = f_2 \, d\rho_2$, then $|f_1| \, d\rho_1 = |f_2| \, d\rho_2$.

Exercise 13.5 Let $(X, \mathcal{A})$ be a measurable space and let μ and ν be two finite measures. We say μ and ν are *equivalent measures* if $\mu \ll \nu$ and $\nu \ll \mu$. Show that μ and ν are equivalent if and only if there exists a μ-integrable function f that is strictly positive a.e. with respect to μ such that $d\nu = f \, d\mu$.

Exercise 13.6 Suppose μ and ν are two finite measures such that ν is absolutely continuous with respect to μ. Let $\rho = \mu + \nu$. Note that $\mu(A) \leq \rho(A)$ and $\nu(A) \leq \rho(A)$ for each measurable A. In particular, $\mu \ll \rho$ and $\nu \ll \rho$. Prove that if $f = d\mu/d\rho$ and $g = d\nu/d\rho$, then f is strictly positive for almost every x with respect to μ, $f + g = 1$, and $d\nu = (g/f)\, d\mu$.

Exercise 13.7 If μ is a signed measure on $(X, \mathcal{A})$ and $|\mu|$ is the total variation measure, prove that there exists a real-valued function f that is measurable with respect to $\mathcal{A}$ such that $|f| = 1$ a.e. with respect to μ and $d\mu = f\, d|\mu|$.

Exercise 13.8 Suppose $\nu \ll \mu$ and $\rho \ll \nu$. Prove that $\rho \ll \mu$ and

$$\frac{d\rho}{d\mu} = \frac{d\rho}{d\nu} \cdot \frac{d\nu}{d\mu}.$$

Exercise 13.9 Suppose λ_n is a sequence of positive measures on a measurable space $(X, \mathcal{A})$ with $\sup_n \lambda_n(X) < \infty$ and μ is another finite positive measure on $(X, \mathcal{A})$. Suppose $\lambda_n = f_n\, d\mu + \nu_n$ is the Lebesgue decomposition of λ_n; in particular, $\nu_n \perp \mu$. If $\lambda = \sum_{n=1}^{\infty} \lambda_n$ is a finite measure, show that

$$\lambda = \left(\sum_{n=1}^{\infty} f_n \right) d\mu + \sum_{n=1}^{\infty} \nu_n$$

is the Lebesgue decomposition of λ.

Exercise 13.10 The point of this exercise is to demonstrate that the Radon-Nikodym derivative can depend on the σ-algebra.

Suppose X is a set and $\mathcal{E} \subset \mathcal{F}$ are two σ-algebras of subsets of X. Let μ, ν be two finite positive measures on $(X, \mathcal{F})$ and suppose $\nu \ll \mu$. Let $\overline{\mu}$ be the restriction of μ to $(X, \mathcal{E})$ and $\overline{\nu}$ the restriction of ν to $\mathcal{E}$. Find an example of the above framework where $d\overline{\nu}/d\overline{\mu} \neq d\nu/d\mu$, that is, where the Radon-Nikodym derivative of $\overline{\nu}$ with respect to $\overline{\mu}$ (in terms of $\mathcal{E}$) is not the same as the Radon-Nikodym derivative of ν with respect to μ (in terms of $\mathcal{F}$).

Exercise 13.11 Let $(X, \mathcal{F}, \mu)$ be a measure space, and suppose $\mathcal{E}$ is a sub-σ-algebra of $\mathcal{F}$, that is, $\mathcal{E}$ is itself a σ-algebra and $\mathcal{E} \subset \mathcal{F}$. Suppose f is a non-negative integrable function that is measurable with respect to $\mathcal{F}$. Define $\nu(A) = \int_A f \, d\mu$ for $A \in \mathcal{E}$ and let $\bar{\mu}$ be the restriction of μ to $\mathcal{E}$.

(1) Prove that $\nu \ll \bar{\mu}$.

(2) Since ν and $\bar{\mu}$ are measures on $\mathcal{E}$, then $g = d\nu/d\bar{\mu}$ is measurable with respect to $\mathcal{E}$. Prove that

$$\int_A g \, d\mu = \int_A f \, d\mu \tag{13.3}$$

whenever $A \in \mathcal{E}$. g is called the *conditional expectation* of f with respect to $\mathcal{E}$ and we write $g = \mathbb{E}\left[f \mid \mathcal{E}\right]$. If f is integrable and real-valued but not necessarily non-negative, we define

$$\mathbb{E}\left[f \mid \mathcal{E}\right] = \mathbb{E}\left[f^+ \mid \mathcal{E}\right] - \mathbb{E}\left[f^- \mid \mathcal{E}\right].$$

(3) Show that $f = g$ if and only if f is measurable with respect to $\mathcal{E}$.

(4) Prove that if h is $\mathcal{E}$ measurable and $\int_A h \, d\mu = \int_A f \, d\mu$ for all $A \in \mathcal{E}$, then $h = g$ a.e. with respect to μ.

Chapter 14

Differentiation

In this chapter we want to look at when a function from $\mathbb{R}$ to $\mathbb{R}$ is differentiable and when the fundamental theorem of calculus holds. Briefly, our results are the following.

(1) The derivative of $\int_a^x f(y)\,dy$ is equal to f a.e. if f is integrable (Theorem 14.5);

(2) Functions of bounded variation, in particular monotone functions, are differentiable (Theorem 14.8);

(3) $\int_a^b f'(y)\,dy = f(b) - f(a)$ if f is absolutely continuous (Theorem 14.15).

Our approach uses what are known as maximal functions and uses the Radon-Nikodym theorem and the Lebesgue decomposition theorem. However, some students and instructors prefer a more elementary proof of the results on differentiation. In Sections 14.5, 14.6, and 14.7 we give an alternative approach that avoids the use of the Radon-Nikodym theorem and Lebesgue decomposition theorem.

The definition of derivative is the same as in elementary calculus. A function f is *differentiable* at x if

$$\lim_{h \to 0} \frac{f(x+h) - f(x)}{h}$$

exists, and the limit is called the *derivative* of f at x and is denoted $f'(x)$. If $f : [a,b] \to \mathbb{R}$, we say f is differentiable on $[a,b]$ if the

107

derivative exists for each $x \in (a, b)$ and both

$$\lim_{h \to 0+} \frac{f(a + h) - f(a)}{h} \qquad \text{and} \qquad \lim_{h \to 0-} \frac{f(b + h) - f(b)}{h}$$

exist.

14.1   Maximal functions

In this section we consider real-valued functions on $\mathbb{R}^n$. Let $B(x, r)$ be the open ball with center x and radius r.

The following is an example of what is known as a *covering lemma*. We use m for Lebesgue measure on $\mathbb{R}^n$ throughout this section.

Proposition 14.1 *Suppose $E \subset \mathbb{R}^n$ is covered by a collection of balls $\{B_\alpha\}$ and there exists a positive real number R such that the diameter of each B_α is bounded by R. Then there exists a disjoint sequence $B_1, B_2, \ldots$ of elements of $\{B_\alpha\}$ such that*

$$m(E) \leq 5^n \sum_k m(B_k).$$

Proof. Let $d(B_\alpha)$ be the diameter of B_α. Choose B_1 such that

$$d(B_1) \geq \tfrac{1}{2} \sup_\alpha d(B_\alpha).$$

Once $B_1, \ldots, B_k$ are chosen, choose B_{k+1} disjoint from $B_1, \ldots, B_k$ such that

$$d(B_{k+1}) \geq \tfrac{1}{2} \sup\{d(B_\alpha) : B_\alpha \text{ is disjoint from } B_1, \ldots, B_k\}.$$

The procedure might terminate after a finite number of steps or it might not.

If $\sum_k m(B_k) = \infty$, we have our result. Suppose $\sum_k m(B_k) < \infty$. Let B_k^* be a ball with the same center as B_k but 5 times the radius. We claim $E \subset \cup_k B_k^*$. Once we have this,

$$m(E) \leq m(\cup_k B_k^*) \leq \sum_k m(B_k^*) = 5^n \sum_k m(B_k).$$

To show $E \subset \cup_k B_k^*$, it suffices to show each $B_\alpha \subset \cup_k B_k^*$, since $\{B_\alpha\}$ is a cover of E. Fix α. If B_α is one of the B_k, we are done.

If $\sum_k m(B_k) < \infty$, then $d(B_k) \to 0$. Let k be the smallest integer such that $d(B_{k+1}) < \frac{1}{2}d(B_\alpha)$. B_α must intersect one of $B_1, \ldots, B_k$, or else we would have chosen it instead of B_{k+1}. Therefore B_α intersects B_{j_0} for some $j_0 \leq k$. We know $\frac{1}{2}d(B_\alpha) \leq d(B_{j_0})$, and some simple geometry shows that $B_\alpha \subset B_{j_0}^*$. In fact, let x_{j_0} be the center of B_{j_0} and y a point in $B_\alpha \cap B_{j_0}$. If $x \in B_\alpha$, then

$$|x - x_{j_0}| \leq |x - y| + |y - x_{j_0}| < d(B_\alpha) + d(B_{j_0})/2 \leq \tfrac{5}{2}d(B_{j_0}),$$

or $x \in B_{j_0}^*$. Therefore $B_\alpha \subset B_{j_0}$, and the proof is complete.   □

We say f is *locally integrable* if $\int_K |f(x)| \, dx$ is finite whenever K is compact. If f is locally integrable, define

$$Mf(x) = \sup_{r>0} \frac{1}{m(B(x,r))} \int_{B(x,r)} |f(y)| \, dy.$$

Note that without the supremum, we are looking at the average of $|f|$ over $B(x,r)$. The function Mf is called the *maximal function* of f.

We now prove a *weak 1-1 inequality*, due to Hardy and Littlewood. It is so named because M does not map integrable functions into integrable functions, but comes close in a certain sense to doing so.

Theorem 14.2 *If f is integrable, then for all $\beta > 0$*

$$m(\{x : Mf(x) > \beta\}) \leq \frac{5^n}{\beta} \int |f(x)| \, dx.$$

Proof. Fix β and let $E_\beta = \{x : Mf(x) > \beta\}$. If $x \in E_\beta$, then there exists a ball B_x centered at x such that $\int_{B_x} |f(x)| \, dx > \beta m(B_x)$ by the definition of $Mf(x)$. Then

$$m(B_x) \leq \frac{\int |f|}{\beta},$$

so $\{B_x\}$ is a cover of E_β by balls whose diameters are bounded by some number independent of x. Extract a disjoint sequence

$B_1, B_2, \ldots$ such that $m(E_\beta) \le 5^n \sum_k m(B_k)$. Then

$$m(E_\beta) \le 5^n \sum_k m(B_k) \le \frac{5^n}{\beta} \sum_k \int_{B_k} |f|$$
$$= \frac{5^n}{\beta} \int_{\cup_k B_k} |f| \le \frac{5^n}{\beta} \int |f|,$$

as desired.   □

If we look at the function $f(x) = \chi_B$, where B is the unit ball, note that $Mf(x)$ is approximately a constant times $|x|^{-n}$ for x large, so Mf is not integrable. Hence M does not map the class of integrable functions into the class of integrable functions.

Theorem 14.3 *Let*

$$f_r(x) = \frac{1}{m(B(x,r))} \int_{B(x,r)} f(y) \, dy. \tag{14.1}$$

If f is locally integrable, then $f_r(x) \to f(x)$ a.e. as $r \to 0$.

Proof. It suffices to prove that for each N, $f_r(x) \to f(x)$ for almost every $x \in B(0, N)$. Fix N. We may suppose without loss of generality that f is 0 outside of $B(0, 2N)$, and thus we may suppose f is integrable.

Fix $\beta > 0$. Let $\varepsilon > 0$. Using Theorem 8.4, take g continuous with compact support such that $\int |f - g| \, dm < \varepsilon$. If g_r is defined analogously to f_r using (14.1),

$$|g_r(x) - g(x)| = \left| \frac{1}{m(B(x,r))} \int_{B(x,r)} [g(y) - g(x)] \, dy \right| \tag{14.2}$$
$$\le \frac{1}{m(B(x,r))} \int_{B(x,r)} |g(y) - g(x)| \, dy \to 0$$

as $r \to 0$ by the continuity of g. We have

$$\limsup_{r \to 0} |f_r(x) - f(x)| \le \limsup_{r \to 0} |f_r(x) - g_r(x)|$$
$$+ \limsup_{r \to 0} |g_r(x) - g(x)|$$
$$+ |g(x) - f(x)|.$$

The second term on the right is 0 by (14.2). We now use Theorem 14.2 and Lemma 10.4 to write

$$m(\{x : \limsup_{r \to 0} |f_r(x) - f(x)| > \beta\})$$

$$\leq m(\{x : \limsup_{r \to 0} |f_r(x) - g_r(x)| > \beta/2\})$$

$$+ m(\{x : |f(x) - g(x)| > \beta/2\})$$

$$\leq m(\{x : M(f - g)(x) > \beta/2\}) + \frac{\int |f - g|}{\beta/2}$$

$$\leq \frac{2(5^n + 1)}{\beta} \int |f - g|$$

$$< \frac{2(5^n + 1)\varepsilon}{\beta},$$

where we use the definition of the maximal function to see that

$$|f_r(x) - g_r(x)| \leq M(f - g)(x)$$

for all r. This is true for every ε, so

$$m(\{x : \limsup_{r \to 0} |f_r(x) - f(x)| > \beta\}) = 0.$$

We apply this with $\beta = 1/j$ for each positive integer j, and we conclude

$$m(\{x : \limsup_{r \to 0} |f_r(x) - f(x)| > 0\})$$

$$\leq \sum_{j=1}^{\infty} m(\{x : \limsup_{r \to 0} |f_r(x) - f(x)| > 1/j\}) = 0.$$

This is the result we seek. $\qquad\square$

We can get a stronger statement:

Theorem 14.4 *For almost every x*

$$\frac{1}{m(B(x,r))} \int_{B(x,r)} |f(y) - f(x)| \, dy \to 0$$

as $r \to 0$.

Proof. For each rational c there exists a set N_c of measure 0 such that

$$\frac{1}{m(B(x,r))} \int_{B(x,r)} |f(y) - c|\, dy \to |f(x) - c|$$

for $x \notin N_c$; we see this by applying Theorem 14.3 to the function $|f(x) - c|$. Let $N = \cup_{c \in \mathbb{Q}} N_c$ and suppose $x \notin N$. Let $\varepsilon > 0$ and choose c rational such that $|f(x) - c| < \varepsilon$. Then

$$\frac{1}{m(B(x,r))} \int_{B(x,r)} |f(y) - f(x)|\, dy$$

$$\leq \frac{1}{m(B(x,r))} \int_{B(x,r)} |f(y) - c|\, dy$$

$$+ \frac{1}{m(B(x,r))} \int_{B(x,r)} |f(x) - c|\, dy$$

$$= \frac{1}{m(B(x,r))} \int_{B(x,r)} |f(y) - c|\, dy + |f(x) - c|$$

and hence

$$\limsup_{r \to 0} \frac{1}{m(B(x,r))} \int_{B(x,r)} |f(y) - f(x)|\, dy \leq 2|f(x) - c| < 2\varepsilon.$$

Since ε is arbitrary, our result follows.   □

If we apply the above to the function $f = \chi_E$, then for almost all $x \in E$

$$\frac{1}{m(B(x,r))} \int_{B(x,r)} \chi_E \to 1,$$

or

$$\frac{m(E \cap B(x,r))}{m(B(x,r))} \to 1,$$

and similarly, for almost all $x \notin E$, the ratio tends to 0. The points where the ratio tends to 1 are called *points of density* for E.

14.2   Antiderivatives

For the remainder of the chapter we consider real-valued functions on the real line $\mathbb{R}$. We can use the results on maximal functions to show that the derivative of the antiderivative of an integrable function is the function itself. A ball $B(x, h)$ in $\mathbb{R}$ is merely the interval $(x - h, x + h)$. We use m for Lebesgue measure throughout.

Define the *indefinite integral* or *antiderivative* of an integrable function f by

$$F(x) = \int_a^x f(t) \, dt.$$

Recall by Exercise 7.6 that F is continuous.

Theorem 14.5 *Suppose $f : \mathbb{R} \to \mathbb{R}$ is integrable and $a \in \mathbb{R}$. Define*

$$F(x) = \int_a^x f(y) \, dy.$$

Then F is differentiable almost everywhere and $F'(x) = f(x)$ a.e.

Proof. If $h > 0$, we have

$$F(x + h) - F(x) = \int_x^{x+h} f(y) \, dy,$$

so

$$\left| \frac{F(x + h) - F(x)}{h} - f(x) \right| = \frac{1}{h} \left| \int_x^{x+h} (f(y) - f(x)) \, dy \right|$$

$$\leq 2 \frac{1}{m(B(x, h))} \int_{x-h}^{x+h} |f(y) - f(x)| \, dy.$$

By Theorem 14.4, the right hand side goes to 0 as $h \to 0$ for almost every x, and we conclude the right hand derivative of F exists and equals f for almost every x. The left hand derivative is handled similarly. □

14.3 Bounded variation

In this section we show that functions of bounded variation are differentiable almost everywhere. We start with right continuous increasing functions.

Lemma 14.6 *Suppose $H : \mathbb{R} \to \mathbb{R}$ is increasing, right continuous, and constant for $x \geq 1$ and $x \leq 0$. Let λ be the Lebesgue-Stieltjes measure defined using the function H and suppose λ and m are mutually singular. Then*

$$\lim_{r \to 0} \frac{\lambda(B(x,r))}{m(B(x,r))} = 0$$

for almost every x.

Proof. This is clear if $x < 0$ or $x > 1$. Since $\lambda \perp m$, there exist measurable sets E and F such that $\lambda(F) = 0$, $m(E) = 0$, and $F = E^c$. Let $\varepsilon > 0$.

Step 1. The first step of the proof is to find a bounded open set G such that $F \subset G$ and $\lambda(G) < \varepsilon$. By the definition of Lebesgue-Stieltjes measure, there exist $a_i < b_i$ such that $F \subset \cup_{i=1}^{\infty}(a_i, b_i]$ and

$$\sum_{i=1}^{\infty}[H(b_i) - H(a_i)] < \varepsilon/2.$$

Since H is right continuous, for each i there exists $b_i' > b_i$ such that

$$H(b_i') - H(b_i) < \varepsilon/2^{i+1}.$$

If $G' = \cup_{i=1}^{\infty}(a_i, b_i')$, then G' is open, G' contains F, and

$$\lambda(G') \leq \sum_{i=1}^{\infty} \lambda((a_i, b_i')) \leq \sum_{i=1}^{\infty} \lambda((a_i, b_i']) = \sum_{i=1}^{\infty}[H(b_i') - H(a_i)] < \varepsilon.$$

Since H is constant on $(-\infty, 0]$ and $[1, \infty)$, we can take G to be the set $G = G' \cap (-1, 2)$.

Step 2. If $\beta > 0$, let

$$A_\beta = \left\{ x \in F \cap [0,1] : \limsup_{r \to 0} \frac{\lambda((x,r))}{m(B(x,r))} > \beta \right\}.$$

The second step is to show that $m(A_\beta) = 0$. If $x \in A_\beta$, then $x \in F \subset G$, and there exists an open ball B_x centered at x and contained in G such that $\lambda(B_x)/m(B_x) > \beta$. Use Proposition 14.1 to find a disjoint subsequence $B_1, B_2, \ldots$ such that

$$m(A_\beta) \leq 5 \sum_{i=1}^{\infty} m(B_i).$$

Then

$$m(A_\beta) \leq 5 \sum_{i=1}^{\infty} m(B_i) \leq \frac{5}{\beta} \sum_{i=1}^{\infty} \lambda(B_i) \leq \frac{5}{\beta} \lambda(G) \leq \frac{5}{\beta} \varepsilon.$$

Since ε is arbitrary, and our construction of G did not depend on β, then $m(A_\beta) = 0$.

Since $m(A_{1/k}) = 0$ for each k, then

$$m(\{x \in F \cap [0,1] : \limsup_{r \to 0} \lambda(B(x,r))/m(B(x,r)) > 0\}) = 0.$$

Since $m(E) = 0$, this completes the proof. □

Proposition 14.7 *Let $F : \mathbb{R} \to \mathbb{R}$ be an increasing and right continuous function. Then F' exists a.e. Moreover, F' is locally integrable and for every $a < b$, $\int_a^b F'(x)\,dx \leq F(b) - F(a)$.*

Proof. We will show F is differentiable a.e. on $[0,1]$. Once we have that, the same argument can be used to show that F is differentiable a.e. on $[-N, N]$ for each N, and that proves that F is differentiable a.e. on $\mathbb{R}$. If we redefine F so that $F(x) = \lim_{y \to 0+} F(y)$ if $x \leq 0$ and $F(x) = F(1)$ if $x > 1$, then F is still right continuous and increasing, and we have not affected the differentiability of F on $[0,1]$ except possibly at the points 0 and 1.

Let ν be the Lebesgue-Stieltjes measure defined in terms of F. By the Lebesgue decomposition theorem, we can write $\nu = \lambda + \rho$, where $\lambda \perp m$ and $\rho \ll m$. Note

$$\rho([0,1]) \leq \nu([0,1]) = F(1) - F(0).$$

By the Radon-Nikodym theorem there exists a non-negative integrable function f such that $\rho(A) = \int_A f\,dm$ for each measurable A.

Let

$$H(x) = \lambda((0,x]) = \nu((0,x]) - \rho((0,x]) = F(x) - F(0) - \int_0^x f(y)\,dy.$$

By Exercise 7.6, the function $x \to \int_0^x f(y)\,dy$ is continuous, so H is right continuous, increasing, and λ is the Lebesgue-Stieltjes measure defined in terms of H. By Lemma 14.6,

$$\limsup_{h \to 0+} \frac{H(x+h) - H(x)}{h} \leq \limsup_{h \to 0+} \frac{H(x+h) - H(x-h)}{h}$$

$$= \limsup_{h \to 0+} \frac{\lambda((x-h, x+h])}{h}$$

$$\leq 4 \limsup_{h \to 0+} \frac{\lambda(B(x, 2h))}{4h} = 0$$

for almost every x. The same is true for the left hand derivative, so H' exists and equals 0 for almost every x. We saw by Theorem 14.5 that the function $x \to \int_0^x f(y) \, dy$ is differentiable almost everywhere, and we conclude that F is differentiable a.e.

We have shown that $F' = f$ a.e. If $a < b$,

$$\int_a^b F'(x) \, dx = \int_a^b f(x) \, dx = \rho((a, b]) \leq \nu((a, b]) = F(b) - F(a).$$

This completes the proof. $\square$

Here is the main theorem on the differentiation of increasing functions.

Theorem 14.8 *If $F : \mathbb{R} \to \mathbb{R}$ is increasing, then F' exists a.e. and*

$$\int_a^b F'(x) \, dx \leq F(b) - F(a) \tag{14.3}$$

whenever $a < b$.

Proof. Let $G(x) = \lim_{y \to x+} F(y)$. Since F is increasing, there are at most countably many values of x where F is not continuous (Proposition 1.6), so $F(x) = G(x)$ a.e. Since G is increasing and right continuous, G is differentiable a.e. by Proposition 14.7. We will show that if x is a point where G is differentiable and at the same time $F(x) = G(x)$, then $F'(x)$ exists and is equal to $G'(x)$.

Let x be such a point, let $L = G'(x)$ and let $\varepsilon > 0$. Because F and G are increasing, for any $h > 0$ there exists a point x_h strictly between $x + h$ and $x + (1 + \varepsilon)h$ where F and G agree, and so

$$F(x+h) \leq F(x_h) = G(x_h) \leq G(x + (1 + \varepsilon)h).$$

Then

$$\limsup_{h\to 0+}\frac{F(x+h)-F(x)}{h} \le \limsup_{h\to 0+}\frac{G(x+(1+\varepsilon)h)-G(x)}{h}$$

$$= (1+\varepsilon)\limsup_{h\to 0+}\frac{G(x+(1+\varepsilon)h)-G(x)}{(1+\varepsilon)h}$$

$$= (1+\varepsilon)L.$$

Similarly, $\liminf_{h\to 0+}[F(x+h)-F(x)]/h \ge (1-\varepsilon)L$. Since ε is arbitrary, the right hand derivative of F exists at x and is equal to L. That the left hand derivative equals L is proved similarly.

Since $F' = G'$ a.e., then F' is locally integrable. If $a < b$, take $a_n \downarrow a$ and $b_n \uparrow b$ such that F and G agree on a_n and b_n. Then using Proposition 14.7,

$$F(b) - F(a) \ge F(b_n) - F(a_n)$$

$$= G(b_n) - G(a_n) \ge \int_{a_n}^{b_n} G'(x)\,dx$$

$$= \int_{a_n}^{b_n} F'(x)\,dx.$$

Now let $n \to \infty$ and use the monotone convergence theorem. □

Remark 14.9 Note that if F is the Cantor-Lebesgue function, then $F'(x) = 0$ a.e., in fact at every point of C^c, where C is the Cantor set. Thus

$$1 = F(1) - F(0) > 0 = \int_0^1 F'(x)\,dx,$$

and we do not in general have equality in (14.3).

A real-valued function f is of *bounded variation* on $[a,b]$ if

$$\sup\left\{\sum_{i=1}^k |f(x_i) - f(x_{i-1})|\right\}$$

is finite, where the supremum is over all partitions $a = x_0 < x_1 < \cdots < x_k = b$ of $[a,b]$.

Lemma 14.10 *If f is of bounded variation on $[a, b]$, then f can be written as $f = f_1 - f_2$, the difference of two increasing functions on $[a, b]$.*

Proof. Define

$$f_1(y) = \sup\left\{\sum_{i=1}^{k}[f(x_i) - f(x_{i-1})]^+\right\}$$

and

$$f_2(y) = \sup\left\{\sum_{i=1}^{k}[f(x_i) - f(x_{i-1})]^-\right\},$$

where the supremum is over all partitions $a = x_0 < x_1 < \cdots < x_k = y$ for $y \in [a, b]$. f_1 and f_2 are measurable since they are both increasing. Since

$$\sum_{i=1}^{k}[f(x_i) - f(x_{i-1})]^+ = \sum_{i=1}^{k}[f(x_i) - f(x_{i-1})]^- + f(y) - f(a),$$

taking the supremum over all partitions of $[a, y]$ yields

$$f_1(y) = f_2(y) + f(y) - f(a).$$

Clearly f_1 and f_2 are increasing in y, and the result follows by solving for $f(y)$.   □

Using this lemma and Theorem 14.8, we see that functions of bounded variation are differentiable a.e. Note that the converse is not true: the function $\sin(1/x)$ defined on $(0, 1]$ is differentiable everywhere, but is not of bounded variation.

Remark 14.11 If we write a function f of bounded variation as the difference of two increasing functions f_1 and f_2 as in Lemma 14.10, then the quantity $(f_1(b) + f_2(b)) - (f_1(a) + f_2(a))$ is called the *total variation* of f on the interval $[a, b]$. We make the observation that if f is of bounded variation on the interval $[a, b]$ and on the interval $[b, c]$, then it is of bounded variation on the interval $[a, c]$.

Remark 14.12 If f is an increasing function $[a, b]$ that is continuous from the right, we can write $f = f_1 - f_2$, where f_1 is continuous and

$$f_2(x) = \sum_{a < t \leq x}(f(t) - f(t-)).$$

Only countably many of the summands in the definition of f_2 can be non-zero in view of Proposition 1.6, each summand is non-negative, and the sum is finite because it is bounded by $f(x) - f(a)$. Using Lemma 14.10, we can similarly decompose any function of bounded variation that is continuous from the right.

14.4 Absolutely continuous functions

A real-valued function f is *absolutely continuous* on $[a, b]$ if given ε there exists δ such that $\sum_{i=1}^{k} |f(b_i) - f(a_i)| < \varepsilon$ whenever $\{(a_i, b_i)\}$ is a finite collection of disjoint intervals with $\sum_{i=1}^{k} |b_i - a_i| < \delta$.

It is easy to see that absolutely continuous functions are continuous and that the Cantor-Lebesgue function is not absolutely continuous.

Lemma 14.13 *If f is absolutely continuous, then it is of bounded variation.*

Proof. By the definition of absolutely continuous function with $\varepsilon = 1$, there exists δ such that $\sum_{i=1}^{k} |f(b_i) - f(a_i)| < 1$ whenever $\sum_{i=1}^{k} (b_i - a_i) \leq \delta$ and the (a_i, b_i) are disjoint open intervals. Hence for each j the total variation of f on $[a + j\delta, a + (j + 1)\delta]$ is less than or equal to 1. Using Remark 14.11, we see the total variation of f on $[a, b]$ is finite. $\square$

Lemma 14.14 *Suppose f is of bounded variation and we decompose f as $f = f_1 - f_2$ as in Lemma 14.10, where f_1 and f_2 are increasing functions. If f is absolutely continuous, then so are f_1 and f_2.*

Proof. Given ε there exists δ such that $\sum_{i=1}^{k} |f(b_i) - f(a_i)| < \varepsilon$ whenever $\sum_{i=1}^{k} (b_i - a_i) \leq \delta$ and the (a_i, b_i) are disjoint open intervals. Partitioning each interval (a_i, b_i) into subintervals with $a_i = s_{i0} < s_{i1} < \cdots < s_{iJ_i} = b_i$, then

$$\sum_{i=1}^{k} \sum_{j=0}^{J_i-1} (s_{i,j+1} - s_{ij}) = \sum_{i=1}^{k} (b_i - a_i) \leq \delta.$$

Hence

$$\sum_{i=1}^{k} \sum_{j=0}^{J_i-1} |f(s_{i,j+1}) - f(s_{ij})| \le \varepsilon.$$

Taking the supremum over all such partitions,

$$\sum_{i=1}^{k} |(f_1 + f_2)(b_i) - (f_1 + f_2)(a_i)| \le \varepsilon,$$

and our conclusion follows.   □

Here is the main theorem on absolutely continuous functions.

Theorem 14.15 *If F is absolutely continuous, then F' exists a.e., and*

$$\int_a^b F'(x)\,dx = F(b) - F(a).$$

Proof. By Lemma 14.14 it suffices to suppose F is increasing and absolutely continuous. Let ν be the Lebesgue-Stieltjes measure defined in terms of F. Since F is continuous, $F(d) - F(c) = \nu((c,d))$.

Taking a limit as $k \to \infty$, we see that given ε there exists δ such that $\sum_{i=1}^{\infty} |F(b_i) - F(a_i)| \le \varepsilon$ whenever $\{(a_i, b_i)\}$ is a collection of disjoint intervals with $\sum_{i=1}^{\infty}(b_i - a_i) < \delta$. Since any open set G can be written as the union of disjoint intervals $\{(a_i, b_i)\}$, this can be rephrased as saying, given ε there exists δ such that

$$\nu(G) = \sum_{i=1}^{\infty} \nu((a_i, b_i)) = \sum_{i=1}^{\infty}(F(b_i) - F(a_i)) \le \varepsilon$$

whenever G is open and $m(G) < \delta$. If $m(A) < \delta$ and A is Borel measurable, then there exists an open set G containing A such that $m(G) < \delta$, and then $\nu(A) \le \nu(G) \le \varepsilon$. We conclude that $\nu \ll m$.

Hence there exists a non-negative integrable function f such that

$$\nu(A) = \int_A f\,dm$$

for all Borel measurable sets A. In particular, for each $x \in [a, b]$,

$$F(x) - F(a) = \nu((a, x)) = \int_a^x f(y)\,dy.$$

By Theorem 14.5, F' exists and is equal to f a.e. Setting $x = b$ we obtain

$$F(b) - F(a) = \int_a^b F'(y)\,dy$$

as desired. □

14.5 Approach 2 – differentiability

In this and the following two sections we give an alternative approach to Theorems 14.5, 14.8, and 14.15 that avoids the use of the Radon-Nikodym theorem, and instead proceeds via a covering lemma due to Vitali.

Let m be Lebesgue measure. Let $E \subset \mathbb{R}$ be a measurable set and let $\mathcal{G}$ be a collection of intervals. We say $\mathcal{G}$ is a *Vitali cover* of E if for each $x \in E$ and each $\varepsilon > 0$ there exists an interval $G \in \mathcal{G}$ containing x whose length is less than ε.

The following is known as the *Vitali covering lemma*, and is a refinement of Proposition 14.1.

Lemma 14.16 *Suppose E has finite measure and let $\mathcal{G}$ be a Vitali cover of E. Given $\varepsilon > 0$ there exists a finite subcollection of disjoint intervals $I_1, \ldots, I_n \in \mathcal{G}$ such that $m(E - \cup_{i=1}^n I_n) < \varepsilon$.*

Proof. We may replace each interval in $\mathcal{G}$ by a closed one, since the set of endpoints of a finite subcollection will have measure 0.

Let G be an open set of finite measure containing E. Since $\mathcal{G}$ is a Vitali cover, we may suppose without loss of generality that each set of $\mathcal{G}$ is contained in G. Let

$$a_0 = \sup\{m(I) : I \in \mathcal{G}\}.$$

Let I_1 be any element of $\mathcal{G}$ with $m(I_1) \geq a_0/2$. Let

$$a_1 = \sup\{m(I) : I \in \mathcal{G}, I \text{ disjoint from } I_1\},$$

and choose $I_2 \in \mathcal{G}$ disjoint from I_1 such that $m(I_2) \geq a_1/2$. Continue in this way, choosing I_{n+1} disjoint from $I_1, \ldots, I_n$ and in $\mathcal{G}$

with length at least one half as large as any other such interval in $\mathcal{G}$ that is disjoint from $I_1, \ldots, I_n$.

If the process stops at some finite stage, we are done. If not, we generate a sequence of disjoint intervals $I_1, I_2, \ldots$ Since they are disjoint and all contained in G, then $\sum_{i=1}^{\infty} m(I_i) \leq m(G) < \infty$. Therefore there exists N such that $\sum_{i=N+1}^{\infty} m(I_i) < \varepsilon/5$.

Let $R = E - \cup_{i=1}^{N} I_i$. We claim $m(R) < \varepsilon$. Let I_n^* be the interval with the same center as I_n but five times the length. Let $x \in R$. Since we supposed each interval in $\mathcal{G}$ was to be modified so as to include its endpoints, then $\cup_{i=1}^{n} I_i$ is closed. Hence there exists an interval $I \in \mathcal{G}$ containing x with I disjoint from $I_1, \ldots, I_N$. Since $\sum_n m(I_n) < \infty$, then $\sum_n a_n \leq 2 \sum_n m(I_n) < \infty$, and $a_n \to 0$. Hence I must either be one of the I_n for some $n > N$ or at least intersect it, for otherwise we would have chosen I at some stage. Let n be the smallest integer such that I intersects I_n; note $n > N$. We have $m(I) \leq a_{n-1} \leq 2m(I_n)$. Since x is in I and I intersects I_n, the distance from x to the midpoint of I_n is at most $m(I) + m(I_n)/2 \leq (5/2)m(I_n)$. Therefore $x \in I_n^*$.

Thus we have $R \subset \cup_{i=N+1}^{\infty} I_i^*$, so

$$m(R) \leq \sum_{i=N+1}^{\infty} m(I_i^*) = 5 \sum_{i=N+1}^{\infty} m(I_i) < \varepsilon.$$

This completes the proof.   □

Given a real-valued function f, we define the *derivates* of f at x by

$$D^+ f(x) = \limsup_{h \to 0+} \frac{f(x+h) - f(x)}{h},$$

$$D^- f(x) = \limsup_{h \to 0-} \frac{f(x+h) - f(x)}{h},$$

$$D_+ f(x) = \liminf_{h \to 0+} \frac{f(x+h) - f(x)}{h},$$

$$D_- f(x) = \liminf_{h \to 0-} \frac{f(x+h) - f(x)}{h}.$$

If all the derivates are equal, then f is differentiable at x and $f'(x)$ is the common value.

Theorem 14.17 *Suppose f is increasing on $[a, b]$. Then f is differentiable almost everywhere, f' is integrable, and*

$$\int_a^b f'(x)\,dx \le f(b) - f(a). \tag{14.4}$$

Proof. We will show that the set where any two derivates are unequal has measure zero. Let us consider the set

$$E = \{x : D^+ f(x) > D_- f(x)\},$$

the other sets being similar. Let

$$E_{uv} = \{x : D^+ f(x) > v > u > D_- f(x)\}.$$

If we show $m(E_{uv}) = 0$, then observing that $E \subset \cup_{u,v \in \mathbb{Q}, u < v} E_{uv}$ will show that $m(E) = 0$.

Let $s = m(E_{uv})$, let $\varepsilon > 0$, and choose an open set G such that $E_{uv} \subset G$ and $m(G) < s + \varepsilon$. For each $x \in E_{uv}$ there exists an arbitrarily small interval $[x - h, x]$ contained in G such that $f(x) - f(x - h) < uh$. Use Lemma 14.16 to choose $I_1, \ldots, I_N$ which are disjoint and whose interiors cover a subset A of E_{uv} of measure greater than $s - \varepsilon$. Write $I_n = [x_n - h_n, x_n]$. Thus

$$A = E_{uv} \cap (\cap_{n=1}^N (x_n - h_n, x_n)).$$

Taking a sum,

$$\sum_{n=1}^N [f(x_n) - f(x_n - h_n)] < u \sum_{n=1}^n h_n < um(G) < u(s + \varepsilon).$$

Each point y in the subset A is the left endpoint of an arbitrarily small interval $(y, y + k)$ that is contained in some I_n and for which $f(y + k) - f(y) > vk$. Using Lemma 14.16 again, we pick out a finite collection $J_1, \ldots, J_M$ whose union contains a subset of A of measure larger than $s - 2\varepsilon$. Summing over these intervals yields

$$\sum_{i=1}^M [f(y_i + k_i) - f(y_i)] > v \sum_{i=1}^M k_i > v(s - 2\varepsilon).$$

Each interval J_i is contained in some interval I_n, and if we sum over those i for which $J_i \subset I_n$ we find

$$\sum_{\{i : J_i \subset I_n\}} [f(y_i + k_i) - f(y_i)] \le f(x_n) - f(x_n - h_n),$$

since f is increasing. Thus

$$\sum_{n=1}^{N}[f(x_n) - f(x_n - h_n)] \geq \sum_{i=1}^{M}[f(y_i + k_i) - f(y_i)],$$

and so $u(s + \varepsilon) > v(s - 2\varepsilon)$. This is true for each ε, so $us \geq vs$. Since $v > u$, this implies $s = 0$.

We prove similarly that $D^+ f = D_+ f$ a.e. and $D^- f = D_- f$ a.e. and conclude that

$$g(x) = \lim_{h \to 0} \frac{f(x + h) - f(x)}{h}$$

is defined almost everywhere and that f is differentiable wherever g is finite.

Define $f(x) = f(b)$ if $x \geq b$. Let

$$g_n(x) = n[f(x + 1/n) - f(x)].$$

Then $g_n(x) \to g(x)$ for almost all x, and so g is measurable. Since f is increasing, $g_n \geq 0$. By Fatou's lemma and the fact that f is increasing,

$$\int_a^b g \leq \liminf_n \int_a^b g_n$$

$$= \liminf_n n \int_a^b [f(x + 1/n) - f(x)]$$

$$= \liminf_n \left[n \int_b^{b+1/n} f - n \int_a^{a+1/n} f \right]$$

$$= \liminf_n \left[f(b) - n \int_a^{a+1/n} f \right]$$

$$\leq f(b) - f(a).$$

We used a change of variables in the second equality. This shows that g is integrable and hence finite almost everywhere.   □

We refer the reader to Lemma 14.10 to see that a function of bounded variation is differentiable almost everywhere.

14.6 Approach 2 – antiderivatives

Continuing the alternative approach, we look at when the derivative of

$$F(x) = \int_a^x f(t)\,dt \qquad (14.5)$$

is equal to $f(x)$ a.e.

Theorem 14.18 *If f is integrable and F is defined by (14.5), then $F'(x) = f(x)$ for almost every x.*

Proof. By writing $f = f^+ - f^-$, it suffices to consider the case where f is non-negative. In this case F is increasing, and so F' exists a.e. By Exercise 7.6 we know F is continuous.

Suppose for the moment that f is bounded by K. Then

$$\left| \frac{F(x + 1/n) - F(x)}{1/n} \right| = \left| n \int_x^{x+1/n} f(t)\,dt \right|$$

is also bounded by K. By the dominated convergence theorem,

$$\int_a^c F'(x)\,dx = \lim_n n \int_a^c [F(x + 1/n) - F(x)]\,dx$$

$$= \lim_n \left[n \int_c^{c+1/n} F(x)\,dx - n \int_a^{a+c} F(x)\,dx \right]$$

$$= F(c) - F(a) = \int_a^c f(x)\,dx.$$

We used a change of variables for the second equality and the fact that F is continuous for the third equality. Therefore

$$\int_a^c [F'(x) - f(x)]\,dx = 0$$

for all c, which implies $F' = f$ a.e. by Corollary 8.3.

We continue to assume f is non-negative but now allow f to be unbounded. Since $f - (f \wedge K) \geq 0$, then

$$G_K(x) = \int_a^x [f - (f \wedge K)]\,dx$$

is increasing, and hence has a derivative almost everywhere. Moreover,

$$G'_K(x) = \lim_{n \to \infty} \frac{G_K(x + 1/n) - G_K(x)}{1/n} \geq 0$$

at points x where G' exists since G is increasing. By the preceding paragraph, we know the derivative of

$$H_K(x) = \int_a^x (f \wedge K)\, dx$$

is equal to $f \wedge K$ almost everywhere. Therefore

$$F'(x) = G'_K(x) + H'_K(x) \geq (f \wedge K)(x), \qquad \text{a.e.}$$

Since K is arbitrary, $F' \geq f$ a.e., and so

$$\int_a^b F' \geq \int_a^b f = F(b) - F(a).$$

Combining with (14.4) we conclude that $\int_a^b [F' - f] = 0$. Since $F' - f \geq 0$ a.e., this tells us that $F' = f$ a.e. $\square$

14.7   Approach 2 − absolute continuity

Finally, we continue the alternative approach to look at when $\int_a^b F'(y)\, dy = F(b) - F(a)$.

We refer the reader to Lemma 14.13 for the proof that if f is absolutely continuous, then it is of bounded variation.

Lemma 14.19 *If f is absolutely continuous on $[a, b]$ and $f'(x) = 0$ a.e., then f is constant.*

The Cantor-Lebesgue function is an example to show that we need the absolute continuity.

Proof.   Let $c \in [a, b]$, let $E = \{x \in [a, c] : f'(x) = 0\}$, and let $\varepsilon > 0$. Choose δ such that $\sum_{i=1}^{K} |f(b_i) - f(a_i)| < \varepsilon$ whenever $\sum_{i=1}^{K} |b_i - a_i| \leq \delta$ and the (a_i, b_i) are disjoint intervals. For each point $x \in E \cap [a, c)$ there exist arbitrarily small intervals $[x, x+h] \subset$

$[a, c]$ such that $|f(x + h) - f(x)| < \varepsilon h$. By Lemma 14.16 we can find a finite disjoint collection of such intervals that cover all of E except for a set of measure less than δ. We label the intervals $[a_i, b_i]$ so that $a_i < b_i \leq a_{i+1}$. Except for a set of measure less than δ, E is covered by $\cup_i(a_i, b_i)$. This implies that $\cup_i(b_i, a_{i+1})$ has measure less than δ, or $\sum_i |a_{i+1} - b_i| \leq \delta$. By our choice of δ and the definition of absolute continuity,

$$\sum_i |f(a_{i+1}) - f(b_i)| < \varepsilon.$$

On the other hand, by our choice of the intervals (a_i, b_i),

$$\sum_i |f(b_i) - f(a_i)| < \varepsilon \sum_i (b_i - a_i) \leq \varepsilon(c - a).$$

Adding these two inequalities together,

$$|f(c) - f(a)| = \left| \sum_i [f(a_{i+1}) - f(b_i)] + \sum_i [f(b_i) - f(a_i)] \right|$$

$$\leq \varepsilon + \varepsilon(c - a).$$

Since ε is arbitrary, then $f(c) = f(a)$, which implies that f is constant. $\square$

Theorem 14.20 *If F is absolutely continuous, then*

$$F(b) - F(a) = \int_a^b F'(y)\, dy.$$

Proof. Suppose F is absolutely continuous on $[a, b]$. Then F is of bounded variation, so $F = F_1 - F_2$ where F_1 and F_2 are increasing, and F' exists a.e. Since $|F'(x)| \leq F_1'(x) + F_2'(x)$, then

$$\int |F'(x)|\, dx \leq (F_1(b) + F_2(b)) - (F_1(a) - F_2(a)),$$

and hence F' is integrable. If

$$G(x) = \int_a^x F'(t)\, dt,$$

then G is absolutely continuous by Exercise 14.2, and hence $F - G$ is absolutely continuous. Then $(F - G)' = F' - G' = F' - F' = 0$

a.e., using Theorem 14.18 for the second equality. By Lemma 14.19, $F - G$ is constant, and thus $F(x) - G(x) = F(a) - G(a)$. We conclude

$$F(x) = \int_a^x F'(t)\, dt + F(a).$$

If we set $x = b$, we get our result.   □

14.8   Exercises

Exercise 14.1 (1) Show that if f and g are absolutely continuous on an interval $[a, b]$, then the product fg is also.
(2) Prove the *integration by parts* formula:

$$fb)g(b) - f(a)g(a) = \int_a^b f(x)g'(x)\, dx + \int_a^b f'(x)g(x)\, dx.$$

Exercise 14.2 If f is integrable and real-valued, $a \in \mathbb{R}$, and

$$F(x) = \int_a^x f(y)\, dy,$$

prove that F is of bounded variation and is absolutely continuous.

Exercise 14.3 Suppose that f is a real-valued continuous function on $[0, 1]$ and that $\varepsilon > 0$. Prove that there exists a continuous function g such that $g'(x)$ exists and equals 0 for a.e. x and

$$\sup_{x \in [0,1]} |f(x) - g(x)| < \varepsilon.$$

Exercise 14.4 Suppose f is a real-valued continuous function on $[0, 1]$ and f is absolutely continuous on $(a, 1]$ for every $a \in (0, 1)$. Is f necessarily absolutely continuous on $[0, 1]$? If f is also of bounded variation on $[0, 1]$, is f absolutely continuous on $[0, 1]$? If not, give counterexamples.

Exercise 14.5 A real-valued function f is *Lipschitz* with constant M if

$$|f(x) - f(y)| \leq M|x - y|$$

for all $x, y \in \mathbb{R}$. Prove that f is Lipschitz with constant M if and only if f is absolutely continuous and $|f'| \leq M$ a.e.

Exercise 14.6 Suppose F_n is a sequence of increasing non-negative right continuous functions on $[0, 1]$ such that $\sup_n F_n(1) < \infty$. Let $F = \sum_{n=1}^{\infty} F_n$ and suppose $F(1) < \infty$. Prove that

$$F'(x) = \sum_{n=1}^{\infty} F_n'(x)$$

for almost every x.

Exercise 14.7 Suppose f is absolutely continuous on $[0, 1]$ and for $A \subset [0, 1]$ we let $f(A) = \{f(x) : x \in A\}$. Prove that if A has Lebesgue measure 0, then $f(A)$ has Lebesgue measure 0.

Exercise 14.8 If f is real-valued and differentiable at each point of $[0, 1]$, is f necessarily absolutely continuous on $[0, 1]$? If not, find a counterexample.

Exercise 14.9 Find an increasing function f such that $f' = 0$ a.e. but f is not constant on any open interval.

Exercise 14.10 If $f : [a, b] \to \mathbb{R}$ is continuous, let $M(y)$ be the number of points x in $[a, b]$ such that $f(x) = y$. $M(y)$ may be finite or infinite. Prove that M is Borel measurable and $\int M(y)\, dy$ equals the total variation of f on $[a, b]$.

Exercise 14.11 Let $\alpha \in (0, 1)$. Find a Borel subset E of $[-1, 1]$ such that

$$\lim_{r \to 0+} \frac{m(E \cap [-r, r])}{2r} = \alpha.$$

Exercise 14.12 Suppose f is a real-valued continuous function on $[a, b]$ and the derivate $D^+ f$ is non-negative on $[a, b]$. Prove that $f(b) \geq f(a)$. What if instead we have that $D_+ f$ is non-negative on $[a, b]$?

Exercise 14.13 Let

$$f(x) = \int_{-\infty}^{\infty} \frac{e^{-xy^2}}{1 + y^2}\, dy.$$

(1) Find the derivative of f.
(2) Find an ordinary differential equation that f solves. Find the solution to this ordinary differential equation to determine an explicit value for $f(x)$.

Exercise 14.14 Let $(X, \mathcal{A}, \mu)$ be a measure space where $\mu(X) > 0$ and let f be a real-valued integrable function. Define

$$g(x) = \int |f(y) - x| \, \mu(dy)$$

for $x \in \mathbb{R}$.
(1) Prove that g is absolutely continuous.
(2) Prove that $\lim_{x \to \infty} g(x) = \infty$ and $\lim_{x \to -\infty} g(x) = \infty$.
(3) Find $g'(x)$ and prove that $g(x_0) = \inf_{x \in \mathbb{R}} g(x)$ if and only if

$$\mu(\{y : f(y) > x_0\}) = \mu(\{y : f(y) < x_0\}).$$

Exercise 14.15 Suppose $A \subset [0, 1]$ has Lebesgue measure zero. Find an increasing function $f : [0, 1] \to \mathbb{R}$ that is absolutely continuous, but

$$\lim_{h \to 0} \frac{f(x + h) - f(x)}{h} = \infty$$

for each $x \in A$.

Exercise 14.16 Suppose that μ is a measure on the Borel σ-algebra on $[0, 1]$ and for every f that is real-valued and continuously differentiable we have

$$\left| \int f'(x) \, \mu(dx) \right| \leq \left(\int_0^1 f(x)^2 \, dx \right)^{1/2}.$$

(1) Show that μ is absolutely continuous with respect to Lebesgue measure on $[0, 1]$.
(2) If g is the Radon-Nikodym derivative of μ with respect to Lebesgue measure, prove that there exists a constant $c > 0$ such that

$$|g(x) - g(y)| \leq c|x - y|^{1/2}, \qquad x, y \in [0, 1].$$

Exercise 14.17 Let $p > 1$ and $f, g \in L^p(\mathbb{R})$. Define

$$H(t) = \int_{-\infty}^{\infty} |f(x) + tg(x)|^p \, dx$$

for $t \in \mathbb{R}$. Prove that H is a differentiable function and find its derivative.

Chapter 15

L^p spaces

We introduce some spaces of functions, called the L^p spaces. We define the L^p norm of a function, prove completeness of the norm, discuss convolutions, and consider the bounded linear functionals on L^p. We assume throughout this chapter that the measure μ is σ-finite.

15.1 Norms

Let $(X, \mathcal{A}, \mu)$ be a σ-finite measure space. For $1 \leq p < \infty$, define the L^p norm of f by

$$\|f\|_p = \left(\int |f(x)|^p \, d\mu \right)^{1/p}. \tag{15.1}$$

For $p = \infty$, define the L^∞ norm of f by

$$\|f\|_\infty = \inf\{M : \mu(\{x : |f(x)| \geq M\}) = 0\}. \tag{15.2}$$

Thus the L^∞ norm of a function f is the smallest number M such that $|f| \leq M$ a.e.

For $1 \leq p \leq \infty$ the space L^p is the set $\{f : \|f\|_p < \infty\}$. One can also write $L^p(X)$ or $L^p(\mu)$ if one wants to emphasize the space or the measure. It is clear that $\|f\|_p = 0$ if and only if $f = 0$ a.e.

If $1 < p < \infty$, we define q by

$$\frac{1}{p} + \frac{1}{q} = 1$$

131

and call q the *conjugate exponent* of p.

Basic to the study of L^p spaces is *Hölder's inequality*. Note that when $p = q = 2$, this is the Cauchy-Schwarz inequality.

Proposition 15.1 *If* $1 < p, q < \infty$ *and* $p^{-1} + q^{-1} = 1$, *then*

$$\int |fg| \, d\mu \leq \|f\|_p \|g\|_q.$$

This also holds if $p = \infty$ *and* $q = 1$.

Proof. If $M = \|f\|_\infty$, then $|f| \leq M$ a.e. and $\int |fg| \leq M \int |g|$. The case $p = \infty$ and $q = 1$ follows.

Now let us assume $1 < p, q < \infty$. If $\|f\|_p = 0$, then $f = 0$ a.e. and $\int |fg| = 0$, so the result is clear if $\|f\|_p = 0$ and similarly if $\|g\|_q = 0$. Let $F(x) = |f(x)|/\|f\|_p$ and $G(x) = |g(x)|/\|g\|_q$. Note $\|F\|_p = 1$ and $\|G\|_q = 1$, and it suffices to show that $\int FG \, d\mu \leq 1$.

The second derivative of the function e^x is again e^x, which is everywhere positive. Any function whose second derivative is everywhere non-negative is convex, so if $0 \leq \lambda \leq 1$, we have

$$e^{\lambda a + (1-\lambda)b} \leq \lambda e^a + (1 - \lambda)e^b \tag{15.3}$$

for every pair of reals $a \leq b$. If $F(x), G(x) \neq 0$, let $a = p \log F(x)$, $b = q \log G(x)$, $\lambda = 1/p$, and $1 - \lambda = 1/q$. We then obtain from (15.3) that

$$F(x)G(x) \leq \frac{F(x)^p}{p} + \frac{G(x)^q}{q}.$$

Clearly this inequality also holds if $F(x) = 0$ or $G(x) = 0$. Integrating,

$$\int FG \, d\mu \leq \frac{\|F\|_p^p}{p} + \frac{\|G\|_q^q}{q} = \frac{1}{p} + \frac{1}{q} = 1.$$

This completes the proof. □

One application of Hölder's inequality is to prove Minkowski's inequality, which is simply the triangle inequality for L^p.

We first need the following lemma:

Lemma 15.2 *If* $a, b \geq 0$ *and* $1 \leq p < \infty$, *then*

$$(a + b)^p \leq 2^{p-1}a^p + 2^{p-1}b^p. \tag{15.4}$$

Proof. The case $a = 0$ is obvious, so we assume $a > 0$. Dividing both sides by a^p, letting $x = b/a$, and setting

$$f(x) = 2^{p-1} + 2^{p-1}x^p - (1+x)^p,$$

the inequality we want to prove is equivalent to showing $f(x) \geq 0$ for $x \geq 0$. Note $f(0) > 0$, $f(1) = 0$, $\lim_{x \to \infty} f(x) = \infty$, and the only solution to $f'(x) = 0$ on $(0, \infty)$ is $x = 1$. We conclude that f takes its minimum at $x = 1$ and hence $f(x) \geq 0$ for $x \geq 0$. $\square$

Here is *Minkowski's inequality*.

Proposition 15.3 *If* $1 \leq p \leq \infty$, *then*

$$\|f + g\|_p \leq \|f\|_p + \|g\|_p.$$

Proof. Since $|(f + g)(x)| \leq |f(x)| + |g(x)|$, integrating gives the case when $p = 1$. The case $p = \infty$ is also easy. Now let us suppose $1 < p < \infty$. If $\|f\|_p$ or $\|g\|_p$ is infinite, the result is obvious, so we may assume both are finite. The inequality (15.4) with $a = |f(x)|$ and $b = |g(x)|$ yields, after an integration,

$$\int |(f+g)(x)|^p \, d\mu \leq 2^{p-1} \int |f(x)|^p \, d\mu + 2^{p-1} \int |g(x)|^p \, d\mu.$$

We therefore have $\|f+g\|_p < \infty$. Clearly we may assume $\|f+g\|_p > 0$.

Now write

$$|f + g|^p \leq |f|\,|f + g|^{p-1} + |g|\,|f + g|^{p-1}$$

and apply Hölder's inequality with $q = (1 - \frac{1}{p})^{-1}$. We obtain

$$\int |f+g|^p \leq \|f\|_p \left(\int |f+g|^{(p-1)q} \right)^{1/q} + \|g\|_p \left(\int |f+g|^{(p-1)q} \right)^{1/q}.$$

Since $p^{-1} + q^{-1} = 1$, then $(p-1)q = p$, so we have

$$\|f + g\|_p^p \leq \left(\|f\|_p + \|g\|_p \right) \|f + g\|_p^{p/q}.$$

Dividing both sides by $\|f+g\|_p^{p/q}$ and using the fact that $p - (p/q) = 1$ gives us our result. $\square$

Recall the definition of normed linear space from Chapter 1. We would like to say that by virtue of Minkowski's inequality, L^p is a normed linear space. This is not quite right. The L^p norm of a function satisfies all the properties of a norm except that $\|f\|_p = 0$ does not imply that f is the zero function, only that $f = 0$ a.e. The procedure we follow to circumvent this is to say two functions are equivalent if they differ on a set of measure 0. This is an equivalence relation for functions. We then define the space L^p to be the set of equivalence classes with respect to this equivalence relation, and define $\|f\|_p$ to be the L^p norm of any function in the same equivalence class as f. We then have that $\|\cdot\|_p$ is a norm on L^p. We henceforth keep this interpretation in the back of our minds when we talk about a function being in L^p; the understanding is that we identify functions that are equal a.e.

Recall Definition 10.1: f_n converges to f in L^p if $\int |f_n - f|^p \to 0$ as $n \to \infty$. In terms of L^p norms, this is equivalent to $\|f_n - f\|_p^p \to 0$ as $n \to \infty$.

Related to the definition of L^∞ is the following terminology. Given a real-valued measurable function f, the *essential supremum* and *essential infimum* are defined by

$$\text{ess sup } f = \inf\{M : \mu(\{x : f(x) > M\}) = 0\}$$

and

$$\text{ess inf } f = \sup\{m : \mu(\{x : f(x) < m\}) = 0\}.$$

15.2   Completeness

We show that the space L^p viewed as a metric space is complete.

Theorem 15.4 *If $1 \le p \le \infty$, then L^p is complete.*

Proof. We will do only the case $p < \infty$ and leave the case $p = \infty$ as Exercise 15.1.

Step 1. Suppose f_n is a Cauchy sequence in L^p. Our first step is to find a certain subsequence. Given $\varepsilon = 2^{-(j+1)}$, there exists n_j such that if $n, m \ge n_j$, then $\|f_n - f_m\|_p \le 2^{-(j+1)}$. Without loss of generality we may assume $n_j \ge n_{j-1}$ for each j.

Step 2. Set $n_0 = 0$ and define f_0 to be identically 0. Our candidate for the limit function will be $\sum_m (f_{n_m} - f_{n_{m-1}})$. In this step we show absolute convergence of this series.

Set $g_j(x) = \sum_{m=1}^{j} |f_{n_m}(x) - f_{n_{m-1}}(x)|$. Of course, $g_j(x)$ increases in j for each x. Let $g(x)$, which might be infinite, be the limit. By Minkowski's inequality

$$\|g_j\|_p \leq \sum_{m=1}^{j} \|f_{n_m} - f_{n_{m-1}}\|_p$$

$$\leq \|f_{n_1} - f_{n_0}\|_p + \sum_{m=2}^{j} 2^{-m}$$

$$\leq \|f_{n_1}\|_p + \tfrac{1}{2}.$$

By Fatou's lemma,

$$\int |g(x)|^p \, \mu(dx) \leq \lim_{j\to\infty} \int |g_j(x)|^p \, \mu(dx)$$

$$= \lim_{j\to\infty} \|g_j\|_p^p$$

$$\leq \tfrac{1}{2} + \|f_{n_1}\|_p.$$

Hence g is finite a.e. This proves the absolute convergence for almost every x.

Step 3. We define our function f. Set

$$f(x) = \sum_{m=1}^{\infty} [f_{n_m}(x) - f_{n_{m-1}}(x)].$$

We showed in Step 2 that this series is absolutely convergent for almost every x, so f is well defined for a.e. x. Set $f(x) = 0$ for any x where absolute convergence does not hold. We have

$$f(x) = \lim_{K\to\infty} \sum_{m=1}^{K} [f_{n_m}(x) - f_{n_{m-1}}(x)] = \lim_{K\to\infty} f_{n_K}(x)$$

since we have a telescoping series. By Fatou's lemma,

$$\|f - f_{n_j}\|_p^p = \int |f - f_{n_j}|^p \leq \liminf_{K\to\infty} \int |f_{n_K} - f_{n_j}|^p$$

$$= \liminf_{K\to\infty} \|f_{n_K} - f_{n_j}\|_p^p \leq 2^{(-j+1)p}.$$

Step 4. We have thus shown that $\|f - f_{n_j}\|_p \to 0$ as $j \to \infty$. It is standard that a Cauchy sequence with a convergent subsequence itself converges. Here is the proof in our case. Given $\varepsilon > 0$, there exists N such that $\|f_n - f_m\|_p < \varepsilon$ if $m, n \geq N$. In particular, $\|f_{n_j} - f_m\|_p < \varepsilon$ if j is large enough. By Fatou's lemma,

$$\|f - f_m\|_p^p \leq \liminf_{j \to \infty} \|f_{n_j} - f_m\|_p^p \leq \varepsilon^p$$

if $m \geq N$. This shows that f_m converges to f in L^p norm.   □

Next we show:

Proposition 15.5 *The set of continuous functions with compact support is dense in $L^p(\mathbb{R})$.*

Proof. Suppose $f \in L^p$. We have $\int |f - f\chi_{[-n,n]}|^p \to 0$ as $n \to \infty$ by the dominated convergence theorem, the dominating function being $|f|^p$. Hence it suffices to approximate functions in L^p that have compact support. By writing $f = f^+ - f^-$ we may suppose $f \geq 0$. Consider simple functions s_m increasing to f; then we have $\int |f - s_m|^p \to 0$ by the dominated convergence theorem, so it suffices to approximate simple functions with compact support. By linearity, it suffices to approximate characteristic functions with compact support. Given E, a Borel measurable set contained in a bounded interval, and $\varepsilon > 0$, we showed in Proposition 8.4 that there exists g continuous with compact support and with values in $[0, 1]$ such that $\int |g - \chi_E| < \varepsilon$. Since $|g - \chi_E| \leq 1$, then $\int |g - \chi_E|^p \leq \int |g - \chi_E| < \varepsilon$. This completes the proof.   □

The same proof shows the following corollary.

Corollary 15.6 *The set of continuous functions on $[a, b]$ are dense in the space $L^2([a, b])$ with respect to $L^2([a, b])$ norm.*

15.3 Convolutions

The *convolution* of two measurable functions f and g is defined by

$$f * g(x) = \int f(x - y)g(y)\, dy,$$

provided the integral exists. By a change of variables, this is the same as $\int f(y)g(x-y)\,dy$, so $f * g = g * f$.

Proposition 15.7 *If $f, g \in L^1$, then $f * g$ is in L^1 and*

$$\|f * g\|_1 \le \|f\|_1 \|g\|_1. \tag{15.5}$$

Proof. We have

$$\int |f * g(x)|\,dx \le \int \int |f(x-y)|\,|g(y)|\,dy\,dx. \tag{15.6}$$

Since the integrand on the right is non-negative, we can apply the Fubini theorem to see that the right hand side is equal to

$$\int \int |f(x-y)|\,dx\,|g(y)|\,dy = \int \int |f(x)|\,dx\,|g(y)|\,dy \tag{15.7}$$
$$= \|f\|_1 \|g\|_1.$$

The first equality here follows by a change of variables (see Exercise 8.1). This together with (15.6) proves (15.5). From (15.5) we conclude that $f * g$ is finite a.e. □

15.4 Bounded linear functionals

A *linear functional* on L^p is a map H from L^p to $\mathbb{R}$ satisfying

$$H(f + g) = H(f) + H(g), \qquad H(af) = aH(f)$$

whenever $f, g \in L^p$ and $a \in \mathbb{R}$. (One can also have complex-valued linear functionals, but we do not consider them in this section. See, however, Exercise 15.28.) H is a *bounded linear functional* if

$$\|H\| = \sup\{|Hf| : \|f\|_p \le 1\} \tag{15.8}$$

is finite. The dual space of L^p is the collection of all bounded linear functionals with norm given by (15.8). Our goal in this section is to identify the dual of L^p.

We define the *signum function* or *sign function* by

$$\operatorname{sgn}(x) = \begin{cases} -1, & x < 0; \\ 0, & x = 0; \\ 1, & x > 0. \end{cases}$$

Note $x \operatorname{sgn}(x) = |x|$.

The following is very useful.

Theorem 15.8 *For $1 < p < \infty$ and $p^{-1} + q^{-1} = 1$,*

$$\|f\|_p = \sup\left\{\int fg\, d\mu : \|g\|_q \leq 1\right\}. \tag{15.9}$$

When $p = 1$, (15.9) holds if we take $q = \infty$, and if $p = \infty$, (15.9) holds if we take $q = 1$.

Proof. The right hand side of (15.9) is less than the left hand side by Hölder's inequality. Thus we need only show that the right hand side is greater than the left hand side.

Case 1: $p = 1$. Take $g(x) = \operatorname{sgn} f(x)$. Then $|g|$ is bounded by 1 and $fg = |f|$. This takes care of the case $p = 1$.

Case 2: $p = \infty$. If $\|f\|_\infty = 0$, the result is trivial, so suppose $\|f\|_\infty > 0$. Since μ is σ-finite, there exist sets F_n increasing up to X such that $\mu(F_n) < \infty$ for each n. If $M = \|f\|_\infty$, let a be any finite real less than M. By the definition of L^∞ norm, the measure of $A_n = \{x \in F_n : |f(x)| > a\}$ must be positive if n is sufficiently large. Let

$$g_n(x) = \frac{\operatorname{sgn}(f(x))\chi_{A_n}(x)}{\mu(A_n)}.$$

Then the L^1 norm of g_n is 1 and $\int fg_n = \int_{A_n} |f|/\mu(A_n) \geq a$. Since a is arbitrary, the supremum on the right hand side of (15.9) must be M.

Case 3: $1 < p < \infty$. We may suppose $\|f\|_p > 0$. Let F_n be measurable sets of finite measure increasing to X, q_n a sequence of non-negative simple functions increasing to f^+, r_n a sequence of non-negative simple functions increasing to f^-, and

$$s_n(x) = (q_n(x) - r_n(x))\chi_{F_n}(x).$$

Then $s_n(x) \to f(x)$ for each x, $|s_n(x)|$ increases to $|f(x)|$ for each x, each s_n is a simple function, and $\|s_n\|_p < \infty$ for each n. Then $\|s_n\|_p \to \|f\|_p$ by the monotone convergence theorem, whether or not $\|f\|_p$ is finite. For n sufficiently large, $\|s_n\|_p > 0$.

Let

$$g_n(x) = (\operatorname{sgn} f(x))\frac{|s_n(x)|^{p-1}}{\|s_n\|_p^{p/q}}.$$

g_n is again a simple function. Since $(p-1)q = p$, then

$$\|g_n\|_q = \frac{(\int |s_n|^{(p-1)q})^{1/q}}{\|s_n\|_p^{p/q}} = \frac{\|s_n\|_p^{p/q}}{\|s_n\|_p^{p/q}} = 1.$$

On the other hand, since $|f| \geq |s_n|$,

$$\int f g_n = \frac{\int |f| \, |s_n|^{p-1}}{\|s_n\|_p^{p/q}} \geq \frac{\int |s_n|^p}{\|s_n\|_p^{p/q}} = \|s_n\|_p^{p-(p/q)}.$$

Since $p - (p/q) = 1$, then $\int f g_n \geq \|s_n\|_p$, which tends to $\|f\|_p$. This proves the right hand side of (15.9) is at least as large as the left hand side. $\qquad \square$

The proof of Theorem 15.8 also establishes

Corollary 15.9 *For $1 < p < \infty$ and $p^{-1} + q^{-1} = 1$,*

$$\|f\|_p = \sup\left\{ \int fg : \|g\|_q \leq 1, g \text{ simple}\right\}.$$

Proposition 15.10 *Suppose $1 < p < \infty$, $p^{-1} + q^{-1} = 1$, and $g \in L^q$. If we define $H(f) = \int fg$ for $f \in L^p$, then H is a bounded linear functional on L^p and $\|H\| = \|g\|_q$.*

Proof. The linearity is obvious. That $\|H\| \leq \|g\|_q$ follows by Hölder's inequality. Using Theorem 15.8 and writing

$$\|H\| = \sup_{\|f\|_p \leq 1} |H(f)| = \sup_{\|f\|_p \leq 1} \left| \int fg \right| \geq \sup_{\|f\|_p \leq 1} \int fg = \|g\|_q$$

completes the proof. $\qquad \square$

Theorem 15.11 *Suppose $1 < p < \infty$, $p^{-1} + q^{-1} = 1$, and H is a real-valued bounded linear functional on L^p. Then there exists $g \in L^q$ such that $H(f) = \int fg$ and $\|g\|_q = \|H\|$.*

This theorem together with Proposition 15.10 allows us to identify the dual space of L^p with L^q.

Proof. Suppose we are given a bounded linear functional H on L^p. First suppose $\mu(X) < \infty$. Define $\nu(A) = H(\chi_A)$. We will show that ν is a measure, that $\nu \ll \mu$ and that $g = d\nu/d\mu$ is the function we seek.

If A and B are disjoint, then

$$\nu(A \cup B) = H(\chi_{A \cup B}) = H(\chi_A + \chi_B)$$
$$= H(\chi_A) + H(\chi_B) = \nu(A) + \nu(B).$$

To show ν is countably additive, it suffices to show that if $A_n \uparrow A$, then $\nu(A_n) \to \nu(A)$, and then use Exercise 3.1. But if $A_n \uparrow A$, then $\chi_{A_n} \to \chi_A$ in L^p, and so $\nu(A_n) = H(\chi_{A_n}) \to H(\chi_A) = \nu(A)$; we use here the fact that $\mu(X) < \infty$. We conclude that ν is a countably additive signed measure. Moreover, if $\mu(A) = 0$, then $\chi_A = 0$ a.e., hence $\nu(A) = H(\chi_A) = 0$. Using Exercise 13.1, which is the Radon-Nikodym theorem for signed measures, we see there exists a real-valued integrable function g such that $\nu(A) = \int_A g$ for all sets A.

If $s = \sum_i a_i \chi_{A_i}$ is a simple function, by linearity we have

$$H(s) = \sum_i a_i H(\chi_{A_i}) = \sum_i a_i \nu(A_i) = \sum_i a_i \int g\chi_{A_i} = \int gs.$$
$$(15.10)$$

By Corollary 15.9 and (15.10),

$$\|g\|_q = \sup\left\{ \int gs : \|s\|_p \leq 1, s \text{ simple} \right\}$$
$$= \sup\{H(s) : \|s\|_p \leq 1, s \text{ simple}\} \leq \|H\|.$$

If s_n are simple functions tending to f in L^p (see Exercise 15.2), then $H(s_n) \to H(f)$, while by Hölder's inequality

$$\left| \int s_n g - \int fg \right| = \left| \int (s_n - f)g \right| \leq \|s_n - f\|_p \|g\|_q \to 0,$$

so $\int s_n g \to \int fg$. We thus have $H(f) = \int fg$ for all $f \in L^p$, and $\|g\|_q \leq \|H\|$. By Hölder's inequality, $\|H\| \leq \|g\|_q$.

In the case where μ is σ-finite, but not necessarily finite, let $F_n \uparrow X$ so that $\mu(F_n) < \infty$ for each n. Define functionals H_n by $H_n(f) = H(f\chi_{F_n})$. Clearly each H_n is a bounded linear functional on L^p. Applying the above argument, we see there exist g_n such that $H_n(f) = \int fg_n$ and $\|g_n\|_q = \|H_n\| \leq \|H\|$. It is easy to see that g_n is 0 if $x \notin F_n$. Moreover, by the uniqueness part of the Radon-Nikodym theorem, if $n > m$, then $g_n = g_m$ on F_m. Define g by setting $g(x) = g_n(x)$ if $x \in F_n$. Then g is well defined. By Fatou's lemma, g is in L^q with a norm bounded by $\|H\|$. Note $f \chi_{F_n} \to f$ in L^p by the dominated convergence theorem. Since H is a bounded linear functional on L^p, we have $H_n(f) = H(f \chi_{F_n}) \to H(f)$. On the other hand

$$H_n(f) = \int_{F_n} f g_n = \int_{F_n} f g \to \int f g$$

by the dominated convergence theorem. Thus $H(f) = \int f g$. Again by Hölder's inequality $\|H\| \leq \|g\|_q$. □

15.5 Exercises

Exercise 15.1 Show that L^∞ is complete.

Exercise 15.2 Prove that the collection of simple functions is dense in L^p.

Exercise 15.3 Prove the equality

$$\int |f(x)|^p \, dx = \int_0^\infty p t^{p-1} m(\{x : |f(x)| \geq t\}) \, dt$$

for $p \geq 1$.

Exercise 15.4 Consider the measure space $([0,1], \mathcal{B}, m)$, where $\mathcal{B}$ is the Borel σ-algebra and m is Lebesgue measure, and suppose f is a measurable function. Prove that $\|f\|_p \to \|f\|_\infty$ as $p \to \infty$.

Exercise 15.5 When does equality hold in Hölder's inequality? When does equality hold in the Minkowski inequality?

Exercise 15.6 Give an example to show that $L^p \not\subset L^q$ in general if $1 < p < q < \infty$. Give an example to show that $L^q \not\subset L^p$ in general if $1 < p < q < \infty$.

Exercise 15.7 Define

$$g_n(x) = n\chi_{[0,n^{-3}]}(x).$$

(1) Show that if $f \in L^2([0,1])$, then

$$\int_0^1 f(x)g_n(x)\,dx \to 0$$

as $n \to \infty$.

(2) Show that there exists $f \in L^1([0,1])$ such that $\int_0^1 f(x)g_n(x)\,dx \not\to 0$.

Exercise 15.8 Suppose μ is a finite measure on the Borel subsets of $\mathbb{R}$ such that

$$f(x) = \int_{\mathbb{R}} f(x+t)\,\mu(dt), \qquad \text{a.e.},$$

whenever f is real-valued, bounded, and integrable. Prove that $\mu(\{0\}) = 1$.

Exercise 15.9 Suppose μ is a measure with $\mu(X) = 1$ and $f \in L^r$ for some $r > 0$, where we define L^r for $r < 1$ exactly as in (15.1). Prove that

$$\lim_{p \to 0} \|f\|_p = \exp\left(\int \log|f|\,d\mu\right),$$

where we use the convention that $\exp(-\infty) = 0$.

Exercise 15.10 Suppose $1 < p < \infty$ and q is the conjugate exponent to p. Suppose $f_n \to f$ a.e. and $\sup_n \|f_n\|_p < \infty$. Prove that if $g \in L^q$, then

$$\lim_{n \to \infty} \int f_n g = \int f g.$$

Does this extend to the case where $p = 1$ and $q = \infty$? If not, give a counterexample.

Exercise 15.11 If $f \in L^1(\mathbb{R})$ and $g \in L^p(\mathbb{R})$ for some $p \in [1, \infty)$, prove that

$$\|f * g\|_p \le \|f\|_1 \|g\|_p.$$

Exercise 15.12 Suppose $p \in (1, \infty)$ and q is its conjugate exponent. Prove that if $f \in L^p(\mathbb{R})$ and $g \in L^q(\mathbb{R})$, then $f * g$ is uniformly continuous and $f * g(x) \to 0$ as $x \to \infty$ and as $x \to -\infty$.

Exercise 15.13 Show that if f and g are continuous with compact support, then $f * g$ is continuous with compact support.

Exercise 15.14 Suppose $f \in L^\infty(\mathbb{R})$, $f_h(x) = f(x + h)$, and

$$\lim_{h \to 0} \|f_h - f\|_\infty = 0.$$

Prove that there exists a uniformly continuous function g on $\mathbb{R}$ such that $f = g$ a.e.

Exercise 15.15 Let $p \in [1, \infty)$. Prove that $f \in L^p(\mu)$ if and only if

$$\sum_{n=1}^{\infty} (2^n)^p \mu(\{x : |f(x)| > 2^n\}) < \infty.$$

Exercise 15.16 Suppose $\mu(X) = 1$ and f and g are non-negative functions such that $fg \geq 1$ a.e. Prove that

$$\left(\int f \, d\mu \right) \left(\int g \, d\mu \right) \geq 1.$$

Exercise 15.17 Suppose $f : [1, \infty) \to \mathbb{R}$, $f(1) = 0$, f' exists and is continuous and bounded, and $f' \in L^2([1, \infty))$. Let $g(x) = f(x)/x$. Show $g \in L^2([1, \infty))$.

Exercise 15.18 Find an example of a measurable $f : [1, \infty) \to \mathbb{R}$ such that $f(1) = 0$, f' exists and is continuous and bounded, $f' \in L^1([1, \infty))$, but the function $g(x) = f(x)/x$ is not in L^1.

Exercise 15.19 Prove the *generalized Minkowski inequality*: If $(X, \mathcal{A}, \mu)$ and $(Y, \mathcal{B}, \nu)$ are measure spaces, f is measurable with respect to $\mathcal{A} \times \mathcal{B}$, and $1 < p < \infty$, then

$$\left(\int_X \left(\int_Y |f(x,y)| \, \nu(dy) \right)^p \mu(dx) \right)^{1/p}$$
$$\leq \int_Y \left(\int_X |f(x,y)|^p \, \mu(dx) \right)^{1/p} \nu(dy).$$

This could be rephrased as

$$\left\| \|f\|_{L^1(\nu)} \right\|_{L^p(\mu)} \leq \left\| \|f\|_{L^p(\mu)} \right\|_{L^1(\nu)}.$$

Does this extend to the cases where $p = 1$ or $p = \infty$? If not, give counterexamples.

If $Y = \{1, 2\}$, $\nu(dy) = \delta_1(dy) + \delta_2(dy)$, where δ_1 and δ_2 are point masses at 1 and 2, resp., and we let $g_1(x) = f(x, 1)$, $g_2(x) = f(x, 2)$, we recover the usual Minkowski inequality, Proposition 15.3.

Exercise 15.20 Let $\alpha \in (0, 1)$ and $K(x) = |x|^{-\alpha}$ for $x \in \mathbb{R}$. Note that K is not in L^p for any $p \geq 1$. Prove that if f is non-negative, real-valued, and integrable on $\mathbb{R}$ and

$$g(x) = \int f(x - t) K(t) \, dt,$$

then g is finite a.e.

Exercise 15.21 Suppose $p > 1$ and q is its conjugate exponent, f is an absolutely continuous function on $[0, 1]$ with $f' \in L^p$, and $f(0) = 0$. Prove that if $g \in L^q$, then

$$\int_0^1 |fg| \, dx \leq \left(\frac{1}{p}\right)^{1/p} \|f'\|_p \|g\|_q.$$

Exercise 15.22 Suppose $f : \mathbb{R} \to \mathbb{R}$ is in L^p for some $p > 1$ and also in L^1. Prove there exist constants $c > 0$ and $\alpha \in (0, 1)$ such that

$$\int_A |f(x)| \, dx \leq cm(A)^\alpha$$

for every Borel measurable set $A \subset \mathbb{R}$, where m is Lebesgue measure.

Exercise 15.23 Suppose $f : \mathbb{R} \to \mathbb{R}$ is integrable and there exist constants $c > 0$ and $\alpha \in (0, 1)$ such that

$$\int_A |f(x)| \, dx \leq cm(A)^\alpha$$

for every Borel measurable set $A \subset \mathbb{R}$, where m is Lebesgue measure. Prove there exists $p > 1$ such that $f \in L^p$.

Exercise 15.24 Suppose $1 < p < \infty$, $f : (0, \infty) \to \mathbb{R}$, and $f \in L^p$ with respect to Lebesgue measure. Define

$$g(x) = \frac{1}{x} \int_0^x f(y) \, dy.$$

Prove that

$$\|g\|_p \leq \frac{p}{p-1}\|f\|_p.$$

This is known as *Hardy's inequality*.

Exercise 15.25 Suppose $(X, \mathcal{A}, \mu)$ is a measure space and suppose $K : X \times X \to \mathbb{R}$ is measurable with respect to $\mathcal{A} \times \mathcal{A}$. Suppose there exists $M < \infty$ such that

$$\int_X |K(x,y)|\, \mu(dy) \leq M$$

for each x and

$$\int_X |K(x,y)|\, \mu(dx) \leq M$$

for each y. If f is measurable and real-valued, define

$$Tf(x) = \int_X K(x,y)f(y)\, \mu(dy)$$

if the integral exists.
(1) Show that $\|Tf\|_1 \leq M\|f\|_1$.
(2) If $1 < p < \infty$, show that $\|Tf\|_p \leq M\|f\|_p$.

Exercise 15.26 Suppose A and B are two Borel measurable subsets of $\mathbb{R}$, each with finite strictly positive Lebesgue measure. Show that $\chi_A * \chi_B$ is a continuous non-negative function that is not identically equal to 0.

Exercise 15.27 Suppose A and B are two Borel measurable subsets of $\mathbb{R}$ with strictly positive Lebesgue measure. Show that

$$C = \{x + y : x \in A, y \in B\}$$

contains a non-empty open interval.

Exercise 15.28 Suppose $1 < p < \infty$ and q is the conjugate exponent of p. Prove that if H is a bounded complex-valued linear functional on L^p, then there exists a complex-valued measurable function $g \in L^q$ such that $H(f) = \int fg$ for all $f \in L^p$ and $\|H\| = \|g\|_q$.

.

Chapter 16

Fourier transforms

Fourier transforms give a representation of a function in terms of frequencies. There is a great deal known about Fourier transforms and their applications. We give an introduction here.

16.1 Basic properties

If f is a complex-valued function and $f \in L^1(\mathbb{R}^n)$, define the Fourier transform $\widehat{f}$ to be the function with domain $\mathbb{R}^n$ and range $\mathbb{C}$ given by

$$\widehat{f}(u) = \int_{\mathbb{R}^n} e^{iu \cdot x} f(x)\, dx, \qquad u \in \mathbb{R}^n. \tag{16.1}$$

We are using $u \cdot x$ for the standard inner product in $\mathbb{R}^n$. Various books have slightly different definitions. Some put a negative sign and/or 2π before the $iu \cdot x$, some have a $(2\pi)^{-1}$ or a $(2\pi)^{-1/2}$ in front of the integral. The basic theory is the same in any case.

Some basic properties of the Fourier transform are given by

Proposition 16.1 *Suppose f and g are in L^1. Then*
(1) $\widehat{f}$ is bounded and continuous;
(2) $\widehat{(f + g)}(u) = \widehat{f}(u) + \widehat{g}(u)$;
(3) $\widehat{(af)}(u) = a\widehat{f}(u)$ if $a \in \mathbb{C}$;
(4) if $a \in \mathbb{R}^n$ and $f_a(x) = f(x + a)$, then $\widehat{f_a}(u) = e^{-iu \cdot a}\widehat{f}(u)$;

147

(5) if $a \in \mathbb{R}^n$ and $g_a(x) = e^{ia \cdot x} g(x)$, then $\widehat{g}_a(u) = \widehat{g}(u + a)$;
(6) if a is a non-zero real number and $h_a(x) = f(ax)$, then $\widehat{h}_a(u) = a^{-n} \widehat{f}(u/a)$.

Proof. (1) $\widehat{f}$ is bounded because $f \in L^1$ and $|e^{iu \cdot x}| = 1$. We have

$$\widehat{f}(u + h) - \widehat{f}(u) = \int \left(e^{i(u+h) \cdot x} - e^{iu \cdot x} \right) f(x) \, dx.$$

Then

$$|\widehat{f}(u + h) - \widehat{f}(u)| \leq \int \left| e^{iu \cdot x} \right| \cdot \left| e^{ih \cdot x} - 1 \right| |f(x)| \, dx.$$

The integrand is bounded by $2|f(x)|$, which is integrable, and $e^{ih \cdot x} - 1 \to 0$ as $h \to 0$. Thus the continuity follows by the dominated convergence theorem.

(2) and (3) are easy by a change of variables. (4) holds because

$$\widehat{f}_a(u) = \int e^{iu \cdot x} f(x + a) \, dx = \int e^{iu \cdot (x-a)} f(x) \, dx = e^{-iu \cdot a} \widehat{f}(u)$$

by a change of variables. For (5),

$$\widehat{g}_a(u) = \int e^{iu \cdot x} e^{ia \cdot x} f(x) \, dx = \int e^{i(u+a) \cdot x} f(x) \, dx = \widehat{f}(u + a).$$

Finally for (6), by a change of variables,

$$\widehat{h}_a(u) = \int e^{iu \cdot x} f(ax) \, dx = a^{-n} \int e^{iu \cdot (y/a)} f(y) \, dy$$

$$= a^{-n} \int e^{i(u/a) \cdot y} f(y) \, dy = a^{-n} \widehat{f}(u/a),$$

as required. $\qquad \square$

One reason for the usefulness of Fourier transforms is that they relate derivatives and multiplication.

Proposition 16.2 *Suppose $f \in L^1$ and $x_j f(x) \in L^1$, where x_j is the jth coordinate of x. Then*

$$\frac{\partial \widehat{f}}{\partial u_j}(u) = i \int e^{iu \cdot x} x_j f(x) \, dx.$$

Proof. Let e_j be the unit vector in the jth direction. Then

$$\frac{\widehat{f}(u + he_j) - \widehat{f}(u)}{h} = \frac{1}{h} \int \left(e^{i(u+he_j)\cdot x} - e^{iu\cdot x} \right) f(x)\, dx$$

$$= \int e^{iu\cdot x} \left(\frac{e^{ihx_j} - 1}{h} \right) f(x)\, dx.$$

Since

$$\left| \frac{1}{h} \left(e^{ihx_j} - 1 \right) \right| \leq |x_j|$$

and $x_j f(x) \in L^1$, the right hand side converges to $\int e^{iu\cdot x} i x_j f(x)\, dx$ by the dominated convergence theorem. Therefore the left hand side converges. The limit of the left hand side is $\partial \widehat{f}/\partial u_j$. $\qquad \square$

Proposition 16.3 *Suppose $f : \mathbb{R} \to \mathbb{R}$ is integrable, f is absolutely continuous, and f' is integrable. Then the Fourier transform of f' is $-iu\widehat{f}(u)$.*

The higher dimensional version of this is left as Exercise 16.4.

Proof. Since f' is integrable,

$$|f(y) - f(x)| \leq \int_x^y |f'(z)|\, dz \to 0$$

as $x, y \to \infty$ by the dominated convergence theorem. This implies that $f(y_n)$ is a Cauchy sequence whenever $y_n \to \infty$, and we conclude that $f(y)$ converges as $y \to \infty$. Since f is integrable, the only possible value for the limit is 0. The same is true for the limit as $y \to -\infty$.

By integration by parts (use Exercise 14.1 and a limit argument),

$$\widehat{f'}(u) = \int_{-\infty}^{\infty} e^{iux} f'(x)\, dx = -\int_{-\infty}^{\infty} iue^{iux} f(x)\, dx$$

$$= -iu\widehat{f}(u),$$

as desired. $\qquad \square$

Recall the definition of convolution given in Section 15.3. Recall also (15.7), which says that

$$\int \int |f(x - y)|\, |g(y)|\, dx\, dy = \|f\|_1 \|g\|_1. \tag{16.2}$$

Proposition 16.4 *If $f, g \in L^1$, then the Fourier transform of $f * g$ is $\widehat{f}(u)\widehat{g}(u)$.*

Proof. We have

$$\widehat{f * g}(u) = \int e^{iu \cdot x} \int f(x - y)g(y) \, dy \, dx$$

$$= \int \int e^{iu \cdot (x-y)} f(x - y) \, e^{iu \cdot y} g(y) \, dx \, dy$$

$$= \int \widehat{f}(u)e^{iu \cdot y} g(y) \, dy = \widehat{f}(u)\widehat{g}(u).$$

We applied the Fubini theorem in the second equality; this is valid because as we see from (16.2), the absolute value of the integrand is integrable. We used a change of variables to obtain the third equality.   □

16.2   The inversion theorem

We want to give a formula for recovering f from $\widehat{f}$. First we need to calculate the Fourier transform of a particular function.

Proposition 16.5 *(1) Suppose $f_1 : \mathbb{R} \to \mathbb{R}$ is defined by*

$$f_1(x) = \frac{1}{\sqrt{2\pi}} e^{-x^2/2}.$$

Then $\widehat{f_1}(u) = e^{-u^2/2}$.

(2) Suppose $f_n : \mathbb{R}^n \to \mathbb{R}$ is given by

$$f_n(x) = \frac{1}{(2\pi)^{n/2}} e^{-|x|^2/2}.$$

Then $\widehat{f_n}(u) = e^{-|u|^2/2}$.

Proof. (1) may also be proved using contour integration, but let's give a (mostly) real variable proof. Let $g(u) = \int e^{iux} e^{-x^2/2} \, dx$. Differentiate with respect to u. We may differentiate under the integral sign because $(e^{i(u+h)x} - e^{iux})/h$ is bounded in absolute value by $|x|$ and $|x|e^{-x^2/2}$ is integrable; therefore the dominated convergence theorem applies. We then obtain

$$g'(u) = i \int e^{iux} x e^{-x^2/2} \, dx.$$

By integration by parts (see Exercise 14.1) this is equal to

$$-u \int e^{iux} e^{-x^2/2} \, dx = -ug(u).$$

Solving the differential equation $g'(u) = -ug(u)$, we have

$$[\log g(u)]' = \frac{g'(u)}{g(u)} = -u,$$

so $\log g(u) = -u^2/2 + c_1$, and then

$$g(u) = c_2 e^{-u^2/2}. \tag{16.3}$$

By Exercise 11.18, $g(0) = \int e^{-x^2/2} \, dx = \sqrt{2\pi}$, so $c_2 = \sqrt{2\pi}$. Substituting this value of c_2 in (16.3) and dividing both sides by $\sqrt{2\pi}$ proves (1).

For (2), since $f_n(x) = f_1(x_1) \cdots f_1(x_n)$ if $x = (x_1, \ldots, x_n)$, then

$$\widehat{f_n}(u) = \int \cdots \int e^{i \sum_j u_j x_j} f_1(x_1) \cdots f_1(x_n) \, dx_1 \cdots dx_n$$

$$= \widehat{f_1}(u_1) \cdots \widehat{f_1}(u_n) = e^{-|u|^2/2}.$$

This completes the proof.   □

One more preliminary is needed before proving the inversion theorem.

Proposition 16.6 *Suppose φ is in L^1 and $\int \varphi(x) \, dx = 1$. Let $\varphi_\delta(x) = \delta^{-n} \varphi(x/\delta)$.*

(1) If g is continuous with compact support, then $g * \varphi_\delta$ converges to g pointwise as $\delta \to 0$.

(2) If g is continuous with compact support, then $g * \varphi_\delta$ converges to g in L^1 as $\delta \to 0$.

(3) If $f \in L^1$, then $\|f * \varphi_\delta - f\|_1 \to 0$ as $\delta \to 0$.

Proof. (1) We have by a change of variables (Exercise 8.1) that $\int \varphi_\delta(y)\, dy = 1$. Then

$$|g * \varphi_\delta(x) - g(x)| = \left| \int (g(x-y) - g(x))\, \varphi_\delta(y)\, dy \right|$$

$$= \left| \int (g(x - \delta y) - g(x))\, \varphi(y)\, dy \right|$$

$$\leq \int |g(x - \delta y) - g(x)|\, |\varphi(y)|\, dy.$$

Since g is continuous with compact support and hence bounded and φ is integrable, the right hand side goes to zero by the dominated convergence theorem, the dominating function being $2\|g\|_\infty \varphi$.

(2) We now use the Fubini theorem to write

$$\int |g * \varphi_\delta(x) - g(x)|\, dx = \int \left| \int (g(x-y) - g(x))\, \varphi_\delta(y)\, dy \right|\, dx$$

$$= \int \left| \int (g(x - \delta y) - g(x))\, \varphi(y)\, dy \right|\, dx$$

$$\leq \int \int |g(x - \delta y) - g(x)|\, |\varphi(y)|\, dy\, dx$$

$$= \int \int |g(x - \delta y) - g(x)|\, dx\, |\varphi(y)|\, dy.$$

Let

$$G_\delta(y) = \int |g(x - \delta y) - g(x)|\, dx.$$

By the dominated convergence theorem, for each y, $G_\delta(y)$ tends to 0 as $\delta \to 0$, since g is continuous with compact support. Moreover G_δ is bounded in absolute value by $2\|g\|_1$. Using the dominated convergence theorem again and the fact that φ is integrable, we see that $\int G_\delta(y)\, |\varphi(y)|\, dy$ tends to 0 as $\delta \to 0$.

(3) Let $\varepsilon > 0$. Let g be a continuous function with compact support so that $\|f - g\|_1 < \varepsilon$. Let $h = f - g$. A change of variables shows that $\|\varphi_\delta\|_1 = \|\varphi\|_1$. Observe

$$\|f * \varphi_\delta - f\|_1 \leq \|g * \varphi_\delta - g\|_1 + \|h * \varphi_\delta - h\|_1.$$

Also

$$\|h * \varphi_\delta - h\|_1 \leq \|h\|_1 + \|h * \varphi_\delta\|_1 \leq \|h\|_1 + \|h\|_1 \|\varphi_\delta\|_1 < \varepsilon(1 + \|\varphi\|_1)$$

by Proposition 15.7. Therefore, using (2),

$$\limsup_{\delta \to 0} \|f * \varphi_\delta - f\|_1 \le \limsup_{\delta \to 0} \|h * \varphi_\delta - h\|_1 \le \varepsilon(1 + \|\varphi\|_1).$$

Since ε is arbitrary, we have our conclusion. $\qquad\square$

Now we are ready to give the inversion formula. The proof seems longer than one might expect it to be, but there is no avoiding the introduction of the function H_a or some similar function.

Theorem 16.7 *Suppose f and $\widehat{f}$ are both in L^1. Then*

$$f(y) = \frac{1}{(2\pi)^n} \int e^{-iu \cdot y} \widehat{f}(u) \, du, \qquad \text{a.e.}$$

Proof. If $g(x) = a^{-n}k(x/a)$, then the Fourier transform of g is $\widehat{k}(au)$. Hence the Fourier transform of

$$\frac{1}{a^n} \frac{1}{(2\pi)^{n/2}} e^{-x^2/2a^2}$$

is $e^{-a^2 u^2/2}$. If we let

$$H_a(x) = \frac{1}{(2\pi)^n} e^{-|x|^2/2a^2},$$

we have

$$\widehat{H}_a(u) = (2\pi)^{-n/2} a^n e^{-a^2|u|^2/2}.$$

We write

$$\int \widehat{f}(u) e^{-iu \cdot y} H_a(u) \, du = \int \int e^{iu \cdot x} f(x) e^{-iu \cdot y} H_a(u) \, dx \, du$$

$$= \int \int e^{iu \cdot (x-y)} H_a(u) \, du \, f(x) \, dx$$

$$= \int \widehat{H}_a(x - y) f(x) \, dx. \qquad (16.4)$$

We can interchange the order of integration because

$$\int \int |f(x)| \, |H_a(u)| \, dx \, du < \infty$$

and $|e^{iu \cdot x}| = 1$. The left hand side of the first line of (16.4) converges to

$$(2\pi)^{-n} \int \widehat{f}(u) e^{-iu \cdot y} \, dy$$

as $a \to \infty$ by the dominated convergence theorem since $H_a(u) \to (2\pi)^{-n}$ and $\widehat{f} \in L^1$. The last line of (16.4) is equal to

$$\int \widehat{H}_a(y - x) f(x) \, dx = f * \widehat{H}_a(y), \tag{16.5}$$

using that $\widehat{H}_a$ is symmetric. But by Proposition 16.6, setting $\delta = a^{-1}$, we see that $f * \widehat{H}_a$ converges to f in L^1 as $a \to \infty$. $\qquad \square$

16.3 The Plancherel theorem

The last topic that we consider is the *Plancherel theorem*.

Theorem 16.8 *Suppose f is continuous with compact support. Then $\widehat{f} \in L^2$ and*

$$\|f\|_2 = (2\pi)^{-n/2} \|\widehat{f}\|_2. \tag{16.6}$$

Proof. First note that if we combine (16.4) and (16.5), then

$$\int \widehat{f}(u) e^{iu \cdot y} H_a(u) \, du = f * \widehat{H}_a(y).$$

Now take $y = 0$ and use the symmetry of $\widehat{H}_a$ to obtain

$$\int \widehat{f}(u) H_a(u) \, du = f * \widehat{H}_a(0). \tag{16.7}$$

Let $g(x) = \overline{f(-x)}$, where $\overline{a}$ denotes the complex conjugate of a. Since $\overline{ab} = \overline{a}\overline{b}$,

$$\widehat{g}(u) = \int e^{iu \cdot x} \overline{f(-x)} \, dx = \overline{\int e^{-iu \cdot x} f(-x) \, dx}$$

$$= \overline{\int e^{iu \cdot x} f(x) \, dx} = \overline{\widehat{f}(u)}.$$

The third equality follows by a change of variables. By (16.7) with f replaced by $f * g$,

$$\int \widehat{f * g}(u) H_a(u) \, du = f * g * \widehat{H}_a(0). \qquad (16.8)$$

Observe that $\widehat{f * g}(u) = \widehat{f}(u)\widehat{g}(u) = |\widehat{f}(u)|^2$. Thus the left hand side of (16.8) converges by the monotone convergence theorem to

$$(2\pi)^{-n} \int |\widehat{f}(u)|^2 \, du$$

as $a \to \infty$. Since f and g are continuous with compact support, then by Exercise 15.13, $f * g$ is also, and so the right hand side of (16.8) converges to $f * g(0) = \int f(y)g(-y) \, dy = \int |f(y)|^2 \, dy$ by Proposition 16.6(2). $\qquad \square$

Remark 16.9 We can use Theorem 16.8 to define $\widehat{f}$ when $f \in L^2$ so that (16.6) will continue to hold. The set of continuous functions with compact support is dense in L^2 by Proposition 15.5. Given a function f in L^2, choose a sequence $\{f_m\}$ of continuous functions with compact support such that $f_m \to f$ in L^2. Then $\|f_m - f_n\|_2 \to 0$ as $m, n \to \infty$. By (16.6), $\{\widehat{f}_m\}$ is a Cauchy sequence in L^2, and therefore converges to a function in L^2, which we call $\widehat{f}$.

Let us check that the limit does not depend on the choice of the sequence. If $\{f'_m\}$ is another sequence of continuous functions with compact support converging to f in L^2, then $\{f_m - f'_m\}$ is a sequence of continuous functions with compact support converging to 0 in L^2. By (16.6), $\widehat{f}_m - \widehat{f}'_m$ converges to 0 in L^2, and therefore $\widehat{f}'_m$ has the same limit as $\widehat{f}_m$. Thus $\widehat{f}$ is defined uniquely up to almost everywhere equivalence. By passing to the limit in L^2 on both sides of (16.6), we see that (16.6) holds for $f \in L^2$.

16.4 Exercises

Exercise 16.1 Find the Fourier transform of $\chi_{[a,b]}$ and in particular, find the Fourier transform of $\chi_{[-n,n]}$.

Exercise 16.2 Find a real-valued function $f \in L^1$ such that $\widehat{f} \notin L^1$.

Exercise 16.3 Show that if $f \in L^1$ and f is everywhere strictly positive, then $|\widehat{f}(y)| < \widehat{f}(0)$ for $y \neq 0$.

Exercise 16.4 If f is integrable, real-valued, and all the partial derivatives $f_j = \partial f / \partial x_j$ are integrable, prove that the Fourier transform of f_j is given by $\widehat{f_j}(u) = -iu_j \widehat{f}(u)$.

Exercise 16.5 Let $\mathcal{S}$ be the class of real-valued functions f on $\mathbb{R}$ such that for every $k \geq 0$ and $m \geq 0$, $|x|^m |f^{(k)}(x)| \to 0$ as $|x| \to \infty$, where $f^{(k)}$ is the kth derivative of f when $k \geq 1$ and $f^{(0)} = f$. The collection $\mathcal{S}$ is called the *Schwartz class*. Prove that if $f \in \mathcal{S}$, then $\widehat{f} \in \mathcal{S}$.

Exercise 16.6 The Fourier transform of a finite signed measure μ on $\mathbb{R}^n$ is defined by

$$\widehat{\mu}(u) = \int e^{iu \cdot x} \, \mu(dx).$$

Prove that if μ and ν are two finite signed measures on $\mathbb{R}^n$ (with respect to the completion of $\mathcal{L} \times \mathcal{L}$, where $\mathcal{L}$ is the Lebesgue σ-algebra on $\mathbb{R}$) such that $\widehat{\mu}(u) = \widehat{\nu}(u)$ for all $u \in \mathbb{R}^n$, then $\mu = \nu$.

Exercise 16.7 If f is real-valued and continuously differentiable on $\mathbb{R}$, prove that

$$\left(\int |f|^2 \, dx \right)^2 \leq 4 \left(\int |xf(x)|^2 \, dx \right) \left(\int |f'|^2 \, dx \right).$$

Exercise 16.8 Prove *Heisenberg's inequality* (which is very useful in quantum mechanics): there exists $c > 0$ such that if $a, b \in \mathbb{R}$ and f is in L^2, then

$$\left(\int (x-a)^2 |f(x)|^2 \, dx \right) \left(\int (u-b)^2 |\widehat{f}(u)|^2 \, du \right) \geq c \left(\int |f(x)|^2 \, dx \right)^2.$$

Find the best constant c.

Chapter 17

Riesz representation

In Chapter 4 we constructed measures on $\mathbb{R}$. In this chapter we will discuss how to construct measures on more general topological spaces X.

If X is a topological space, let $\mathcal{B}$ be the Borel σ-algebra and suppose μ is a σ-finite measure on $(X, \mathcal{B})$. Throughout this chapter we will restrict our attention to real-valued functions. If f is continuous on X, let us define

$$L(f) = \int_X f \, d\mu.$$

Clearly L is linear, and if $f \geq 0$, then $L(f) \geq 0$. The main topic of this chapter is to prove a converse, the Riesz representation theorem.

We need more hypotheses on X than just that it is a topological space. For simplicity, throughout this chapter we suppose X is a compact metric space. In fact, with almost no changes in the proof, we could let X be a compact Hausdorff space, and with only relatively minor changes, we could even let X be a locally compact Hausdorff metric space. See Remark 17.1. But here we stick to compact metric spaces.

We let $\mathcal{C}(X)$ be the collection of continuous functions from X to $\mathbb{R}$. Recall that the support of a function f is the closure of $\{x : f(x) \neq 0\}$. We write supp (f) for the support of f. If G is an open subset of X, we define $\mathcal{F}_G$ by

$$\mathcal{F}_G = \{f \in \mathcal{C}(X) : 0 \le f \le 1, \operatorname{supp}(f) \subset G\}.$$

Observe that if $f \in \mathcal{F}_G$, then $0 \le f \le \chi_G$, but the converse does not hold. For example, if $X = [-2, 2]$, $G = (-1, 1)$, and $f(x) = (1 - x^2)^+$, then $0 \le f \le \chi_G$, but the support of f, which is $[-1, 1]$, is not contained in G.

17.1   Partitions of unity

The reason we take our set X to be a metric space is that if $K \subset G \subset X$, where K is compact and G is open, then there exists $f \in \mathcal{F}_G$ such that f is 1 on K. If we let

$$f(x) = \left(1 - \frac{d(x, K)}{\delta/2}\right)^+,$$

where $d(x, K) = \inf\{d(x, y) : y \in K\}$ is the distance from x to K and $\delta = \inf\{d(x, y) : x \in K, y \in G^c\}$, then this f will do the job.

Remark 17.1 If X is a compact Hausdorff space instead of a compact metric one, we can still find such an f, that is, $f \in \mathcal{F}_G$ with $f \ge \chi_K$ when $K \subset G$, K is compact, and G is open. Urysohn's lemma is the result from topology that guarantees such an f exists; see Section 20.6. (A Hausdorff space X is one where if $x, y \in X$, $x \ne y$, there exist disjoint open sets G_x and G_y with $x \in G_x$ and $y \in G_y$. An example of a compact Hausdorff space that is not a metric space and cannot be made into a metric space is $[0, 1]^{\mathbb{R}}$ with the product topology.) See Chapter 20 for details.

We will need the following proposition.

Proposition 17.2 *Suppose K is compact and $K \subset G_1 \cup \cdots \cup G_n$, where the G_i are open sets. There exist $g_i \in \mathcal{F}_{G_i}$ for $i = 1, 2, \ldots, n$ such that $\sum_{i=1}^n g_i(x) = 1$ if $x \in K$.*

The collection $\{g_i\}$ is called a *partition of unity* on K, subordinate to the cover $\{G_i\}$.

Proof. Let $x \in K$. Then x will be in at least one G_i. Single points are always compact, so there exists $h_x \in \mathcal{F}_{G_i}$ such that $h_x(x) = 1$.

Let $N_x = \{y : h_x(y) > 0\}$. Since h_x is continuous, then N_x is open, $x \in N_x$, and $\overline{N_x} \subset G_i$.

The collection $\{N_x\}$ is an open cover for the compact set K, so there exists a finite subcover $\{N_{x_1}, \ldots, N_{x_m}\}$. For each i, let

$$F_i = \cup \{\overline{N_{x_j}} : \overline{N_{x_j}} \subset G_i\}.$$

Each F_i is closed, and since X is compact, F_i is compact. We have $F_i \subset G_i$. Let us choose $f_i \in \mathcal{F}_{G_i}$ such that f_i is 1 on F_i.

Now define

$$g_1 = f_1,$$
$$g_2 = (1 - f_1)f_2,$$
$$\ldots$$
$$g_n = (1 - f_1)(1 - f_2) \cdots (1 - f_{n-1})f_n.$$

Clearly $g_i \in \mathcal{F}_{G_i}$. Note $g_1 + g_2 = 1 - (1 - f_1)(1 - f_2)$, and an induction argument shows that

$$g_1 + \cdots + g_n = 1 - (1 - f_1)(1 - f_2) \cdots (1 - f_n).$$

If $x \in K$, then $x \in N_{x_j}$ for some j, so $x \in F_i$ for some i. Then $f_i(x) = 1$, which implies $\sum_{k=1}^{n} g_k(x) = 1$. □

17.2   The representation theorem

Let L be a linear functional mapping $\mathcal{C}(X)$ to $\mathbb{R}$. Thus $L(f + g) = L(f) + L(g)$ and $L(af) = aL(f)$ if $f, g \in \mathcal{C}(X)$ and $a \in \mathbb{R}$. L is a *positive linear functional* if $L(f) \geq 0$ whenever $f \geq 0$ on X.

Here is the *Riesz representation theorem*. $\mathcal{B}$ is the Borel σ-algebra on X, that is, the smallest σ-algebra that contains all the open subsets of X.

Theorem 17.3 *Let X be a compact metric space and L a positive linear functional on $\mathcal{C}(X)$. Then there exists a measure μ on $(X, \mathcal{B})$ such that*

$$L(f) = \int f(y)\,\mu(dy), \qquad f \in \mathcal{C}(X). \tag{17.1}$$

We often write Lf for $L(f)$. Since X is compact, taking f identically equal to 1 in (17.1) shows that μ is a finite measure.

Proof. If G is open, let

$$\ell(G) = \sup\{Lf : f \in \mathcal{F}_G\}$$

and for $E \subset X$, let

$$\mu^*(E) = \inf\{\ell(G) : E \subset G, G \text{ open}\}.$$

Step 1 of the proof will be to show μ^* is an outer measure. Step 2 is to show that every open set is μ^*-measurable. Step 3 is to apply Theorem 4.6 to obtain a measure μ. Step 4 establishes some regularity of μ and Step 5 shows that (17.1) holds.

Step 1. We show μ^* is an outer measure. The only function in $\mathcal{F}_\emptyset$ is the zero function, so $\ell(\emptyset) = 0$, and therefore $\mu^*(\emptyset) = 0$. Clearly $\mu^*(A) \leq \mu^*(B)$ if $A \subset B$.

To show the countable subadditivity of μ^*, first let $G_1, G_2, \ldots$ be open sets. For any open set H we see that $\mu^*(H) = \ell(H)$. Let $G = \cup_i G_i$ and let f be any element of $\mathcal{F}_G$. Let K be the support of f. Then K is compact, $\{G_i\}$ is an open cover for K, and therefore there exists n such that $K \subset \cup_{i=1}^n G_i$. Let $\{g_i\}$ be a partition of unity for K subordinate to $\{G_i\}_{i=1}^n$. Since K is the support of f, we have $f = \sum_{i=1}^n f g_i$. Since $g_i \in \mathcal{F}_{G_i}$ and f is bounded by 1, then $f g_i \in \mathcal{F}_{G_i}$. Therefore

$$Lf = \sum_{i=1}^n L(f g_i) \leq \sum_{i=1}^n \mu^*(G_i) \leq \sum_{i=1}^\infty \mu^*(G_i).$$

Taking the supremum over $f \in \mathcal{F}_G$,

$$\mu^*(G) = \ell(G) \leq \sum_{i=1}^\infty \mu^*(G_i).$$

If $A_1, A_2, \ldots$ are subsets of X, let $\varepsilon > 0$, and choose G_i open such that $\ell(G_i) \leq \mu^*(G_i) + \varepsilon 2^{-i}$. Then

$$\mu^*(\cup_{i=1}^\infty A_i) \leq \mu^*(\cup_{i=1}^\infty G_i) \leq \sum_{i=1}^\infty \mu^*(G_i) \leq \sum_{i=1}^\infty \mu^*(A_i) + \varepsilon.$$

Since ε is arbitrary, countable subadditivity is proved, and we conclude that μ^* is an outer measure.

Step 2. We show that every open set is μ^*-measurable. Suppose G is open and $E \subset X$. It suffices to show

$$\mu^*(E) \geq \mu^*(E \cap G) + \mu^*(E \cap G^c), \qquad (17.2)$$

since the opposite inequality is true by the countable subadditivity of μ^*.

First suppose E is open. Choose $f \in \mathcal{F}_{E \cap G}$ such that

$$L(f) > \ell(E \cap G) - \varepsilon/2.$$

Let K be the support of f. Since K^c is open, we can choose $g \in \mathcal{F}_{E \cap K^c}$ such that $L(g) > \ell(E \cap K^c) - \varepsilon/2$. Then $f + g \in \mathcal{F}_E$, and

$$\begin{aligned}
\ell(E) \geq L(f + g) = Lf + Lg &\geq \ell(E \cap G) + \ell(E \cap K^c) - \varepsilon \\
&= \mu^*(E \cap G) + \mu^*(E \cap K^c) - \varepsilon \\
&\geq \mu^*(E \cap G) + \mu^*(E \cap G^c) - \varepsilon.
\end{aligned}$$

Since ε is arbitrary, (17.2) holds when E is open.

If $E \subset X$ is not necessarily open, let $\varepsilon > 0$ and choose H open such that $E \subset H$ and $\ell(H) \leq \mu^*(E) + \varepsilon$. Then

$$\begin{aligned}
\mu^*(E) + \varepsilon \geq \ell(H) = \mu^*(H) &\geq \mu^*(H \cap G) + \mu^*(H \cap G^c) \\
&\geq \mu^*(E \cap G) + \mu^*(E \cap G^c).
\end{aligned}$$

Since ε is arbitrary, (17.2) holds.

Step 3. Let $\mathcal{B}$ be the Borel σ-algebra on X. By Theorem 4.6, the restriction of μ^* to $\mathcal{B}$, which we call μ, is a measure on $\mathcal{B}$. In particular, if G is open, $\mu(G) = \mu^*(G) = \ell(G)$.

Step 4. In this step we show that if K is compact, $f \in \mathcal{C}(X)$, and $f \geq \chi_K$, then $L(f) \geq \mu(K)$. Let $\varepsilon > 0$ and define

$$G = \{x : f(x) > 1 - \varepsilon\},$$

which is open. If $g \in \mathcal{F}_G$, then $g \leq \chi_G \leq f/(1 - \varepsilon)$, so

$$(1 - \varepsilon)^{-1} f - g \geq 0.$$

Because L is a positive linear functional,

$$L((1 - \varepsilon)^{-1} f - g) \geq 0,$$

which leads to $Lg \le Lf/(1-\varepsilon)$. This is true for all $g \in \mathcal{F}_G$, hence

$$\mu(K) \le \mu(G) \le \frac{Lf}{1-\varepsilon}.$$

Since ε is arbitrary, $\mu(K) \le Lf$.

Step 5. We now establish (17.1). By writing $f = f^+ - f^-$ and using the linearity of L, to show (17.1) for continuous functions we may suppose $f \ge 0$. Since X is compact, then f is bounded, and multiplying by a constant and using linearity, we may suppose $0 \le f \le 1$.

Let $n \ge 1$ and let $K_i = \{x : f(x) \ge i/n\}$. Since f is continuous, each K_i is a closed set, hence compact. K_0 is all of X. Define

$$f_i(x) = \begin{cases} 0, & x \in K_{i-1}^c; \\ f(x) - \frac{i-1}{n}, & x \in K_{i-1} - K_i; \\ \frac{1}{n}, & x \in K_i. \end{cases}$$

Note $f = \sum_{i=1}^n f_i$ and $\chi_{K_i} \le nf_i \le \chi_{K_{i-1}}$. Therefore

$$\frac{\mu(K_i)}{n} \le \int f_i \, d\mu \le \frac{\mu(K_{i-1})}{n},$$

and so

$$\frac{1}{n} \sum_{i=1}^n \mu(K_i) \le \int f \, d\mu \le \frac{1}{n} \sum_{i=0}^{n-1} \mu(K_i). \qquad (17.3)$$

Let $\varepsilon > 0$ and let G be an open set containing K_{i-1} such that $\mu(G) < \mu(K_{i-1}) + \varepsilon$. Then $nf_i \in \mathcal{F}_G$, so

$$L(nf_i) \le \mu(G) \le \mu(K_{i-1}) + \varepsilon.$$

Since ε is arbitrary, $L(f_i) \le \mu(K_{i-1})/n$. By Step 4, $L(nf_i) \ge \mu(K_i)$, and hence

$$\frac{1}{n} \sum_{i=1}^n \mu(K_i) \le L(f) \le \frac{1}{n} \sum_{i=0}^{n-1} \mu(K_i). \qquad (17.4)$$

Comparing (17.3) and (17.4) we see that

$$\left| L(f) - \int f \, d\mu \right| \le \frac{\mu(K_0) - \mu(K_n)}{n} \le \frac{\mu(X)}{n}.$$

Since, as we saw above, $\mu(X) = L(1) < \infty$ and n is arbitrary, then (17.1) is established. $\qquad \square$

Example 17.4 If f is continuous on $[a, b]$, let $L(f)$ be the Riemann integral of f on the interval $[a, b]$. Then L is a positive linear functional on $\mathcal{C}([a, b])$. In this case, the measure whose existence is given by the Riesz representation theorem is Lebesgue measure.

Remark 17.5 Let X be a metric space, not necessarily compact. A continuous function f *vanishes at infinity* if given $\varepsilon > 0$ there exists a compact set K such that $|f(x)| < \varepsilon$ if $x \notin K$. $\mathcal{C}_0(X)$ is the usual notation for the set of continuous functions vanishing at infinity. There is a version of the Riesz representation theorem for $\mathcal{C}_0(X)$. See [4] for details.

17.3 Regularity

We establish the following regularity property of measures on compact metric spaces.

Proposition 17.6 *Suppose X is a compact measure space, $\mathcal{B}$ is the Borel σ-algebra, and μ is a finite measure on the measurable space $(X, \mathcal{B})$. If $E \in \mathcal{B}$ and $\varepsilon > 0$, there exists $K \subset E \subset G$ such that K is compact, G is open, $\mu(G - E) < \varepsilon$, and $\mu(E - K) < \varepsilon$. (K and G depend on ε as well as on E.)*

Proof. Let us say that a subset $E \in \mathcal{B}$ is approximable if given $\varepsilon > 0$ there exists $K \subset E \subset G$ with K compact, G open, $\mu(G - E) < \varepsilon$, and $\mu(E - F) < \varepsilon$. Let $\mathcal{H}$ be the collection of approximable subsets. We will show $\mathcal{H}$ contains all the compact sets and $\mathcal{H}$ is a σ-algebra, which will prove that $\mathcal{H} = \mathcal{B}$, and thus establish the proposition.

If K is compact, let $G_n = \{x : d(x, K) < 1/n\}$. Then the G_n are open sets decreasing to K, and if n is large enough, $\mu(G_n - K) < \varepsilon$. Thus every compact set is in $\mathcal{H}$.

If E is in $\mathcal{H}$ and $\varepsilon > 0$, then choose $K \subset E \subset G$ with K compact, G open, $\mu(E - K) < \varepsilon$, and $\mu(G - E) < \varepsilon$. Then $G^c \subset E^c \subset K^c$, G^c is closed, hence compact, K^c is open, $\mu(K^c - E^c) = \mu(E - K) < \varepsilon$, and $\mu(E^c - G^c) = \mu(G - E) < \varepsilon$. Therefore $\mathcal{H}$ is closed under the operation of taking complements.

Suppose $E_1, E_2, \ldots \in \mathcal{H}$. For each i choose K_i compact and G_i open such that $K_i \subset E_i \subset G_i$, $\mu(G_i - E_i) < \varepsilon 2^{-i}$, and $\mu(E_i - K_i) <$

$\varepsilon 2^{-(i+1)}$. Then $\cup_{i=1}^{\infty} G_i$ is open, contains $\cup_{i=1}^{\infty} E_i$, and

$$\mu(\cup_i G_i - \cup_i E_i) \leq \sum_{i=1}^{\infty} \mu(G_i - E_i) < \varepsilon.$$

We see that $\cup_{i=1}^{\infty} K_i$ is contained in $\cup_{i=1}^{\infty} E_i$ and similarly,

$$\mu(\cup_{i=1}^{\infty} E_i - \cup_{i=1}^{\infty} K_i) \leq \sum_{i=1}^{\infty} \mu(E_i - K_i) < \varepsilon/2.$$

Since $\cup_{i=1}^{n} K_i$ increases to $\cup_{i=1}^{\infty} K_i$, we can choose n large so that

$$\mu(\cup_{i=n+1}^{\infty} K_i) < \varepsilon/2.$$

Then $\cup_{i=1}^{n} K_i$, being the finite union of compact sets, is compact, is contained in $\cup_{i=1}^{\infty} E_i$, and

$$\mu(\cup_{i=1}^{\infty} E_i - \cup_{i=1}^{n} K_i) < \varepsilon.$$

This proves that $\cup_i E_i$ is in $\mathcal{H}$.

Since $\mathcal{H}$ is closed under the operations of taking complements and countable unions and $\cap_i E_i = (\cup_i E_i^c)^c$, then $\mathcal{H}$ is also closed under the operation of taking countable intersections. Therefore $\mathcal{H}$ is a σ-algebra. $\square$

A measure is called *regular* if

$$\mu(E) = \inf\{\mu(G) : G \text{ open}, E \subset G\}$$

and

$$\mu(E) = \sup\{\mu(K) : K \text{ compact}, K \subset E\}$$

for all measurable E. An immediate consequence of what we just proved is that finite measures on $(X, \mathcal{B})$ are regular when X is a compact metric space.

17.4   Bounded linear functionals

We have proved the Riesz representation theorem for positive linear functionals on $\mathcal{C}(X)$. In Chapter 25 we will need a version for complex-valued bounded linear functionals. We do the real-valued

case in this section; the complex-valued case follows relatively easily
and is Exercise 17.10.

The following proposition is key. We set $\|f\| = \sup_{x \in X} |f(x)|$
for $f \in \mathcal{C}(X)$ and if I is a bounded linear functional on $\mathcal{C}(X)$, we
let $\|I\| = \sup_{\|f\|=1} |I(f)|$.

Proposition 17.7 *Suppose I is a bounded linear functional on*
$\mathcal{C}(X)$. *Then there exist positive bounded linear functionals J and*
K *such that* $I = J - K$.

Proof. For $g \in \mathcal{C}(X)$ with $g \geq 0$, define

$$J(g) = \sup\{I(f) : f \in \mathcal{C}(X), 0 \leq f \leq g\}.$$

Since $I(0) = 0$, then $J(g) \geq 0$. Since $|I(f)| \leq \|I\| \, \|f\| \leq \|I\| \, \|g\|$ if
$0 \leq f \leq g$, then $|J(g)| \leq \|I\| \, \|g\|$. Clearly $J(cg) = cJ(g)$ if $c \geq 0$.

We prove that

$$J(g_1 + g_2) = J(g_1) + J(g_2) \tag{17.5}$$

if $g_1, g_2 \in \mathcal{C}(X)$ are non-negative. If $0 \leq f_1 \leq g_1$ and $0 \leq f_2 \leq g_2$
with each of the four functions in $\mathcal{C}(X)$, we have $0 \leq f_1 + f_2 \leq$
$g_1 + g_2$, so

$$J(g_1 + g_2) \geq I(f_1 + f_2) = I(f_1) + I(f_2).$$

Taking the supremum over all such f_1 and f_2,

$$J(g_1 + g_2) \geq J(g_1) + J(g_2). \tag{17.6}$$

To get the opposite inequality, suppose $0 \leq f \leq g_1 + g_2$ with
each function non-negative and in $\mathcal{C}(X)$. Let $f_1 = f \wedge g_1$ and
$f_2 = f - f_1$. Note $f_1, f_2 \in \mathcal{C}(X)$. Since $f_1 \leq f$, then $f_2 \geq$
0. If $f(x) \leq g_1(x)$, we have $f(x) = f_1(x) \leq f_1(x) + g_2(x)$. If
$f(x) > g_1(x)$, we have $f(x) \leq g_1(x) + g_2(x) = f_1(x) + g_2(x)$. Hence
$f \leq f_1 + g_2$, so $f_2 = f - f_1 \leq g_2$. Thus

$$I(f) = I(f_1) + I(f_2) \leq J(g_1) + J(g_2).$$

Taking the supremum over $f \in \mathcal{C}(X)$ with $0 \leq f \leq g_1 + g_2$,

$$J(g_1 + g_2) \leq J(g_1) + J(g_2).$$

Combining with (17.6) proves (17.5).

If $f \in \mathcal{C}(X)$, define $J(f) = J(f^+) - J(f^-)$. We have

$$f_1^+ - f_1^- + f_2^+ - f_2^- = f_1 + f_2 = (f_1 + f_2)^+ - (f_1 + f_2)^-,$$

so

$$f_1^+ + f_2^+ + (f_1 + f_2)^- = (f_1 + f_2)^+ + f_1^- + f_2^-.$$

Hence

$$J(f_1^+) + J(f_2^+) + J((f_1 + f_2)^-) = J((f_1 + f_2)^+) + J(f_1^-) + J(f_2^-).$$

Rearranging,

$$J(f_1 + f_2) = J(f_1) + J(f_2).$$

Showing $J(cf) = cJ(f)$ is easier, and we conclude J is a linear functional on $\mathcal{C}(X)$.

We write

$$\begin{aligned}
|J(f)| = |J(f^+) - J(f^-)| &\leq J(f^+) \vee J(f^-) \\
&\leq (\|I\| \, \|f^+\|) \vee (\|I\| \, \|f^-\|) \\
&= \|I\| \, (\|f^+\| \vee \|f^-\|) \\
&\leq \|I\| \, \|f\|.
\end{aligned}$$

Thus J is a bounded linear functional.

If $f \geq 0$, then $J(f) \geq 0$. Set $K = J - I$. If $f \geq 0$, then $I(f) \leq J(f)$, so $K(f) \geq 0$, and K is also a positive operator. □

We now state the *Riesz representation theorem* for bounded real-valued linear functionals.

Theorem 17.8 *If X is a compact metric space and I is a bounded linear functional on $\mathcal{C}(X)$, there exists a finite signed measure μ on the Borel σ-algebra such that*

$$I(f) = \int f \, d\mu$$

for each $f \in \mathcal{C}(X)$.

Proof. Write $I = J - K$ as in Proposition 17.7. By the Riesz representation theorem for positive linear functionals there exist positive finite measures μ^+ and μ^- such that $J(f) = \int f \, d\mu^+$ and $K(f) = \int f \, d\mu^-$ for every $f \in \mathcal{C}(X)$. Then $\mu = \mu^+ - \mu^-$ will be the signed measure we seek. □

17.5 Exercises

Exercise 17.1 Suppose F is a closed subset of $[0,1]$ and we define

$$L(f) = \int_0^1 f\chi_F \, dx$$

for real-valued continuous functions f on $[0,1]$. Prove that if μ is the measure whose existence is given by the Riesz representation theorem, then $\mu(A) = m(A \cap F)$, where m is Lebesgue measure.

Exercise 17.2 Suppose X is a compact metric space and μ is a finite regular measure on $(X, \mathcal{B})$, where $\mathcal{B}$ is the Borel σ-algebra. Prove that if f is a real-valued measurable function and $\varepsilon > 0$, there exists a closed set F such that $\mu(F^c) < \varepsilon$ and the restriction of f to F is a continuous function on F.

Exercise 17.3 Let $C^1([0,1])$ be the set of functions whose derivative exists and is continuous on $[0,1]$. Suppose L is a linear functional on $C^1([0,1])$ such that

$$|L(f)| \leq c_1\|f'\| + c_2\|f\|$$

for all $f \in C^1([0,1])$, where c_1 and c_2 are positive constants and the norm is the supremum norm. Show there exists a signed measure μ on the Borel subsets of $[0,1]$ and a constant K such that

$$L(f) = \int f' \, d\mu + Kf(0), \qquad f \in C^1([0,1]).$$

Exercise 17.4 Suppose X and Y are compact metric spaces and $F : X \to Y$ is a continuous map from X onto Y. If ν is a finite measure on the Borel sets of Y, prove that there exists a measure μ on the Borel sets of X such that

$$\int_Y f \, d\nu = \int_X f \circ F \, d\mu$$

for all f that are continuous on Y.

Exercise 17.5 Let X be a compact metric space. Prove that $\mathcal{C}(X)$ has a countable dense subset.

Exercise 17.6 Let X be a compact metric space and let $\mathcal{B}$ be the Borel σ-algebra on X. Let μ_n be a sequence of finite measures on $(X, \mathcal{B})$ and let μ be another finite measure on $(X, \mathcal{B})$. Suppose $\mu_n(X) \to \mu(X)$. Prove that the following are equivalent:
(1) $\int f \, d\mu_n \to \int f \, d\mu$ whenever f is a continuous real-valued function on X;
(2) $\limsup_{n \to \infty} \mu_n(F) \leq \mu(F)$ for all closed subsets F of X;
(3) $\liminf_{n \to \infty} \mu_n(G) \geq \mu(G)$ for all open subsets G of X;
(4) $\lim_{n \to \infty} \mu_n(A) = \mu(A)$ whenever A is a Borel subset of X such that $\mu(\partial A) = 0$, where $\partial A = \overline{A} - A^o$ is the boundary of A.

Exercise 17.7 Let X be a compact metric space and let $\mathcal{B}$ be the Borel σ-algebra on X. Let μ_n be a sequence of finite measures on $(X, \mathcal{B})$ and suppose $\sup_n \mu_n(X) < \infty$.
(1) Prove that if $f \in \mathcal{C}(X)$, there is a subsequence $\{n_j\}$ such that $\int f \, d\mu_{n_j}$ converges.
(2) Let A be a countable dense subset of $\mathcal{C}(X)$. Prove that there is a subsequence $\{n_j\}$ such that $\int f \, d\mu_{n_j}$ converges for all $f \in A$.
(3) With $\{n_j\}$ as in (2), prove that $\int f \, d\mu_{n_j}$ converges for all $f \in \mathcal{C}(X)$.
(4) Let $L(f) = \lim_{n_j \to \infty} \int f \, d\mu_{n_j}$. Prove that $L(f)$ is a positive linear functional on $\mathcal{C}(X)$. Conclude that there exists a measure μ such that

$$\int f \, d\mu_{n_j} \to \int f \, d\mu$$

for all $f \in \mathcal{C}(X)$.

Exercise 17.8 Prove that if X is a compact metric space, $\mathcal{B}$ is the Borel σ-algebra, and μ and ν are two finite positive measures on $(X, \mathcal{B})$ such that

$$\int f \, d\mu = \int f \, d\nu$$

for all $f \in \mathcal{C}(X)$, then $\mu = \nu$.

Exercise 17.9 Prove that if X is a compact metric space, $\mathcal{B}$ is the Borel σ-algebra, and μ and ν are two finite signed measures on $(X, \mathcal{B})$ such that

$$\int f \, d\mu = \int f \, d\nu$$

for all $f \in \mathcal{C}(X)$, then $\mu = \nu$.

Exercise 17.10 State and prove a version of the Riesz representation theorem for complex measures.

Exercise 17.11 Prove that if X is a compact metric space, $\mathcal{B}$ is the Borel σ-algebra, and μ is a complex measure on $(X, \mathcal{B})$, then the total variation of μ, defined in Exercise 13.4, equals

$$\sup_{f \in \mathcal{C}(X)} \left| \int f \, d\mu \right|.$$

Chapter 18

Banach spaces

Banach spaces are normed linear spaces that are complete. We will give the definitions, discuss the existence of bounded linear functionals, prove the Baire category theorem, and derive some consequences such as the uniform boundedness theorem and the open mapping theorem.

18.1   Definitions

The definition of normed linear space X over a field of scalars F, where F is either the real numbers or the complex numbers, was given in Chapter 1. Recall that a normed linear space is a metric space if we use the metric $d(x, y) = \|x - y\|$.

Definition 18.1 We define a *Banach space* to be a normed linear space that is complete, that is, where every Cauchy sequence converges.

A *linear map* is a map L from a normed linear space X to a normed linear space Y satisfying $L(x + y) = L(x) + L(y)$ for all $x, y \in X$ and $L(\alpha x) = \alpha L(x)$ for all $x \in X$ and $\alpha \in F$. We will sometimes write Lx for $L(x)$. Since $L(0) = L(0+0) = L(0) + L(0)$, then $L(0) = 0$.

Definition 18.2 A linear map f from X to $\mathbb{R}$ is a *real linear functional*, a linear map from X to $\mathbb{C}$ a *complex linear functional*. f is a *bounded linear functional* if

$$\|f\| = \sup\{|f(x)| : x \in X, \|x\| \le 1\} < \infty.$$

Proposition 18.3 *The following are equivalent.*
(1) The linear functional f is bounded.
(2) The linear functional f is continuous.
(3) The linear functional f is continuous at 0.

Proof. $|f(x) - f(y)| = |f(x-y)| \le \|f\| \|x - y\|$, so (1) implies (2). That (2) implies (3) is obvious. To show (3) implies (1), if f is not bounded, there exists a sequence $x_n \in X$ such that $\|x_n\| = 1$ for each n, but $|f(x_n)| \to \infty$. If we let $y_n = x_n/|f(x_n)|$, then $y_n \to 0$ but $|f(y_n)| = 1 \not\to 0$, contradicting (3). $\square$

18.2 The Hahn-Banach theorem

We want to prove that there are plenty of linear functionals. First we state Zorn's lemma, which is equivalent to the axiom of choice.

If we have a set Y with a partial order "$\le$" (defined in Chapter 1), a linear ordered subset $X \subset Y$ is one such that if $x, y \in X$, then either $x \le y$ or $y \le x$ (or both) holds. A linearly ordered subset $X \subset Y$ has an upper bound if there exists an element z of Y (but it is not necessary that $z \in X$) such that $x \le z$ for all $x \in X$. An element z of Y is maximal if $z \le y$ for $y \in Y$ implies $y = z$.

Here is *Zorn's lemma*.

Lemma 18.4 *If Y is a partially ordered set and every linearly ordered subset of Y has an upper bound, then Y has a maximal element.*

A subspace of a normed linear space X is a subset $M \subset X$ such that M is itself a normed linear space.

We now give the *Hahn-Banach theorem* for real linear functionals.

Theorem 18.5 *If M is a subspace of a normed linear space X and f is a bounded real linear functional on M, then f can be extended to a bounded linear functional F on X such that $\|F\| = \|f\|$.*

Saying that F is an extension of f means that the domain of F contains the domain of f and $F(x) = f(x)$ if x is in the domain of f.

Proof. If $\|f\| = 0$, then we take F to be identically 0, so we may assume that $\|f\| \neq 0$, and then by multiplying by a constant, that $\|f\| = 1$. We first show that we can extend f by at least one dimension.

Choose $x_0 \in X - M$ and let M_1 be the vector space spanned by M and x_0. Thus M_1 consists of all vectors of the form $x + \lambda x_0$, where $x \in X$ and λ is real.

We have for all $x, y \in M$

$$f(x) - f(y) = f(x - y) \leq \|x - y\| \leq \|x - x_0\| + \|y - x_0\|.$$

Hence

$$f(x) - \|x - x_0\| \leq f(y) + \|y - x_0\|$$

for all $x, y \in M$. Choose $\alpha \in \mathbb{R}$ such that

$$f(x) - \|x - x_0\| \leq \alpha \leq f(y) + \|y - x_0\|$$

for all $x, y \in M$. Define $f_1(x + \lambda x_0) = f(x) + \lambda \alpha$. This is clearly an extension of f to M_1.

We need to verify that the norm of f_1 is less than or equal to 1. Let $x \in M$ and $\lambda \in \mathbb{R}$. By our choice of α, $f(x) - \|x - x_0\| \leq \alpha$, or $f(x) - \alpha \leq \|x - x_0\|$, and $\alpha \leq f(x) + \|x - x_0\|$, or $f(x) - \alpha \geq -\|x - x_0\|$. Thus

$$|f(x) - \alpha| \leq \|x - x_0\|.$$

Replacing x by $-x/\lambda$ and multiplying by $|\lambda|$, we get

$$|\lambda| \, | - f(x)/\lambda - \alpha| \leq |\lambda| \, \| - x/\lambda - x_0\|,$$

or

$$|f_1(x + \lambda x_0)| = |f(x) + \lambda \alpha| \leq \|x + \lambda x_0\|,$$

which is what we wanted to prove.

We now establish the existence of an extension of f to all of X. Let $\mathcal{F}$ be the collection of all linear extensions F of f satisfying

$\|F\| \leq 1$. This collection is partially ordered by inclusion. That is, if f_1 is an extension of f to a subspace M_1 and f_2 is an extension of f to a subspace M_2, we say $f_1 \leq f_2$ if $M_1 \subset M_2$. Since the union of any increasing family of subspaces of X is again a subspace, then the union of a linearly ordered subfamily of $\mathcal{F}$ lies in $\mathcal{F}$. By Zorn's lemma, $\mathcal{F}$ has a maximal element, say, F_1. By the construction of the preceding two paragraphs, if the domain of F_1 is not all of X, we can find an extension, which would be a contradiction to F_1 being maximal. Therefore F_1 is the desired extension.   □

To get a version for complex valued linear functionals is quite easy. Note that if $f(x) = u(x) + iv(x)$, then the real part of f, namely, $u = \operatorname{Re} f$, is a real valued linear functional. Also, $u(ix) = \operatorname{Re} f(ix) = \operatorname{Re} if(x) = -v(x)$, so that $v(x) = -u(ix)$, and hence $f(x) = u(x) - iu(ix)$.

Theorem 18.6 *If M is a subspace of a normed linear space X and f is a bounded complex linear functional on M, then f can be extended to a bounded linear functional F on X such that $\|F\| = \|f\|$.*

Proof.   Assume without loss of generality that $\|f\| = 1$. Let $u = \operatorname{Re} f$. Note $|u(x)| \leq |f(x)| \leq \|x\|$. Now use the version of the Hahn-Banach theorem for real linear functionals to find a linear functional U that is an extension of u to X such that $\|U\| \leq 1$. Let $F(x) = U(x) - iU(ix)$.

It only remains to show that the norm of F is at most 1. Fix x, and write $F(x) = re^{i\theta}$. Then

$$|F(x)| = r = e^{-i\theta}F(x) = F(e^{-i\theta}x).$$

Since this quantity is real and non-negative,

$$|F(x)| = U(e^{-i\theta}x) \leq \|U\| \, \|e^{-i\theta}x\| \leq \|x\|.$$

This holds for all x, so $\|F\| \leq 1$.   □

As an application of the Hahn-Banach theorem, given a subspace M and an element x_0 not in M, we can define $f(x + \lambda x_0) = \lambda x_0$ for $x \in M$, and then extend this linear functional to all of X. Then f will be 0 on M but non-zero at x_0.

Another application is to fix $x_0 \neq 0$, let $f(\lambda x_0) = \lambda \|x_0\|$, and then extend f to all of X. Thus there exists a linear functional f such that $f(x_0) = \|x_0\|$ and $\|f\| = 1$.

18.3   Baire's theorem and consequences

We turn now to the Baire category theorem and some of its consequences. Recall that if A is a set, we use $\overline{A}$ for the closure of A and A^o for the interior of A. A set A is dense in X if $\overline{A} = X$ and A is nowhere dense if $(\overline{A})^o = \emptyset$.

The *Baire category theorem* is the following. Completeness of the metric space is crucial to the proof.

Theorem 18.7 *Let X be a complete metric space.*
(1) If G_n are open sets dense in X, then $\cap_n G_n$ is dense in X.
(2) X cannot be written as the countable union of nowhere dense sets.

Proof. We first show that (1) implies (2). Suppose we can write X as a countable union of nowhere dense sets, that is, $X = \cup_n E_n$ where $(\overline{E_n})^o = \emptyset$. We let $F_n = \overline{E_n}$, which is a closed set, and then $F_n^o = \emptyset$ and $X = \cup_n F_n$. Let $G_n = F_n^c$, which is open. Since $F_n^o = \emptyset$, then $\overline{G_n} = X$. Starting with $X = \cup_n F_n$ and taking complements, we see that $\emptyset = \cap_n G_n$, a contradiction to (1).

We must prove (1). Suppose $G_1, G_2, \ldots$ are open and dense in X. Let H be any non-empty open set in X. We need to show there exists a point in $H \cap (\cap_n G_n)$. We will construct a certain Cauchy sequence $\{x_n\}$ and the limit point, x, will be the point we seek.

Let $B(z, r) = \{y \in X : d(z, y) < r\}$, where d is the metric. Since G_1 is dense in X, $H \cap G_1$ is non-empty and open, and we can find x_1 and r_1 such that $\overline{B(x_1, r_1)} \subset H \cap G_1$ and $0 < r_1 < 1$. Suppose we have chosen x_{n-1} and r_{n-1} for some $n \geq 2$. Since G_n is dense, then $G_n \cap B(x_{n-1}, r_{n-1})$ is open and non-empty, so there exists x_n and r_n such that $\overline{B(x_n, r_n)} \subset G_n \cap B(x_{n-1}, r_{n-1})$ and $0 < r_n < 2^{-n}$. We continue and get a sequence x_n in X. If $m, n > N$, then x_m and x_m both lie on $B(x_N, r_N)$, and so $d(x_m, x_n) < 2r_N < 2^{-N+1}$. Therefore x_n is a Cauchy sequence, and since X is complete, x_n converges to a point $x \in X$.

It remains to show that $x \in H \cap (\cap_n G_n)$. Since x_n lies in $\overline{B(x_N, r_N)}$ if $n > N$, then x lies in each $\overline{B(x_N, r_N)}$, and hence in each G_N. Therefore $x \in \cap_n G_n$. Also,

$$x \in \overline{B(x_n, r_n)} \subset B(x_{n-1}, r_{n-1}) \subset \cdots \subset B(x_1, r_1) \subset H.$$

Thus we have found a point x in $H \cap (\cap_n G_n)$.   □

A set $A \subset X$ is called *meager* or of the *first category* if it is the countable union of nowhere dense sets; otherwise it is of the *second category.*

A linear map L from a normed linear space X into a normed linear space Y is a *bounded linear map* if

$$\|L\| = \sup\{\|Lx\| : \|x\| = 1\} \tag{18.1}$$

is finite.

An important application of the Baire category theorem is the *Banach-Steinhaus theorem*, also called the *uniform boundedness theorem.*

Theorem 18.8 *Suppose X is a Banach space and Y is a normed linear space. Let A be an index set and let $\{L_\alpha : \alpha \in A\}$ be a collection of bounded linear maps from X into Y. Then either there exists a positive real number $M < \infty$ such that $\|L_\alpha\| \leq M$ for all $\alpha \in A$ or else $\sup_\alpha \|L_\alpha x\| = \infty$ for some x.*

Proof. Let $\ell(x) = \sup_{\alpha \in A} \|L_\alpha x\|$. Let $G_n = \{x : \ell(x) > n\}$. We argue that G_n is open. The map $x \to \|L_\alpha x\|$ is a continuous function for each α since L_α is a bounded linear functional. This implies that for each α, the set $\{x : \|L_\alpha x\| > n\}$ is open. Since $x \in G_n$ if and only if for some $\alpha \in A$ we have $\|L_\alpha x\| > n$, we conclude G_n is the union of open sets, hence is open.

Suppose there exists N such that G_N is not dense in X. Then there exists x_0 and r such that $\overline{B(x_0, r)} \cap G_N = \emptyset$. This can be rephrased as saying that if $\|x - x_0\| \leq r$, then $\|L_\alpha(x)\| \leq N$ for all $\alpha \in A$. If $\|y\| \leq r$, we have $y = (x_0 + y) - x_0$. Then $\|(x_0 + y) - x_0\| = \|y\| \leq r$, and hence $\|L_\alpha(x_0 + y)\| \leq N$ for all α. Also, of course, $\|x_0 - x_0\| = 0 \leq r$, and thus $\|L_\alpha(x_0)\| \leq N$ for all α. We conclude that if $\|y\| \leq r$ and $\alpha \in A$,

$$\|L_\alpha y\| = \|L_\alpha((x_0 + y) - x_0)\| \leq \|L_\alpha(x_0 + y)\| + \|L_\alpha x_0\| \leq 2N.$$

Consequently, $\sup_\alpha \|L_\alpha\| \leq M$ with $M = 2N/r$.

The other possibility, by the Baire category theorem, is that every G_n is dense in X, and in this case $\cap_n G_n$ is dense in X. But $\ell(x) = \infty$ for every $x \in \cap_n G_n$. □

The following theorem is called the *open mapping theorem*. It is important that L be onto. A mapping $L : X \to Y$ is *open* if $L(U)$ is open in Y whenever U is open in X. For a measurable set A, we let $L(A) = \{Lx : x \in A\}$.

Theorem 18.9 *Let X and Y be Banach spaces. A bounded linear map L from X onto Y is open.*

Proof. We need to show that if $B(x, r) \subset X$, then $L(B(x, r))$ contains a ball in Y. We will show $L(B(0, r))$ contains a ball centered at 0 in Y. Then using the linearity of L, $L(B(x, r))$ will contain a ball centered at Lx in Y. By linearity, to show that $L(B(0, r))$ contains a ball centered at 0, it suffices to show that $L(B(0, 1))$ contains a ball centered at 0 in Y.

Step 1. We show that there exists r such that $B(0, r2^{-n}) \subset \overline{L(B(0, 2^{-n}))}$ for each n. Since L is onto, $Y = \cup_{n=1}^\infty L(B(0, n))$. The Baire category theorem tells us that at least one of the sets $L(B(0, n))$ cannot be nowhere dense. Since L is linear, $L(B(0, 1))$ cannot be nowhere dense. Thus there exist y_0 and r such that $B(y_0, 4r) \subset \overline{L(B(0, 1))}$.

Pick $y_1 \in L(B(0, 1))$ such that $\|y_1 - y_0\| < 2r$ and let $z_1 \in B(0, 1)$ be such that $y_1 = Lz_1$. Then $B(y_1, 2r) \subset B(y_0, 4r) \subset \overline{L(B(0, 1))}$. Thus if $\|y\| < 2r$, then $y + y_1 \in B(y_1, 2r)$, and so

$$y = -Lz_1 + (y + y_1) \in \overline{L(-z_1 + B(0, 1))}.$$

Since $z_1 \in B(0, 1)$, then $-z_1 + B(0, 1) \subset B(0, 2)$, hence

$$y \in \overline{L(-z_1 + B(0, 1))} \subset \overline{L(B(0, 2))}.$$

By the linearity of L, if $\|y\| < r$, then $y \in \overline{L(B(0, 1))}$. It follows by linearity that if $\|y\| < r2^{-n}$, then $y \in \overline{L(B(0, 2^{-n}))}$. This can be rephrased as saying that if $\|y\| < r2^{-n}$ and $\varepsilon > 0$, then there exists x such that $\|x\| < 2^{-n}$ and $\|y - Lx\| < \varepsilon$.

Step 2. Suppose $\|y\| < r/2$. We will construct a sequence $\{x_j\}$ by induction such that $y = L(\sum_{j=1}^\infty x_j)$. By Step 1 with $\varepsilon = r/4$, we can find $x_1 \in B(0, 1/2)$ such that $\|y - Lx_1\| < r/4$. Suppose we have chosen $x_1, \ldots, x_{n-1}$ such that

$$\left\| y - \sum_{j=1}^{n-1} Lx_j \right\| < r2^{-n}.$$

Let $\varepsilon = r2^{-(n+1)}$. By Step 1, we can find x_n such that $\|x_n\| < 2^{-n}$ and

$$\left\| y - \sum_{j=1}^{n} Lx_j \right\| = \left\| \left(y - \sum_{j=1}^{n-1} Lx_j \right) - Lx_n \right\| < r2^{-(n+1)}.$$

We continue by induction to construct the sequence $\{x_j\}$. Let $w_n = \sum_{j=1}^{n} x_j$. Since $\|x_j\| < 2^{-j}$, then w_n is a Cauchy sequence. Since X is complete, w_n converges, say, to x. But then $\|x\| < \sum_{j=1}^{\infty} 2^{-j} = 1$, and since L is continuous, $y = Lx$. That is, if $y \in B(0, r/2)$, then $y \in L(B(0, 1))$. $\qquad \square$

Remark 18.10 Suppose X and Y are Banach spaces and $\mathcal{L}$ is the collection of bounded linear maps from X into Y. If we define $(L + M)x = Lx + Mx$ and $(cL)x = c(Lx)$ for $L, M \in \mathcal{L}$, $x \in X$, and $c \in F$, and if we define $\|L\|$ by (18.1), then $\mathcal{L}$ itself is a normed linear space. Exercise 18.7 asks you to prove that $\mathcal{L}$ is a Banach space, i.e., that $\mathcal{L}$ with the norm given by (18.1) is complete.

When $Y = F$, either the set of real numbers or the set of complex numbers, then $\mathcal{L}$ is the set of bounded linear functionals on X. In this case we write X^* instead of $\mathcal{L}$ and call X^* the *dual space* of X.

18.4   Exercises

Exercise 18.1 Find a measure space $(X, \mathcal{A}, \mu)$, a subspace Y of $L^1(\mu)$, and a bounded linear functional f on Y with norm 1 such that f has two distinct extensions to $L^1(\mu)$ and each of the extensions has norm equal to 1.

Exercise 18.2 Show that $L^p([0, 1])$ is *separable*, that is, there is a countable dense subset, if $1 \le p < \infty$. Show that $L^\infty([0, 1])$ is not separable.

Exercise 18.3 For $k \geq 1$ and functions $f : [0,1] \to \mathbb{R}$ that are k times differentiable, define

$$\|f\|_{C^k} = \|f\|_\infty + \|f'\|_\infty + \cdots + \|f^{(k)}\|_\infty,$$

where $f^{(k)}$ is the kth derivative of f. Let $C^k([0,1])$ be the collection of k times differentiable functions f with $\|f\|_{C^k} < \infty$. Is $C^k(0,1])$ complete with respect to the norm $\|\cdot\|_{C^k}$?

Exercise 18.4 Let $\alpha \in (0,1)$. For f a real-valued continuous function on $[0,1]$ define

$$\|f\|_{C^\alpha} = \sup_{x \in [0,1]} |f(x)| + \sup_{x,y \in [0,1], x \neq y} \frac{|f(x) - f(y)|}{|x - y|^\alpha}.$$

Let $C^\alpha([0,1])$ be the set of functions f with $\|f\|_{C^\alpha} < \infty$. Is $C^\alpha([0,1])$ complete with respect to the norm $\|\cdot\|_{C^\alpha}$?

Exercise 18.5 For positive integers n let

$$A_n = \left\{ f \in L^1([0,1]) : \int_0^1 |f(x)|^2 \, dx \leq n \right\}.$$

Show that each A_n is a closed subset of $L^1([0,1])$ with empty interior.

Exercise 18.6 Suppose L is a linear functional on a normed linear space X. Prove that L is a bounded linear functional if and only if the set $\{x \in X : L(x) = 0\}$ is closed.

Exercise 18.7 Prove that $\mathcal{L}$ as defined in Remark 18.10 is a Banach space.

Exercise 18.8 A set A in a normed linear space is *convex* if

$$\lambda x + (1 - \lambda)y \in A$$

whenever $x, y \in A$ and $\lambda \in [0,1]$.
(1) Prove that if A is convex, then the closure of A is convex.
(2) Prove that the open unit ball in a normed linear space is convex. (The open unit ball is the set of x such that $\|x\| < 1$.)

Exercise 18.9 The unit ball in a normed linear space X is *strictly convex* if $\|\lambda x + (1 - \lambda)y\| < 1$ whenever $\|f\| = \|g\| = 1$, $f \neq g$, and $\lambda \in (0,1)$.
(1) Let $(X, \mathcal{A}, \mu)$ be a measure space. Prove that the unit ball in $L^p(\mu)$ is strictly convex.
(2) Prove that the unit balls in $L^1(\mu)$, $L^\infty(\mu)$, and $\mathcal{C}(X)$ are not strictly convex provided X consists of more than one point.

Exercise 18.10 Let f_n be a sequence of continuous functions on $\mathbb{R}$ that converge at every point. Prove there exist an interval and a number M such that $\sup_n |f_n|$ is bounded by M on that interval.

Exercise 18.11 Suppose $\| \cdot \|_1$ and $\| \cdot \|_2$ are two norms such that $\|x\|_1 \leq \|x\|_2$ for all x in a vector space X, and suppose X is complete with respect to both norms. Prove that there exists a positive constant c such that

$$\|x\|_2 \leq c\|x\|_1$$

for all $x \in X$.

Exercise 18.12 Suppose X and Y are Banach spaces.
(1) Let $X \times Y$ be the set of ordered pairs (x, y) with

$$(x_1 + x_2, y_1 + y_2) = (x_1, y_1) + (x_2, y_2)$$

for each $x_1, x_2 \in X$ and $y_1, y_2 \in Y$ and $c(x, y) = (cx, cy)$ if $x \in \mathbb{R}$. Define $\|(x, y)\| = \|x\| + \|y\|$. Prove that $X \times Y$ is a Banach space.
(2) Let L be a linear map from X into Y such that if $x_n \to x$ in X and $Lx_n \to y$ in Y, then $y = Lx$. Such a map is called a *closed map*. Let G be the *graph* of L, defined by $G = \{(x, y) : y = Lx\}$. Prove that G is a closed subset of $X \times Y$, hence is complete.
(3) Prove that the function $(x, Lx) \to x$ is continuous, one-one, linear, and maps G onto X.
(4) Prove the *closed graph theorem*, which says that if L is a linear map from one Banach space to another that is a closed map, then L is a continuous map.

Exercise 18.13 Let $X = C^1([0, 1])$ (defined in Exercise 18.3) and $Y = C([0, 1])$. Define $D : X \to Y$ by $Df = f'$. Show that D is a closed map but not a bounded one.

Exercise 18.14 Let A be the set of real-valued continuous functions on $[0, 1]$ such that

$$\int_0^{1/2} f(x)\, dx - \int_{1/2}^1 f(x)\, dx = 1.$$

Prove that A is a closed convex subset of $C([0, 1])$, but there does not exist $f \in A$ such that

$$\|f\| = \inf_{g \in A} \|g\|.$$

Exercise 18.15 Let A_n be the subset of the real-valued continuous functions on $[0, 1]$ given by

$$A_n = \{f : \text{there exists } x \in [0, 1] \text{ such that}$$
$$|f(x) - f(y)| \le n|x - y| \text{ for all } y \in [0, 1]\}.$$

(1) Prove that A_n is nowhere dense in $C([0, 1])$.
(2) Prove that there exist functions f in $C([0, 1])$ which are *nowhere differentiable* on $[0, 1]$, that is, $f'(x)$ does not exist at any point of $[0, 1]$.

Chapter 19

Hilbert spaces

Hilbert spaces are complete normed linear spaces that have an inner product. This added structure allows one to talk about orthonormal sets. We will give the definitions and basic properties. As an application we briefly discuss Fourier series.

19.1   Inner products

Recall that if a is a complex number, then $\overline{a}$ represents the complex conjugate. When a is real, $\overline{a}$ is just a itself.

Definition 19.1 Let H be a vector space where the set of scalars F is either the real numbers or the complex numbers. H is an *inner product space* if there is a map $\langle \cdot, \cdot \rangle$ from $H \times H$ to F such that
(1) $\langle y, x \rangle = \overline{\langle x, y \rangle}$ for all $x, y \in H$;
(2) $\langle x + y, z \rangle = \langle x, z \rangle + \langle y, z \rangle$ for all $x, y, z \in H$;
(3) $\langle \alpha x, y \rangle = \alpha \langle x, y \rangle$, for $x, y \in H$ and $\alpha \in F$;
(4) $\langle x, x \rangle \geq 0$ for all $x \in H$;
(5) $\langle x, x \rangle = 0$ if and only if $x = 0$.

We define $\|x\| = \langle x, x \rangle^{1/2}$, so that $\langle x, x \rangle = \|x\|^2$. From the definitions it follows easily that $\langle 0, y \rangle = 0$ and $\langle x, \alpha y \rangle = \overline{\alpha} \langle x, y \rangle$.

The following is the *Cauchy-Schwarz inequality*. The proof is the same as the one usually taught in undergraduate linear algebra classes, except for some complications due to the fact that we allow the set of scalars to be the complex numbers.

Theorem 19.2 *For all $x, y \in H$, we have*

$$|\langle x, y \rangle| \leq \|x\| \, \|y\|.$$

Proof. Let $A = \|x\|^2$, $B = |\langle x, y \rangle|$, and $C = \|y\|^2$. If $C = 0$, then $y = 0$, hence $\langle x, y \rangle = 0$, and the inequality holds. If $B = 0$, the inequality is obvious. Therefore we will suppose that $C > 0$ and $B \neq 0$.

If $\langle x, y \rangle = Re^{i\theta}$, let $\alpha = e^{i\theta}$, and then $|\alpha| = 1$ and $\alpha \langle y, x \rangle = |\langle x, y \rangle| = B$. Since B is real, we have that $\overline{\alpha}\langle x, y \rangle$ also equals $|\langle x, y \rangle|$.

We have for real r

$$
\begin{aligned}
0 &\leq \|x - r\alpha y\|^2 \\
&= \langle x - r\alpha y, x - r\alpha y \rangle \\
&= \langle x, x \rangle - r\alpha \langle y, x \rangle - r\overline{\alpha}\langle x, y \rangle + r^2 \langle y, y \rangle \\
&= \|x\|^2 - 2r|\langle x, y \rangle| + r^2 \|y\|^2.
\end{aligned}
$$

Therefore

$$A - 2Br + Cr^2 \geq 0$$

for all real numbers r. Since we are supposing that $C > 0$, we may take $r = B/C$, and we obtain $B^2 \leq AC$. Taking square roots of both sides gives the inequality we wanted. $\qquad \square$

From the Cauchy-Schwarz inequality we get the *triangle inequality*:

Proposition 19.3 *For all $x, y \in H$ we have*

$$\|x + y\| \leq \|x\| + \|y\|.$$

Proof. We write

$$
\begin{aligned}
\|x + y\|^2 &= \langle x + y, x + y \rangle = \langle x, x \rangle + \langle x, y \rangle + \langle y, x \rangle + \langle y, y \rangle \\
&\leq \|x\|^2 + 2\|x\| \, \|y\| + \|y\|^2 = (\|x\| + \|y\|)^2,
\end{aligned}
$$

as desired. □

The triangle inequality implies

$$\|x - z\| \leq \|x - y\| + \|y - z\|.$$

Therefore $\| \cdot \|$ is a norm on H, and so if we define the distance between x and y by $\|x - y\|$, we have a metric space.

Definition 19.4 A *Hilbert space* H is an inner product space that is complete with respect to the metric $d(x, y) = \|x - y\|$.

Example 19.5 Let μ be a positive measure on a set X, let $H = L^2(\mu)$, and define

$$\langle f, g \rangle = \int f\bar{g} \, d\mu.$$

As is usual, we identify functions that are equal a.e. H is easily seen to be a Hilbert space. To show the completeness we use Theorem 15.4.

If we let μ be counting measure on the natural numbers, we get what is known as the space ℓ^2. An element of ℓ^2 is a sequence $a = (a_1, a_2, \ldots)$ such that $\sum_{n=1}^{\infty} |a_n|^2 < \infty$ and if $b = (b_1, b_2, \ldots)$, then

$$\langle a, b \rangle = \sum_{n=1}^{\infty} a_n \bar{b}_n.$$

We get another common Hilbert space, n-dimensional Euclidean space, by letting μ be counting measure on $\{1, 2, \ldots, n\}$.

Proposition 19.6 *Let $y \in H$ be fixed. Then the functions $x \to \langle x, y \rangle$ and $x \to \|x\|$ are continuous.*

Proof. By the Cauchy-Schwarz inequality,

$$|\langle x, y \rangle - \langle x', y \rangle| = |\langle x - x', y \rangle| \leq \|x - x'\| \, \|y\|,$$

which proves that the function $x \to \langle x, y \rangle$ is continuous. By the triangle inequality, $\|x\| \leq \|x - x'\| + \|x'\|$, or

$$\|x\| - \|x'\| \leq \|x - x'\|.$$

The same holds with x and x' reversed, so

$$| \, \|x\| - \|x'\| \, | \leq \|x - x'\|,$$

and thus the function $x \to \|x\|$ is continuous.   □

19.2   Subspaces

Definition 19.7 A subset M of a vector space is a *subspace* if M is itself a vector space with respect to the same operations of addition and scalar multiplication. A *closed subspace* is a subspace that is closed relative to the metric given by $\langle \cdot, \cdot \rangle$.

For an example of a subspace that is not closed, consider ℓ^2 and let M be the collection of sequences for which all but finitely many elements are zero. M is clearly a subspace. Let $x_n = (1, \frac{1}{2}, \ldots, \frac{1}{n}, 0, 0, \ldots)$ and $x = (1, \frac{1}{2}, \frac{1}{3}, \ldots)$. Then each $x_n \in M$, $x \notin M$, and we conclude M is not closed because

$$\|x_n - x\|^2 = \sum_{j=n+1}^{\infty} \frac{1}{j^2} \to 0$$

as $n \to \infty$.

Since $\|x + y\|^2 = \langle x + y, x + y \rangle$ and similarly for $\|x - y\|^2$, $\|x\|^2$, and $\|y\|^2$, a simple calculation yields the *parallelogram law*:

$$\|x + y\|^2 + \|x - y\|^2 = 2\|x\|^2 + \|y\|^2. \qquad (19.1)$$

A set $E \subset H$ is *convex* if $\lambda x + (1 - \lambda x) \in E$ whenever $0 \leq \lambda \leq 1$ and $x, y \in E$.

Proposition 19.8 *Each non-empty closed convex subset E of H has a unique element of smallest norm.*

Proof. Let $\delta = \inf\{\|x\| : x \in E\}$. Dividing (19.1) by 4, if $x, y \in E$, then

$$\tfrac{1}{4}\|x - y\|^2 = \tfrac{1}{2}\|x\|^2 + \tfrac{1}{2}\|y\|^2 - \left\|\frac{x + y}{2}\right\|^2.$$

Since E is convex, if $x, y \in E$, then $(x+y)/2 \in E$, and we have

$$\|x - y\|^2 \le 2\|x\|^2 + 2\|y\|^2 - 4\delta^2. \qquad (19.2)$$

Choose $y_n \in E$ such that $\|y_n\| \to \delta$. Applying (19.2) with x replaced by y_n and y replaced by y_m, we see that

$$\|y_n - y_m\|^2 \le 2\|y_n\|^2 + 2\|y_m\|^2 - 4\delta^2,$$

and the right hand side tends to 0 as m and n tend to infinity. Hence y_n is a Cauchy sequence, and since H is complete, it converges to some $y \in H$. Since $y_n \in E$ and E is closed, $y \in E$. Since the norm is a continuous function, $\|y\| = \lim \|y_n\| = \delta$.

If y' is another point with $\|y'\| = \delta$, then by (19.2) with x replaced by y' we have $\|y - y'\| = 0$, and hence $y = y'$. $\qquad\square$

We say $x \perp y$, or x is *orthogonal* to y, if $\langle x, y \rangle = 0$. Let $x^\perp$, read "x perp," be the set of all y in X that are orthogonal to x. If M is a subspace, let $M^\perp$ be the set of all y that are orthogonal to all points in M. The subspace $M^\perp$ is called the *orthogonal complement* of M. It is clear from the linearity of the inner product that $x^\perp$ is a subspace of H. The subspace $x^\perp$ is closed because it is the same as the set $f^{-1}(\{0\})$, where $f(x) = \langle x, y \rangle$, which is continuous by Proposition 19.6. Also, it is easy to see that $M^\perp$ is a subspace, and since

$$M^\perp = \cap_{x \in M} x^\perp,$$

$M^\perp$ is closed. We make the observation that if $z \in M \cap M^\perp$, then

$$\|z\|^2 = \langle z, z \rangle = 0,$$

so $z = 0$.

The following is sometimes called the *Riesz representation theorem*, although usually that name is reserved for Theorem 17.3. To motivate the theorem, consider the case where H is n-dimensional Euclidean space. Elements of $\mathbb{R}^n$ can be identified with $n \times 1$ matrices and linear maps from $\mathbb{R}^n$ to $\mathbb{R}^m$ can be represented by multiplication on the left by a $m \times n$ matrix A. For bounded linear functionals on H, $m = 1$, so A is $1 \times n$, and the y of the next theorem is the vector associated with the transpose of A.

Theorem 19.9 *If L is a bounded linear functional on H, then there exists a unique $y \in H$ such that $Lx = \langle x, y \rangle$.*

Proof. The uniqueness is easy. If $Lx = \langle x, y \rangle = \langle x, y' \rangle$, then $\langle x, y - y' \rangle = 0$ for all x, and in particular, when $x = y - y'$.

We now prove existence. If $Lx = 0$ for all x, we take $y = 0$. Otherwise, let $M = \{x : Lx = 0\}$, take $z \neq 0$ in $M^\perp$, and let $y = \alpha z$ where $\alpha = \overline{Lz}/\langle z, z \rangle$. Notice $y \in M^\perp$,

$$Ly = \frac{\overline{Lz}}{\langle z, z \rangle} Lz = |Lz|^2/\langle z, z \rangle = \langle y, y \rangle,$$

and $y \neq 0$.

If $x \in H$ and

$$w = x - \frac{Lx}{\langle y, y \rangle} y,$$

then $Lw = 0$, so $w \in M$, and hence $\langle w, y \rangle = 0$. Then

$$\langle x, y \rangle = \langle x - w, y \rangle = Lx$$

as desired. $\square$

19.3   Orthonormal sets

A subset $\{u_\alpha\}_{\alpha \in A}$ of H is *orthonormal* if $\|u_\alpha\| = 1$ for all α and $\langle u_\alpha, u_\beta \rangle = 0$ whenever $\alpha, \beta \in A$ and $\alpha \neq \beta$.

The Gram-Schmidt procedure from linear algebra also works in infinitely many dimensions. Suppose $\{x_n\}_{n=1}^\infty$ is a linearly independent sequence, i.e., no finite linear combination of the x_n is 0. Let $u_1 = x_1/\|x_1\|$ and define inductively

$$v_N = x_N - \sum_{i=1}^{n-1} \langle x_N, u_i \rangle u_i,$$

$$u_N = v_N/\|v_N\|.$$

We have $\langle v_N, u_i \rangle = 0$ if $i < N$, so $u_1, \ldots, u_N$ are orthonormal.

Proposition 19.10 *If $\{u_\alpha\}_{\alpha \in A}$ is an orthonormal set, then for each $x \in H$,*

$$\sum_{\alpha \in A} |\langle x, u_\alpha \rangle|^2 \leq \|x\|^2. \tag{19.3}$$

This is called *Bessel's inequality.* This inequality implies that only finitely many of the summands on the left hand side of (19.3) can be larger than $1/n$ for each n, hence only countably many of the summands can be non-zero.

Proof. Let F be a finite subset of A. Let

$$y = \sum_{\alpha \in F} \langle x, u_\alpha \rangle u_\alpha.$$

Then

$$0 \le \|x - y\|^2 = \|x\|^2 - \langle x, y \rangle - \langle y, x \rangle + \|y\|^2.$$

Now

$$\langle y, x \rangle = \Big\langle \sum_{\alpha \in F} \langle x, u_\alpha \rangle u_\alpha, x \Big\rangle = \sum_{\alpha \in F} \langle x, u_\alpha \rangle \langle u_\alpha, x \rangle = \sum_{\alpha \in F} |\langle x, u_\alpha \rangle|^2.$$

Since this is real, then $\langle x, y \rangle = \langle y, x \rangle$. Also

$$\|y\|^2 = \langle y, y \rangle = \Big\langle \sum_{\alpha \in F} \langle x, u_\alpha \rangle u_\alpha, \sum_{\beta \in F} \langle x, u_\beta \rangle u_\beta \Big\rangle$$

$$= \sum_{\alpha, \beta \in F} \langle x, u_\alpha \rangle \overline{\langle x, u_\beta \rangle} \langle u_\alpha, u_\beta \rangle$$

$$= \sum_{\alpha \in F} |\langle x, u_\alpha \rangle|^2,$$

where we used the fact that $\{u_\alpha\}$ is an orthonormal set. Therefore

$$0 \le \|y - x\|^2 = \|x\|^2 - \sum_{\alpha \in F} |\langle x, u_\alpha \rangle|^2.$$

Rearranging,

$$\sum_{\alpha \in F} |\langle x, u_\alpha \rangle|^2 \le \|x\|^2$$

when F is a finite subset of A. If N is an integer larger than $n\|x\|^2$, it is not possible that $|\langle x, u_\alpha \rangle|^2 > 1/n$ for more than N of the α. Hence $|\langle x, u_\alpha \rangle|^2 \ne 0$ for only countably many α. Label those α's as $\alpha_1, \alpha_2, \dots$. Then

$$\sum_{\alpha \in A} |\langle x, u_\alpha \rangle|^2 = \sum_{j=1}^{\infty} |\langle x, u_{\alpha_j} \rangle|^2 = \lim_{J \to \infty} \sum_{j=1}^{J} |\langle x, u_{\alpha_j} \rangle|^2 \le \|x\|^2,$$

which is what we wanted. $\qquad\square$

Proposition 19.11 *Suppose $\{u_\alpha\}_{\alpha \in A}$ is orthonormal. Then the following are equivalent.*
(1) If $\langle x, u_\alpha \rangle = 0$ for each $\alpha \in A$, then $x = 0$.
(2) $\|x\|^2 = \sum_{\alpha \in A} |\langle x, u_\alpha \rangle|^2$ for all x.
(3) For each $x \in H$, $x = \sum_{\alpha \in A} \langle x, u_\alpha \rangle u_\alpha$.

We make a few remarks. When (1) holds, we say the orthonormal set is *complete*. (2) is called *Parseval's identity*. In (3) the convergence is with respect to the norm of H and implies that only countably many of the terms on the right hand side are non-zero.

Proof. First we show (1) implies (3). Let $x \in H$. By Bessel's inequality and the remarks following the statement of Proposition 19.10 there can be at most countably many α such that $|\langle x, u_\alpha \rangle|^2 \neq 0$. Let $\alpha_1, \alpha_2, \ldots$ be an enumeration of those α. By Bessel's inequality, the series $\sum_i |\langle x, u_{\alpha_i} \rangle|^2$ converges. Using that $\{u_\alpha\}$ is an orthonormal set,

$$\left\| \sum_{j=m}^{n} \langle x, u_{\alpha_j} \rangle u_{\alpha_j} \right\|^2 = \sum_{j,k=m}^{n} \langle x, u_{\alpha_j} \rangle \overline{\langle x, u_{\alpha_k} \rangle} \langle u_{\alpha_j}, u_{\alpha_k} \rangle$$

$$= \sum_{j=m}^{n} |\langle x, u_{\alpha_j} \rangle|^2 \to 0$$

as $m, n \to \infty$. Thus $\sum_{j=1}^{n} \langle x, u_{\alpha_j} \rangle u_{\alpha_j}$ is a Cauchy sequence, and hence converges. Let $z = \sum_{j=1}^{\infty} \langle x, u_{\alpha_j} \rangle u_{\alpha_j}$. Then $\langle z - x, u_{\alpha_j} \rangle = 0$ for each α_j. By (1), this implies $z - x = 0$.

We see that (3) implies (2) because

$$\|x\|^2 - \sum_{j=1}^{n} |\langle x, u_{\alpha_j} \rangle|^2 = \left\| x - \sum_{j=1}^{n} \langle x, u_{\alpha_j} \rangle u_{\alpha_j} \right\|^2 \to 0.$$

That (2) implies (1) is clear.   □

Example 19.12 Take $H = \ell^2 = \{x = (x_1, x_2, \ldots) : \sum |x_i|^2 < \infty\}$ with $\langle x, y \rangle = \sum_i x_i \overline{y_i}$. Then $\{e_i\}$ is a complete orthonormal system, where $e_i = (0, 0, \ldots, 0, 1, 0, \ldots)$, i.e., the only non-zero coordinate of e_i is the ith one.

If K is a subset of a Hilbert space H, the set of finite linear combinations of elements of K is called the *span* of K.

A collection of elements $\{e_\alpha\}$ is a *basis* for H if the set of finite linear combinations of the e_α is dense in H. A basis, then, is an orthonormal subset of H such that the closure of its span is all of H.

Proposition 19.13 *Every Hilbert space has an orthonormal basis.*

This means that (3) in Proposition 19.11 holds.

Proof. If $B = \{u_\alpha\}$ is orthonormal, but not a basis, let V be the closure of the linear span of B, that is, the closure with respect to the norm in H of the set of finite linear combinations of elements of B. Choose $x \in V^\perp$, and if we let $B' = B \cup \{x/\|x\|\}$, then B' is a basis that is strictly bigger than B.

It is easy to see that the union of an increasing sequence of orthonormal sets is an orthonormal set, and so there is a maximal one by Zorn's lemma. By the preceding paragraph, this maximal orthonormal set must be a basis, for otherwise we could find a larger basis. □

19.4 Fourier series

An interesting application of Hilbert space techniques is to *Fourier series*, or equivalently, to *trigonometric series*. For our Hilbert space we take $H = L^2([0, 2\pi))$ and let

$$u_n = \frac{1}{\sqrt{2\pi}} e^{inx}$$

for n an integer. (n can be negative.) Recall that

$$\langle f, g \rangle = \int_0^{2\pi} f(x)\overline{g(x)} \, dx$$

and $\|f\|^2 = \int_0^{2\pi} |f(x)|^2 \, dx$.

It is easy to see that $\{u_n\}$ is an orthonormal set:

$$\int_0^{2\pi} e^{inx} e^{-imx} \, dx = \int_0^{2\pi} e^{i(n-m)x} \, dx = 0$$

if $n \neq m$ and equals 2π if $n = m$.

Let $\mathcal{F}$ be the set of finite linear combinations of the u_n, i.e., the span of $\{u_n\}$. We want to show that $\mathcal{F}$ is a dense subset of $L^2([0, 2\pi))$. The first step is to show that the closure of $\mathcal{F}$ with respect to the supremum norm is equal to the set of continuous functions f on $[0, 2\pi)$ with $f(0) = f(2\pi)$. We will accomplish this by using the Stone-Weierstrass theorem, Theorem 1.7.

We identify the set of continuous functions on $[0, 2\pi)$ that take the same value at 0 and 2π with the continuous functions on the circle. To do this, let $S = \{e^{i\theta} : 0 \leq \theta < 2\pi\}$ be the unit circle in $\mathbb{C}$. If f is continuous on $[0, 2\pi)$ with $f(0) = f(2\pi)$, define $\widetilde{f} : S \to \mathbb{C}$ by $\widetilde{f}(e^{i\theta}) = f(\theta)$. Note $\widetilde{u}_n(e^{i\theta}) = e^{in\theta}$.

Let $\widetilde{\mathcal{F}}$ be the set of finite linear combinations of the $\widetilde{u}_n$. S is a compact metric space. Since the complex conjugate of $\widetilde{u}_n$ is $\widetilde{u}_{-n}$, then $\widetilde{\mathcal{F}}$ is closed under the operation of taking complex conjugates. Since $\widetilde{u}_n \cdot \widetilde{u}_m = \widetilde{u}_{n+m}$, it follows that $\mathcal{F}$ is closed under the operation of multiplication. That it is closed under scalar multiplication and addition is obvious. $\widetilde{u}_0$ is identically equal to 1, so $\widetilde{\mathcal{F}}$ vanishes at no point. If $\theta_1, \theta_2 \in S$ and $\theta_1 \neq \theta_2$, then $\theta_1 - \theta_2$ is not an integer multiple of 2π, so

$$\frac{\widetilde{u}_1(\theta_1)}{\widetilde{u}_1(\theta_2)} = e^{i(\theta_1 - \theta_2)} \neq 1,$$

or $\widetilde{u}_1(\theta_1) \neq \widetilde{u}_1(\theta_2)$. Therefore $\mathcal{F}$ separates points. By the Stone-Weierstrass theorem (Theorem 1.7), the closure of $\mathcal{F}$ with respect to the supremum norm is equal to the set of continuous complex-valued functions on S.

If $f \in L^2([0, 2\pi))$, then

$$\int |f - f\chi_{[1/m, 2\pi - 1/m]}|^2 \to 0$$

by the dominated convergence theorem as $m \to \infty$. By Corollary 15.6 any function in $L^2([1/m, 2\pi - 1/m])$ can be approximated in L^2 by continuous functions which have support in the interval $[1/m, 2\pi - 1/m]$. By what we showed above, a continuous function with support in $[1/m, 2\pi - 1/m]$ can be approximated uniformly on $[0, 2\pi)$ by elements of $\mathcal{F}$. Finally, if g is continuous on $[0, 2\pi)$ and $g_m \to g$ uniformly on $[0, 2\pi)$, then $g_m \to g$ in $L^2([0, 2\pi))$ by the dominated convergence theorem. Putting all this together proves that $\mathcal{F}$ is dense in $L^2([0, 2\pi))$.

It remains to show the completeness of the u_n. If f is orthogonal to each u_n, then it is orthogonal to every finite linear combination, that is, to every element of $\mathcal{F}$. Since $\mathcal{F}$ is dense in $L^2([0, 2\pi))$, we can find $f_n \in \mathcal{F}$ tending to f in L^2. Then

$$\|f\|^2 = |\langle f, \overline{f} \rangle| \leq |\langle f - f_n, \overline{f} \rangle| + |\langle f_n, \overline{f} \rangle|.$$

The second term on the right of the inequality sign is 0. The first term on the right of the inequality sign is bounded by $\|f - f_n\| \, \|f\|$ by the Cauchy-Schwarz inequality, and this tends to 0 as $n \to \infty$. Therefore $\|f\|^2 = 0$, or $f = 0$, hence the $\{u_n\}$ are complete. Therefore $\{u_n\}$ is a complete orthonormal system.

Given f in $L^2([0, 2\pi))$, write

$$c_n = \langle f, u_n \rangle = \int_0^{2\pi} f \overline{u_n} \, dx = \frac{1}{\sqrt{2\pi}} \int_0^{2\pi} f(x) e^{-inx} \, dx,$$

the Fourier coefficients of f. Parseval's identity says that

$$\|f\|^2 = \sum_n |c_n|^2.$$

For any f in L^2 we also have

$$\sum_{|n| \leq N} c_n u_n \to f$$

as $N \to \infty$ in the sense that

$$\left\| f - \sum_{|n| \leq N} c_n u_n \right\|_2 \to 0$$

as $N \to \infty$.

Using $e^{inx} = \cos nx + i \sin nx$, we have

$$\sum_{n=-\infty}^{\infty} c_n e^{inx} = A_0 + \sum_{n=1}^{\infty} B_n \cos nx + \sum_{n=1}^{\infty} C_n \sin nx,$$

where $A_0 = c_0$, $B_n = c_n + c_{-n}$, and $C_n = i(c_n - c_{-n})$. Conversely, using $\cos nx = (e^{inx} + e^{-inx})/2$ and $\sin nx = (e^{inx} - e^{-inx})/2i$,

$$A_0 + \sum_{n=1}^{\infty} B_n \cos nx + \sum_{n=1}^{\infty} C_n \sin nx = \sum_{n=-\infty}^{\infty} c_n e^{inx}$$

if we let $c_0 = A_0$, $c_n = B_n/2 + C_n/2i$ for $n > 0$ and $c_n = B_n/2 - C_n/2i$ for $n < 0$. Thus results involving the u_n can be transferred to results for series of sines and cosines and vice versa.

19.5   Exercises

Exercise 19.1 For $f, g \in L^2([0,1])$, let $\langle f, g \rangle = \int_0^1 f(x)\overline{g(x)}\, dx$. Let $H = C([0,1])$ be the functions that are continuous on $[0,1]$. Is H a Hilbert space with respect to the norm defined in terms of the inner product $\langle \cdot, \cdot \rangle$? Justify your answer.

Exercise 19.2 Suppose H is a Hilbert space with a countable basis. Suppose $\|x_n\| \to \|x\|$ as $n \to \infty$ and $\langle x_n, y \rangle \to \langle x, y \rangle$ as $n \to \infty$ for every $y \in H$. Prove that $\|x_n - x\| \to 0$ as $n \to \infty$.

Exercise 19.3 Prove that if M is a closed subset of a Hilbert space H, then $(M^\perp)^\perp = M$. Is this necessarily true if M is not closed? If not, give a counterexample.

Exercise 19.4 Give an example of a subspace M of a Hilbert space H such that $M \neq H$ but $M^\perp = \{0\}$.

Exercise 19.5 Prove that if M is a closed subspace of H with $M \neq H$, then $M^\perp$ is strictly larger than the set $\{0\}$.

Exercise 19.6 Prove that if H is infinite-dimensional, that is, it has no finite basis, then the closed unit ball in H is not compact.

Exercise 19.7 Suppose a_n is a sequence of real numbers such that

$$\sum_{n=1}^{\infty} a_n b_n < \infty$$

whenever $\sum_{n=1}^{\infty} b_n^2 < \infty$. Prove that $\sum_{n=1}^{\infty} a_n^2 < \infty$.

Exercise 19.8 We say $x_n \to x$ *weakly* if $\langle x_n, y \rangle \to \langle x, y \rangle$ for every y in H. Prove that if x_n is a sequence in H with $\sup_n \|x_n\| \leq 1$, then there is a subsequence $\{n_j\}$ and an element x of H with $\|x\| \leq 1$ such that x_{n_j} converges to x weakly.

Exercise 19.9 If A is a measurable subset of $[0, 2\pi]$, prove that

$$\lim_{n \to \infty} \int_A e^{inx}\, dx = 0.$$

This is special case of the *Riemann-Lebesgue lemma*.

Exercise 19.10 The purpose of Exercise 13.6 was to show that in proving the Radon-Nikodym theorem, we can assume that $\nu(A) \leq \mu(A)$ for all measurable A. Assume for the current problem that this is the case and that μ and ν are finite measures. We use this to give an alternative proof of the Radon-Nikodym theorem.

For f real-valued and in L^2 with respect to μ, define $L(f) = \int f \, d\nu$.
(1) Show that L is a bounded linear functional on $L^2(\mu)$.
(2) Conclude by Theorem 19.9 that there exists a real-valued measurable function g in $L^2(\mu)$ such that $L(f) = \int fg \, d\mu$ for all $f \in L^2(\mu)$. Prove that $d\nu = g \, d\mu$.

Exercise 19.11 Suppose f is a continuous real-valued function on $\mathbb{R}$ such that $f(x+1) = f(x)$ for every x. Let γ be an irrational number. Prove that

$$\lim_{n \to \infty} \frac{1}{n} \sum_{j=1}^{n} f(j\gamma) = \int_0^1 f(x) \, dx.$$

Exercise 19.12 If M is a closed subspace of a Hilbert space, let $x + M = \{x + y : y \in M\}$.
(1) Prove that $x + M$ is a closed convex subset of H.
(2) Let Qx be the point of $x + M$ of smallest norm and $Px = x - Qx$. P is called the *projection* of x onto M. Prove that P and Q are mappings of H into M and $M^{\perp}$, respectively.
(3) Prove that P and Q are linear mappings.
(4) Prove that if $x \in M$, then $Px = x$ and $Qx = 0$.
(5) Prove that if $x \in M^{\perp}$, then $Px = 0$ and $Qx = x$.
(6) Prove that
$$\|x\|^2 = \|Px\|^2 + \|Qx\|^2.$$

Exercise 19.13 Suppose $\{e_n\}$ is an orthonormal basis for a separable Hilbert space and $\{f_n\}$ is an orthonormal set such that $\sum \|e_n - f_n\| < 1$. Prove that $\{f_n\}$ is a basis.

Chapter 20

Topology

I have assumed up until now that you are familiar with metric spaces. This chapter studies more general topological spaces. Topics include compactness, connectedness, separation results, embeddings, and approximation results.

20.1   Definitions

Definition 20.1 Let X be an arbitrary set. A *topology* $\mathcal{T}$ is a collection of subsets of X such that
(1) $X, \emptyset \in \mathcal{T}$;
(2) if $G_\alpha \in \mathcal{T}$ for each α in a non-empty index set I, then $\cup_{\alpha \in I} G_\alpha \in \mathcal{T}$;
(3) if $G_1, \ldots, G_n \in \mathcal{T}$, then $\cap_{i=1}^n G_i \in \mathcal{T}$.
A *topological space* is a set X together with a topology $\mathcal{T}$ of subsets of X.

Property (2) says that $\mathcal{T}$ is closed under the operation of arbitrary unions, while (3) says that $\mathcal{T}$ is closed under the operation of finite intersections.

An *open set* G is an element of $\mathcal{T}$. A set F is a *closed set* if F^c is open.

Example 20.2 Let X be a metric space with a metric d. A subset G of X is open in the metric space sense if whenever $x \in G$, there exists r depending on x such that $B(x, r) \subset G$, where $B(x, r) = \{y : d(x, y) < r\}$. A metric space becomes a topological space if we let $\mathcal{T}$ be the collection of open sets. We call $\mathcal{T}$ the *topology generated by the metric d.*

Example 20.3 If X is an arbitrary set and $\mathcal{T}$ is the collection of all subsets of X, then the topology $\mathcal{T}$ is called the *discrete topology.*

Example 20.4 If X is an arbitrary set and $\mathcal{T} = \{\emptyset, X\}$, then the topology $\mathcal{T}$ is called the *trivial topology.*

There are a large number of terms associated with topology. Let us start with some that have a geometric interpretation. Let A be a subset of a topological space $(X, \mathcal{T})$, but not necessarily an element of $\mathcal{T}$. A point x is an *interior point* of A if there exists $G \in \mathcal{T}$ such that $x \subset G \subset A$. The *interior* of A, frequently denoted by A^o, is the set of interior points of A.

A point x, not necessarily an element of A, is a *limit point* of A if every open set that contains x contains a point of A other than x. The set of limit points of A is sometimes denoted A'. Another name for limit point is *accumulation point.* The *closure* of A, frequently denoted $\overline{A}$, is the set $A \cup A'$.

The *boundary* of A, sometimes written ∂A, is $\overline{A} - A^o$. A point $x \in A$ is an *isolated point* of A if $x \in A - A'$, that is, it is a point of A that is not a limit point of A.

If X is the real line, with the topology coming from the usual metric $d(x, y) = |x - y|$ and $A = (0, 1]$, then $A^o = (0, 1)$, $A' = [0, 1]$, $\overline{A} = [0, 1]$, and $\partial A = \{0, 1\}$. A has no isolated points. If $B = \{1, \frac{1}{2}, \frac{1}{3}, \frac{1}{4}, \ldots\}$, then $B^o = \emptyset$, $B' = \{0\}$, $\overline{B} = \{0, 1, \frac{1}{2}, \frac{1}{3}, \ldots\}$, and $\partial B = \overline{B}$. Each point of B is an isolated point of B. If C is the set of rationals in $[0, 1]$, then $\overline{C} = [0, 1]$, $C^o = \emptyset$, and $\partial C = [0, 1]$.

A set A is a *neighborhood* of x if $x \in A^o$, that is, if there exists an open set G such that $x \in G \subset A$. Some authors require a neighborhood to be open, but this is not common usage. We will call A an *open neighborhood* when A is both a neighborhood and an open set.

Let us prove two propositions which will give some practice with the definitions.

Proposition 20.5 *(1) If $F_1, \ldots, F_n$ are closed sets, then $\cup_{i=1}^{n} F_i$ is closed.*
(2) If F_α is a closed set for each α in a non-empty index set I, then $\cap_{\alpha \in I} F_\alpha$ is closed.

Proof. (1) Since each F_i is closed, then each F_i^c is open. Hence

$$(\cup_{i=1}^{n} F_i)^c = \cap_{i=1}^{n} F_i^c$$

is open. Therefore $\cup_{i=1}^{n} F_i$ is closed.

(2) is similar. $\qquad\qquad\qquad\qquad\qquad\qquad\qquad\qquad$ □

Proposition 20.6 *(1) If A is a subset of X, then*

$$\overline{A} = \cap \{F : F \ closed, A \subset F\}. \qquad (20.1)$$

(2) $\overline{A}$ is closed.

Proof. Let B denote the right hand side of (20.1). We first show that $\overline{A} \subset B$ by showing that if $x \notin B$, then $x \notin \overline{A}$. If $x \notin B$, there exists a closed set F containing A such that $x \notin F$. Then F^c is an open set containing x which is disjoint from A. Since $x \in F^c$, then $x \notin A$ and x is not a limit point of A, hence $x \notin \overline{A}$.

We finish the proof of (1) by showing $B \subset \overline{A}$. Let $x \in B$. One possibility is that $x \in A$, in which case $x \in \overline{A}$. The second possibility is that $x \notin A$. Let G be an open set containing x. If G is disjoint from A, then G^c is a closed set containing A that does not contain the point x, a contradiction to the definition of B. Therefore, in this second case where $x \notin A$, every open set containing x intersects A, which says that x is a limit point of A, hence $x \in \overline{A}$.

This proves (1). Since the intersection of closed sets is closed, (2) follows. $\qquad\qquad\qquad\qquad\qquad\qquad\qquad\qquad\qquad\qquad$ □

Next let us discuss some situations where there are several topologies present. Let $(X, \mathcal{T})$ be a topological space and let Y be a subset of a set X. If we define $\mathcal{U} = \{G \cap Y : G \in \mathcal{T}\}$, then it is routine to check that $\mathcal{U}$ is a topology of subsets of Y. The space

$(Y, \mathcal{U})$ is a *subspace* of $(X, \mathcal{T})$. We say an element of $\mathcal{U}$ is *relatively open* and call $\mathcal{U}$ the *relative topology*.

As an example, let $X = [0, 1]$ with the usual metric and let $Y = [1/2, 1]$. The set $A = [1/2, 3/4)$ is relatively open but is not an open subset of X.

Given two topologies $\mathcal{T}$ and $\mathcal{T}'$ on a set X with $\mathcal{T} \subset \mathcal{T}'$, we say $\mathcal{T}$ is *weaker* or *coarser* than $\mathcal{T}'$ and $\mathcal{T}'$ is *stronger* or *finer* than $\mathcal{T}$. A stronger topology has more open sets.

Suppose $(X, \mathcal{T})$ is a topological space and $\sim$ is an equivalence relation for X. Let $\overline{X}$ be the set of equivalence classes and let $E : X \to \overline{X}$ be defined by setting $E(x)$ equal to the equivalence class containing x. Define $\mathcal{U} = \{A \subset \overline{X} : E^{-1}(A) \in \mathcal{T}\}$, where $E^{-1}(A) = \{x : E(x) \subset A\}$. Then $\mathcal{U}$ is called the *quotient topology* on $\overline{X}$.

Next we discuss bases, subbases, and the product topology.

A subcollection $\mathcal{B}$ of $\mathcal{T}$ is an *open base* if every element of $\mathcal{T}$ is a union of sets in $\mathcal{B}$. A subcollection $\mathcal{S}$ of $\mathcal{T}$ is a *subbase* if the collection of finite intersections of elements of $\mathcal{S}$ is an open base for $\mathcal{T}$.

As an example, consider $\mathbb{R}^2$ with the topology generated by the metric

$$d((x_1, y_1), (x_2, y_2)) = (|x_1 - x_2|^2 + |y_1 - y_2|^2)^{1/2},$$

the usual Euclidean metric. If $B(x, r) = \{y \in \mathbb{R}^2 : d(x, y) < r\}$, then the collection of balls $\{B(x, r) : x \in \mathbb{R}^2, r > 0\}$ forms an open base for $\mathcal{T}$. The set of rectangles $\{(x, y) : a < x < b, c < y < d\}$ where $a < b, c < d$ also forms an open base. To give an example of a subbase, let

$$\mathcal{C}_1 = \{(x, y) : a < x < b, y \in \mathbb{R}\}, \quad \mathcal{C}_2 = \{(x, y) : x \in \mathbb{R}, c < y < d\},$$

and then let $\mathcal{S} = \mathcal{C}_1 \cup \mathcal{C}_2$. Every set in $\mathcal{S}$ is open, and any rectangle is the intersection of an element of $\mathcal{C}_1$ with an element of $\mathcal{C}_2$. Therefore the finite intersections of elements in $\mathcal{S}$ form a base, and therefore $\mathcal{S}$ is a subbase.

Any collection $\mathcal{C}$ of subsets of a set X generates a topology $\mathcal{T}$ on X by letting $\mathcal{T}$ be the smallest topology that has $\mathcal{C}$ as a subbase. This means that we first take the collection $\mathcal{B}$ of all finite intersections of elements of $\mathcal{C}$, and then let $\mathcal{T}$ be the collection of arbitrary unions of elements of $\mathcal{B}$. It is easy to see that $\mathcal{T}$ is a topology which has $\mathcal{C}$ as a subbase.

Suppose I is a non-empty index set and for each $\alpha \in I$, $(X_\alpha, \mathcal{T}_\alpha)$ is a topological set. (We will always take our index sets to be non-empty.) Let $X = \prod_{\alpha \in I} X_\alpha$, the product set. Let π_α be the projection of X onto X_α.

When all the X_α are equal to the same space X, we use the notation X^I for $\prod_{\alpha \in I} X_\alpha$. We remark that just as n-tuples $(x_1, \ldots, x_n)$ can be viewed as functions from $\{1, \ldots, n\}$ into a set and sequences can be viewed as functions from $\{1, 2, \ldots\}$ into a set, then elements of X^I can be viewed as functions from I into X.

Let $\{X_\alpha\}$, $\alpha \in I$, be a non-empty collection of topological spaces, let $\mathcal{T}_\alpha$ be the topology on X_α, let $X = \prod_{\alpha \in I} X_\alpha$, and let π_α be the projection of X onto X_α. Set

$$\mathcal{C}_\alpha = \{\pi_\alpha^{-1}(A) : A \in \mathcal{T}_\alpha\}.$$

The *product topology* is the topology generated by $\cup_{\alpha \in I} \mathcal{C}_\alpha$.

This is a bit confusing, so let us look at the special case where $I = \{1, 2, \ldots, n\}$. Then X is the set of n-tuples $\{x_1, x_2, \ldots, x_n\}$, where $x_i \in X_i$. If $x = (x_1, x_2, \ldots, x_n)$, then $\pi_i(x) = x_i$, the ith coordinate. The collection $\mathcal{C}_i$ is the collection of sets of the form

$$\left(\prod_{j=1}^{i-1} X_j\right) \times A \times \left(\prod_{j=i+1}^{n} X_j\right),$$

where A is open in X_i. Let $\mathcal{S} = \cup_{i=1}^n \mathcal{C}_i$ and let $\mathcal{B}$ be the collection of finite intersections of elements of $\mathcal{S}$. A set in $\mathcal{B}$ will be of the form $A_1 \times \cdots \times A_n$, where A_i is open in X_i for each i. (Nothing prevents some of the A_i being all of X_i.) The product topology is then the set of arbitrary unions of sets in $\mathcal{B}$.

A subcollection $\mathcal{B}_x$ of open sets containing the point x is an *open base at the point* x if every open set containing x contains an element of $\mathcal{B}_x$.

We discuss some terms connected to infinite sets. A set $A \subset X$ is *dense* in X if $\overline{A} = X$. The set A is *nowhere dense* if the closure of A has empty interior, that is, $(\overline{A})^\circ = \emptyset$. A space X is *separable* if there exists a countable subset of X that is countable. A space X is *second countable* if it has a countable base. A topological space is *first countable* if every point x has a countable open base at x.

A sequence $\{x_1, x_2, \ldots\}$ converges to a point y if whenever G is an open set containing y, there exists N such that $x_n \in G$ if $n \geq N$. If there exists a subsequence of $\{x_1, x_2, \ldots\}$ that converges to a point y, then y is called a *subsequential limit* of the sequence. Another name for a subsequential limit point is *cluster point*.

Proposition 20.7 *Let X be a metric space. Then X is second countable if and only if it is separable.*

Proof. Suppose X is second countable, and $\mathcal{B} = \{G_1, G_2, \ldots\}$ is a countable base. Pick a point $x_i \in G_i$ for each i. Clearly $A = \{x_i\}$ is countable, and we claim that it is dense in X. If $y \in X$ and H is an open set containing y, then by the definition of base, $y \in G_j \subset H$ for some j. Therefore H contains x_j, and so intersects A. Since H is arbitrary, this shows $y \in \overline{A}$. Since y is arbitrary, $X = \overline{A}$. Note that this part of the proof did not use the fact that X is a metric space.

Now suppose X is a separable metric space with $\{x_i\}$ a countable dense subset of X. Let

$$\mathcal{B} = \{B(x_i, r) : r \text{ rational}, r > 0, i = 1, 2, \ldots\}.$$

Note that $\mathcal{B}$ is countable and we show that $\mathcal{B}$ is a base. It suffices to show that if $y \in X$ and G is an open set of X containing y, then there exists an element of $\mathcal{B}$ containing y and contained in G. Since G is open, there exists s such that $y \in B(y, s) \subset G$. Since $\{x_i\}$ is separable, there exists j such that $B(y, s/4)$ contains x_j. Take r rational with $s/4 < r < s/2$. Then $B(x_j, r) \in \mathcal{B}$. Since $d(x_j, y) < s/4$, then $y \in B(x_j, r)$. Since $d(x_j, y) < s/4$ and $B(y, s) \subset G$, then $B(x_j, r) \subset G$. $\qquad \square$

We next define nets. A set I is a *directed set* if there exists an ordering "$\leq$" satisfying
(1) $\alpha \leq \alpha$ for all $\alpha \in I$;
(2) if $\alpha \leq \beta$ and $\beta \leq \gamma$, then $\alpha \leq \gamma$;
(3) if α and β are in I, there exists $\gamma \in I$ such that $\alpha \leq \gamma$ and $\beta \leq \gamma$.

Here are two examples. For the first, let $I = \{1, 2, \ldots\}$ and say $j \leq k$ if j is less than or equal to k in the usual sense. For the second, let x be a point in a topological space, let I be the collection of open neighborhoods of x, and say $N_1 \leq N_2$ if $N_2 \subset N_1$. Note that this ordering is the reverse of the usual inclusion ordering.

A *net* is a mapping from a directed set I into a topological space X. A net $\langle x_\alpha \rangle$, $\alpha \in I$, *converges* to a point y if for each open set G containing y there is an $\alpha_0 \in I$ such that $x_\alpha \in G$ whenever $\alpha \geq \alpha_0$.

If I is the first example of directed sets, namely, the positive integers, the notion of convergence is the same as that for a sequence to converge.

Proposition 20.8 *Let E be a subset of a topological space. If there is a net consisting of infinitely many different points in E that converges to y, then y is a limit point of E. If y is a limit point of E, then there is a net taking values in E that converges to y.*

Proof. It is easy to see that if there is an infinite net $\{x_\alpha\}$ taking values in E that converges to y, then y is a limit point of E. Suppose y is a limit point of E. We take as a directed set I our second example, the collection of all open neighborhoods of y, ordered by reverse inclusion. For each G_α in this collection, we choose (the axiom of choice is used here) an element x_α of $G_\alpha \cap E$ different than x. Such a point x_α exists because y is a limit point of E.

It now remains to show that $\langle x_\alpha \rangle$ converges to y, and that is a matter of checking the definitions. If G is an open set containing y, then G is equal to G_{α_0} for some $\alpha_0 \in I$. If $\alpha \geq \alpha_0$, then $G_\alpha \subset G_{\alpha_0}$, so $x_\alpha \in G_\alpha \subset G_{\alpha_0} = G$. This is what it means for the net $\langle x_\alpha \rangle$ to converge to the point y. $\square$

Exercise 20.19 shows why the convergence of nets is more useful for general topological spaces than the convergence of sequences.

Remark 20.9 We have talked quite a bit in this book about almost every convergence. One might ask whether one can construct a topology which is in some sense consistent with this type of convergence. The answer is no.

To see this, recall Example 10.7 where we had a sequence of bounded measurable functions $\{f_n\}$ converging to 0 in measure but not almost everywhere. If there were a topology consistent

with almost everywhere convergence, then there would be a neighborhood A of the function 0 such that $f_n \notin A$ for infinitely many
n. We can thus extract a subsequence $\{f_{n_j}\}$ such that no f_{n_j} is in
A. However the subsequence $\{f_{n_j}\}$ still converges in measure to 0,
hence there is a further subsequence $\{f_{n_{j_k}}\}$ which converges almost
everywhere to 0 by Proposition 10.2. This implies that $f_{n_{j_k}} \in A$
for all k sufficiently large, contradicting the fact that no f_{n_j} is in
the neighborhood A of 0.

Finally we talk about continuous functions. Suppose $(X, \mathcal{T})$
and $(Y, \mathcal{U})$ are two topological spaces. A function $f : X \to Y$ is
continuous if $f^{-1}(G) \in \mathcal{T}$ whenever $G \in \mathcal{U}$. The function f is
open if $f(H)$ is in $\mathcal{U}$ whenever $H \in \mathcal{T}$. A *homeomorphism* between
X and Y is a function f that is one-to-one, onto, continuous, and
open. In this case, since f is one-to-one and onto, then f^{-1} exists,
and saying f is open is the same as saying f^{-1} is continuous.

Suppose f is a continuous function from X into Y and F is
closed in Y. Then $(f^{-1}(F))^c = f^{-1}(F^c)$ will be open in X since
F^c is open in Y, and therefore $f^{-1}(F)$ is closed in X. Thus the
inverse image of a closed set under a continuous function is closed.

Conversely, suppose the inverse image of every closed set in Y
is closed in X and G is open in Y. Then $(f^{-1}(G))^c = f^{-1}(G^c)$ will
be closed in X, and so the inverse image of G is open in X. This
implies that f is continuous.

Given a topological space $(Y, \mathcal{U})$ and a non-empty collection of
functions $\{f_\alpha\}$, $\alpha \in I$, from X to Y, the topology on X generated
by the f_α is defined to be the topology generated by

$$\{f_\alpha^{-1}(G) : G \in \mathcal{U}, \alpha \in I\}.$$

Proposition 20.10 *Suppose f is a function from a topological
space $(X, \mathcal{T})$ to a topological space $(Y, \mathcal{U})$. Let $\mathcal{S}$ be a subbase for
Y. If $f^{-1}(G) \in \mathcal{T}$ whenever $G \in \mathcal{S}$, then f is continuous.*

Proof. Let $\mathcal{B}$ be the collection of finite intersections of elements
of $\mathcal{S}$. By the definition of subbase, $\mathcal{B}$ is a base for Y. Suppose
$H = G_1 \cap G_2 \cap \cdots \cap G_n$ with each $G_i \in \mathcal{S}$. Since $f^{-1}(H) =$
$f^{-1}(G_1) \cap \cdots \cap f^{-1}(G_n)$ and $\mathcal{T}$ is closed under the operation of
finite intersections, then $f^{-1}(H) \in \mathcal{T}$. If J is an open subset of
Y, then $J = \cup_{\alpha \in I} H_\alpha$, where I is a non-empty index set and each

$H_\alpha \in \mathcal{B}$. Then $f^{-1}(J) = \cup_{\alpha \in I} f^{-1}(H_\alpha)$, which proves $f^{-1}(J) \in \mathcal{T}$. That is what we needed to show. $\square$

20.2 Compactness

Let X be a topological space. Let A be a subset of X. An *open cover* of A is a non-empty collection $\{G_\alpha\}, \alpha \in I$, of open subsets of X such that $A \subset \cup_{\alpha \in I} G_\alpha$. A *subcover* is a subcollection of $\{G_\alpha\}$ that is also a cover of A. A is *compact* if every open cover of A has a finite subcover.

We will develop several characterizations of compactness. For now, observe that every finite set is compact.

It is easier to give examples of sets that are not compact. If $X = \mathbb{R}$ with the usual metric, then X is not compact. To see this, notice that $\{(n, n+2)\}$, n an integer, covers $\mathbb{R}$, but any finite subcollection can cover at most a bounded set. For another example, let $A = (0, 1/4]$. If we let $G_i = (2^{-i-2}, 2^{-i})$, $i = 1, 2, \ldots$, then $\{G_i\}$ covers A but if $\{G_{i_1}, \ldots G_{i_n}\}$ is any finite subcollection, the interval $(0, 2^{-I-2}]$ will not be covered, where $I = i_1 \vee \cdots \vee i_n$.

Proposition 20.11 *If $A \subset B$, B is compact, and A is closed, then A is compact.*

Proof. Let $\mathcal{G} = \{G_\alpha\}$, $\alpha \in I$, be an open cover for A. Add to this collection the set A^c, which is open. This larger collection, $\mathcal{H} = \mathcal{G} \cup \{A^c\}$, will be an open cover for B, and since B is compact, there is a finite subcover $\mathcal{H}'$. If A^c is in $\mathcal{H}'$, discard it, and let $\mathcal{G}' = \mathcal{H}' - \{A^c\}$. Then $\mathcal{G}'$ is finite, is a subset of $\mathcal{G}$ and covers A. $\square$

Proposition 20.12 *Let X and Y be topological spaces, f a continuous function from X into Y, and A a compact subset of X. Then $f(A) = \{f(x) : x \in A\}$ is a compact subset of Y.*

Proof. Let $\{G_\alpha\}$, $\alpha \in I$, be an open cover for $f(A)$. Then $\{f^{-1}(G_\alpha)\}$, $\alpha \in I$, will be an open cover for A. We used here the fact that since f is continuous, the inverse image of an open set is open. Since A is compact, there exist finitely many sets $\{f^{-1}(G_{\alpha_1}), \ldots, f^{-1}(G_{\alpha_n})\}$ that cover A. Then $\{G_{\alpha_1}, \ldots, G_{\alpha_n}\}$ is a finite subcover for $f(A)$.   □

A collection of closed subsets of X has the *finite intersection property* if every finite subcollection has non-empty intersection.

Proposition 20.13 *A topological space X is compact if and only if any collection of closed sets with the finite intersection property has non-empty intersection.*

Proof. Suppose X is compact and $\{F_\alpha\}$, $\alpha \in I$, is a non-empty collection of closed sets with the finite intersection property. If $\cap_{\alpha \in I} F_\alpha = \emptyset$, then $\{F_\alpha^c\}$ is an open cover for X. Thus there exist finitely many sets $\{F_{\alpha_1}^c, \ldots, F_{\alpha_n}^c\}$ which form a finite subcover for X. This means that $\cap_{i=1}^n F_{\alpha_i} = \emptyset$, which contradicts the finite intersection property.

Conversely, suppose any collection of closed sets with the finite intersection property has non-empty intersection. If $\{G_\alpha\}$, $\alpha \in I$, is an open cover for X, then $\{G_\alpha^c\}$ has empty intersection. Hence there must exist $\{G_{\alpha_1}^c, \ldots, G_{\alpha_n}^c\}$ which has empty intersection. Then $\{G_{\alpha_1}, \ldots, G_{\alpha_n}\}$ is a finite subcover. Therefore X is compact.   □

Here are a few more definitions. A set A is *precompact* if $\overline{A}$ is compact. A set A is *σ-compact* if there exist $K_1, K_2, \ldots$ compact such that $A = \cup_{i=1}^\infty K_i$. A set A is *countably compact* if every countable cover of A has a finite subcover.

A set A is *sequentially compact* if every sequence of elements in A has a subsequence which converges to a point of A. A set A has the *Bolzano-Weierstrass property* if every infinite subset of A has a limit point in A.

20.3   Tychonoff's theorem

Tychonoff's theorem says that the product of compact spaces is compact. We will get to this theorem in stages. We will need Zorn's lemma to prove Tychonoff's theorem; this cannot be avoided, since it is known that Tychonoff's theorem implies the axiom of choice.

Let $(X, \mathcal{T})$ be a topological space, let $\mathcal{B}$ be a basis for $\mathcal{T}$, and let $\mathcal{S}$ be a subbasis. Naturally enough, if A is a subset of X and $\{G_\alpha\}$ is an open cover for A such that each $G_\alpha \in \mathcal{B}$, then $\{G_\alpha\}$ is called a *basic open cover*, while if each $G_\alpha \in \mathcal{S}$, then $\{G_\alpha\}$ is a *subbasic open cover* .

Proposition 20.14 *Suppose A is a subset of X and every basic open cover of A has a finite subcover. Then A is compact.*

Proof. Let $\{G_\alpha\}$ be an open cover for A; we only assume here that $G_\alpha \in \mathcal{T}$. If $x \in A$, there exists α_x such that $x \in G_{\alpha_x}$, and by definition of basis, there exists $B_x \in \mathcal{B}$ such that $x \in B_x \subset G_{\alpha_x}$. Then $\{B_x\}$, $x \in A$, is a basic open cover of A. By hypothesis there is a basic open subcover $\{B_{x_1}, \dots B_{x_n}\}$. Since $B_{x_i} \subset G_{\alpha_{x_i}}$, then $\{G_{\alpha_{x_1}}, \dots, G_{\alpha_{x_n}}\}$ will be a finite subcollection of $\{G_\alpha\}$ that covers A. Thus every open cover of A has a finite subcover, and hence A is compact. □

Much harder is the fact that for A to be compact, it suffices that every subbasic open cover have a finite subcover. First we prove the following lemma, which is where Zorn's lemma is used.

Lemma 20.15 *Let A be a subset of X. Suppose $\mathcal{C} \subset \mathcal{E}$ are two collections of open subsets of X and suppose that no finite subcollection of $\mathcal{C}$ covers A. Then there exists a maximal subset $\mathcal{D}$ of $\mathcal{E}$ that contains $\mathcal{C}$ and such that no finite subcollection of $\mathcal{D}$ covers A.*

Saying that $\mathcal{D}$ is maximal means that if $\mathcal{D} \subset \mathcal{D}' \subset \mathcal{E}$ and no finite subcollection of $\mathcal{D}'$ covers A, then $\mathcal{D}'$ must equal $\mathcal{D}$.

Proof. Let $\mathbb{B}$ be the class of all subcollections $\mathcal{B}$ of $\mathcal{E}$ such that $\mathcal{B}$ contains $\mathcal{C}$ and no subcollection of $\mathcal{B}$ covers A. We order $\mathbb{B}$ by inclusion. If we prove that every totally ordered subset of $\mathbb{B}$ has an upper bound in $\mathbb{B}$, our lemma will follow by Zorn's lemma.

Let $\mathbb{B}'$ be a totally ordered subset of $\mathbb{B}$. Let

$$\mathcal{B} = \cup\{\mathcal{B}_\alpha : \mathcal{B}_\alpha \in \mathbb{B}'\}.$$

Clearly $\mathcal{C} \subset \mathcal{B} \subset \mathcal{E}$, and we must show that no finite subcollection of $\mathcal{B}$ covers A.

Suppose there exist $B_1, \ldots, B_n$ in $\mathcal{B}$ such that $A \subset \cup_{i=1}^n B_i$. For each i, there exists α_i such that $B_i \in \mathcal{B}_{\alpha_i}$ for some $\mathcal{B}_{\alpha_i} \in \mathbb{B}'$. Since $\mathbb{B}'$ is totally ordered, one of the $\mathcal{B}_{\alpha_i}$ contains all the others. Let us suppose it is $\mathcal{B}_{\alpha_n}$, since otherwise we can relabel. But then $B_i \in \mathcal{B}_{\alpha_i} \subset \mathcal{B}_{\alpha_n}$ for each i, contradicting that $\mathcal{B}_{\alpha_n}$ has no finite subcollection that covers A. We conclude that $\mathcal{B}$ is an upper bound for $\mathbb{B}'$. $\qquad\qquad\qquad\qquad\qquad\qquad\qquad\qquad\qquad\qquad\qquad\qquad\quad \square$

Theorem 20.16 *Suppose A is a subset of X and every subbasic open cover of A has a finite subcover. Then A is compact.*

Proof. Let $\mathcal{B}$ be a basis for $\mathcal{T}$, the topology on X, and let $\mathcal{S}$ be a subbasis. We will show that every basic open cover of A has a finite subcover and then apply Proposition 20.14. We will achieve this by supposing that $\mathcal{C}$ is a basic open cover of A having no finite subcover and show that this leads to a contradiction.

Step 1. The first step is to enlarge $\mathcal{C}$. Since $\mathcal{C} \subset \mathcal{B}$ and no finite subcover of $\mathcal{C}$ covers A, by Lemma 20.15 there exists a maximal $\mathcal{D}$ such that $\mathcal{C} \subset \mathcal{D} \subset \mathcal{B}$ and no finite subcover of $\mathcal{D}$ covers A.

Step 2. We write $\mathcal{D} = \{B_\alpha : \alpha \in I\}$, where I is an index set and each $B_\alpha \in \mathcal{B}$. Fix α for now. By the definition of subbase, we can find $n \geq 1$ and $S_1, \ldots, S_n \in \mathcal{S}$ such that $B_\alpha = S_1 \cap \cdots \cap S_n$.

We claim that at least one of the S_i in in $\mathcal{D}$. Suppose not. Let $i \leq n$. Since S_i is a subbasic open set, it is also a basic open set, and therefore $\mathcal{C} \subset \mathcal{D} \cup \{S_i\} \subset \mathcal{B}$. By the maximality property of $\mathcal{D}$, the collection $\mathcal{D} \cup \{S_i\}$ must have a finite subcover of A. Thus there exist $B_{i1}, \ldots, B_{ik_i} \in \mathcal{D}$ such that

$$A \subset S_i \cup B_{i1} \cup \cdots \cup B_{ik_i}. \qquad\qquad (20.2)$$

This holds for each i.

If $x \in A$, one possibility is that $x \in B_{ij}$ for some $i \leq n$, $j \leq k_i$. The other possibility is that $x \notin B_{ij}$ for any $i \leq n$, $j \leq k_i$. In this second case, (20.2) implies that $x \in S_i$ for each i, and hence $x \in S_1 \cap \cdots \cap S_n = B_\alpha$.

Under either possibility we have that $\{B_{ij} : i \leq n, j \leq k_i\} \cup \{B_\alpha\}$ is a finite subcollection of $\mathcal{D}$ that covers A. This is a contradiction. We conclude that at least one of the S_i is in $\mathcal{D}$. We rename S_i as S_α.

Step 3. Now we no longer fix α and we do the above argument for each α, obtaining a collection of subbasic open sets $\{S_\alpha\}$. Since $\{S_\alpha\} \subset \mathcal{D}$, then $\{S_\alpha\}$ has no finite subcover of A. On the other hand, $B_\alpha \subset S_\alpha$, so $A \subset \cup_\alpha B_\alpha \subset \cup_\alpha S_\alpha$. Therefore $\{S_\alpha\}$ is a subbasic open cover of A. By the hypothesis of the theorem, $\{S_\alpha\}$ has a finite subcover. This is our contradiction. We conclude that $\{B_\alpha\}$ must have a finite subcover, and thus A is compact. $\square$

We now state and prove the *Tychonoff theorem.*

Theorem 20.17 *The non-empty product of compact topological spaces is compact.*

Proof. Suppose we have a non-empty family $\{X_\alpha\}$, $\alpha \in I$, of compact topological spaces, and we let $X = \prod_{\alpha \in I} X_\alpha$. A subbase for X is the collection $\{\pi_\alpha^{-1}(G_\alpha)\}$, where $\alpha \in I$, G_α is an open subset of X_α, and π_α is the projection of X onto X_α.

Let $\mathcal{H} = \{H_\beta\}$ be a collection of subbasic open sets for X that covers X. Assume that $\mathcal{H}$ has no finite subcover.

Fix α for the moment. Let $\mathcal{H}_\alpha = \mathcal{H} \cap \mathcal{C}_\alpha$, where

$$\mathcal{C}_\alpha = \{\pi_\alpha^{-1}(G_\alpha) : G_\alpha \text{ is open in } X_\alpha\}.$$

Thus $H_\beta \in \mathcal{H}_\alpha$ if $H_\beta \in \mathcal{H}$ and there exists an open set $G_{\alpha\beta}$ in X_α such that $H_\beta = \pi_\alpha^{-1}(G_{\alpha\beta})$.

If $\{\pi_\alpha(H_\beta) : H_\beta \in \mathcal{H}_\alpha\}$ covers X_α, then since X_α is compact, there exists a finite subcover $\{\pi_\alpha(H_{\beta_1}), \ldots, \pi_\alpha(H_{\beta_n})\}$ of X_α. But then $\{H_{\beta_1}, \ldots, H_{\beta_n}\}$ is a finite cover of X, a contradiction. Therefore there exists $x_\alpha \in X_\alpha$ such that $x_\alpha \notin \cup_{H_\beta \in \mathcal{H}_\alpha} \pi_\alpha(H_\beta)$.

We do this for each α. Let x be the point of X whose α^{th} coordinate is x_α, that is, $\pi_\alpha(x) = x_\alpha$ for each α. If $x \in H_\beta$ for some $H_\beta \in \mathcal{H}$, then $x \in \pi_\alpha^{-1}(G_{\alpha\beta})$ for some $\alpha \in I$ and some $G_{\alpha\beta}$ open in X_α. But then $x_\alpha = \pi_\alpha(x) \in G_{\alpha\beta}$, a contradiction to the definition of x_α. Therefore $x \notin \cup_{H_\beta \in \mathcal{H}} H_\beta$, or $\mathcal{H}$ is not a cover of X, a contradiction.

We conclude that our assumption that $\mathcal{H}$ had no finite subcover is wrong, and a finite subcover does indeed exist. Then X is compact by Theorem 20.16.   □

Remark 20.18 Consider $X = [-1, 1]^{\mathbb{N}}$, where $\mathbb{N} = \{1, 2, \ldots\}$. This means $X = \prod_{i=1}^{\infty} X_i$ where each $X_i = [0, 1]$. By the Tychonoff theorem, X is compact when furnished with the product topology.

This does not contradict Exercise 19.6, which says that the closed unit ball in an infinite-dimensional Hilbert space is never compact. The reason is that there are two different topologies involved. If we consider $[-1, 1]^{\mathbb{N}}$ as a Hilbert space, the metric is given by

$$d(x, y) = \Big(\sum_{i=1}^{\infty} |x_i - y_i|^2 \Big)^{1/2}.$$

Let e_n be the point whose coordinates are zero except for the nth coordinate, which is 1. Then $\{e_n\}$ is a sequence in X that does not converge to 0 when X has the topology inherited as a metric space with the metric d. However, $\pi_i(e_n) \to 0$ as $n \to \infty$ for each i, and so by Exercise 20.18, $e_n \to 0$ in the product topology. Therefore the two topologies are different.

20.4   Compactness and metric spaces

Most undergraduate classes do not deal with compactness in metric spaces in much detail. We will provide that detail here.

Let X be a metric space with metric d. A set A is a *bounded set* if there exists $x_0 \in X$ and $M > 0$ such that $A \subset B(x_0, M)$.

Proposition 20.19 *If A is a compact subset of a metric space X, then A is closed and bounded.*

Proof.   To show boundedness, choose $x_0 \in A$ and notice that $\{B(x_0, n)\}$, $n \geq 1$, is an open cover for X and hence for A. Since A is compact, it has a finite subcover, and boundedness follows.

To show A is closed, let x be a limit point of A and suppose $x \notin A$. For each $y \in A$, let $r_y = d(x,y)/4$. Observe that $\{B(y,r_y)\}$, $y \in A$, is an open cover for A, and hence there is a finite subcover $\{B(y_1, r_{y_1}), \ldots, B(y_n, r_{y_n})\}$. Then

$$F = \cup_{i=1}^n \overline{B(y_i, r_{y_i})}$$

will be a closed set (recall $\overline{C}$ is the closure of C) containing A but not containing x. Therefore F^c is an open set containing x but no point of A, contradicting that x is a limit point of A. □

Recall that A has the Bolzano-Weierstrass property if every infinite set in A has a limit point in A and A is sequentially compact if every sequence in A has a subsequence which converges to a point in A.

Proposition 20.20 *Let X be a metric space. A subset A has the Bolzano-Weierstrass property if and only if it is sequentially compact.*

Proof. First suppose that A has the Bolzano-Weierstrass property. If $\{x_i\}$ is a sequence in A, one possibility is that there are only finitely many distinct points. In that case one of the points must appear in the sequence infinitely often, and those appearances form a subsequence that converge to a point in A. The other possibility is that $\{x_i\}$ is an infinite set. Then there exists y that is a limit point of this set. Let $i_1 = 1$. For each n, choose $i_n > i_{n-1}$ such that $d(x_{i_n}, y) < 2^{-n}$. It is possible to choose such a number i_n because if not, we can find $r < 2^{-n}$ small enough so that $B(y, r)$ does not contain any point of $\{x_i\}$ other than possibly y itself, which contradicts that y is a limit point. The subsequence $\{x_{i_n}\}$ is the subsequence we seek.

Now suppose that A is sequentially compact. Let B be an infinite subset of A. We can choose distinct points $x_1, x_2, \ldots$ in B. If y is a subsequential limit point of this sequence that is in A, then y will be a limit point of B that is in A. □

A bit harder is the following theorem.

Theorem 20.21 *Let X be a metric space and let A be a subset of X. The following are equivalent.*

(1) A is compact;
(2) A is sequentially compact;
(3) A has the Bolzano-Weierstrass property.

Proof. We already know (2) and (3) are equivalent. We first prove that (1) implies (3). Let B be an infinite subset of A. If B has no limit point in A, then for each $y \in A$ we can find r_y such that $B(y, r_y)$ contains no point of B except possibly y itself. Choose a finite subcover $B(y_1, r_{y_1}), \ldots, B(y_n, r_{y_n})$ of A. Then $B \subset A \subset \cup_{i=1}^n B(y_i, r_{y_i})$, but at the same time, the union contains at most n points of B, contradicting that B is infinite.

We next prove (2) implies (1). Let $\{G_\alpha\}$, $\alpha \in I$, be an open cover of A. First we show that there exists $\varepsilon > 0$ with the property that if $x \in A$, then there exists $\alpha_x \in I$ such that $x \in B(x, \varepsilon) \subset G_{\alpha_x}$. If not, for all n large enough there exist x_n such that $B(x_n, 1/n)$ is not contained in any G_α. Let y be a subsequential limit point of $\{x_n\}$. y is in some G_β, and since G_β is open, there exists $\delta > 0$ such that $y \in B(y, \delta) \subset G_\beta$. However y is a subsequential limit point of $\{x_n\}$, and so there exists $m > 2/\delta$ such that $d(x_m, y) < \delta/2$. Then

$$x_m \in B(x_m, 1/m) \subset B(x_m, \delta/2) \subset B(y, \delta) \subset G_\beta,$$

a contradiction to how the x_n were chosen.

Now we can prove that if (2) holds, then A is compact. Let $\{G_\alpha\}$ be an open cover of A and let ε be chosen as in the previous paragraph. Pick $x_1 \in A$. If $B(x_1, \varepsilon)$ covers A, stop. If not, choose $x_2 \in A - B(x_1, \varepsilon)$. If $\{B(x_1, \varepsilon), B(x_2, \varepsilon)\}$ covers A, stop. If not, choose $x_3 \in A - B(x_1, \varepsilon) - B(x_2, \varepsilon)$. Continue. This procedure must stop after finitely many steps, or else we have an infinite sequence $\{x_n\}$ such that $d(x_i, x_j) \geq \varepsilon$ if $i \neq j$, and such a sequence cannot have a convergent subsequence. Therefore we have a collection $\{B(x_1, \varepsilon), \ldots, B(x_n, \varepsilon)\}$ that covers A. For each x_i choose G_{α_i} such that $x_i \in B(x, \varepsilon) \subset G_{\alpha_i}$; this is possible by our choice of ε. Then $\{G_{\alpha_1}, \ldots, G_{\alpha_n}\}$ is the desired subcover. $\square$

As a corollary we get the *Heine-Borel theorem*, although there are easier proofs.

Theorem 20.22 *A subset of $\mathbb{R}^n$ is compact if and only if it is closed and bounded.*

Proof. We have already shown that compact sets are closed and bounded. Since closed subsets of compact sets are compact, to show the converse it suffices to show that $[-M, M]^n$ is compact for each integer M, since any bounded set A will be contained in such a set if M is large enough. By the Tychonoff theorem, it suffices to show $[-M, M]$ is compact in $\mathbb{R}$.

Let B be an infinite subset of $[-M, M]$. Let $J_1 = [a_1, b_1] = [-M, M]$. One of the two intervals $[a_1, (a_1 + b_1)/2]$, $[(a_1 + b_1)/2, b_1]$ must contain infinitely many points of B (perhaps both intervals do). Choose one that has infinitely many points and call it $J_2 = [a_2, b_2]$. At least one of the intervals $[a_2, (a_2 + b_2)/2]$, $[(a_2 + b_2)/2, b_2]$ contains infinitely many points of B. Choose it and call it $J_3 = [a_3, b_3]$. Continue. The sequence $a_1, a_2, a_3, \ldots$ is an increasing sequence of real numbers bounded by M, and so this sequence has a least upper bound z. Let $\varepsilon > 0$. Since $B(z, \varepsilon)$ contains J_n for all n sufficiently large, then $B(z, \varepsilon)$ contains infinitely many points of B, and hence z is a limit point of B. Therefore $[-M, M]$ has the Bolzano-Weierstrass property, and so is compact. $\square$

Given a set A, an *ε-net* for A is a subset $\{x_1, x_2, \ldots\}$ such that $\{B(x_i, \varepsilon)\}$ covers A. A is *totally bounded* if for each ε there exists a finite ε-net. Recall that a set A is complete if every Cauchy sequence in A converges to a point in A. (The notion of Cauchy sequence makes sense only in metric spaces, not in general topological spaces.)

Theorem 20.23 *A subset A of a metric space is compact if and only if it is both complete and totally bounded.*

Proof. First suppose A is compact. If $\varepsilon > 0$, then $\{B(x, \varepsilon)\}$, $x \in A$, is an open cover of A. Choosing a finite subcover shows that A has a finite ε-net. Since ε is arbitrary, A is totally bounded.

Let $\{x_n\}$ be a Cauchy sequence in A. Since A is compact, by Theorem 20.21 there is a subsequence $\{x_{n_j}\}$ that converges, say to $y \in A$. If $\varepsilon > 0$, there exists N such that $d(x_n, x_m) < \varepsilon/2$ if $n, m \geq N$. Choose $n_j > N$ such that $d(x_{n_j}, y) < \varepsilon/2$. Then if $n \geq N$,

$$d(x_n, y) \leq d(x_n, x_{n_j}) + d(x_{n_j}, y) < \varepsilon.$$

Therefore the entire sequence converges to y, which proves that A is complete.

Now suppose that A is totally bounded and complete. We let $B = \{x_1, x_2, \ldots\}$ be a sequence in A. If the set is finite, clearly there is a convergent subsequence, so we suppose there are infinitely many distinct points. Since A is totally bounded, it has a finite $1/2$-net, and there exist balls $B(y_{11}, 1/2)$, ..., $B(y_{1n_1}, 1/2)$ that cover A. Choose one, call it B_1', that contains infinitely many of the x_i and let $C_1 = B_1' \cap B$. Since A is totally bounded, it has a finite $1/4$-net and there exist balls $B(y_{21}, 1/4), \ldots, B(y_{2n_2}, 1/4)$ that cover A. At least one of these, call it B_2', must contain infinitely many points of C_1; let $C_2 = B_2' \cap C_1$. Continue to obtain a sequence $C_1 \supset C_2 \supset \cdots$ so that each C_i contains infinitely many points of B. Choose $n_i > n_{i-1}$ such that $x_{n_i} \in C_i$.

We claim $\{x_{n_i}\}$ is a Cauchy sequence. Let $\varepsilon > 0$ and choose N such that $2^{-N+1} < \varepsilon$. If $N \leq i < j$, then $x_{n_j} \in C_j \subset C_i$, $x_{n_i} \in C_i$, and C_i is contained in a ball of radius 2^{-i}, hence

$$d(x_{n_i}, x_{n_j}) \leq 2 \cdot 2^{-i} \leq 2^{-N+1} < \varepsilon.$$

Therefore $\{x_{n_i}\}$ is a Cauchy sequence.

Since A is complete, then $\{x_{n_i}\}$ converges to a point in A. This implies that B has a subsequence that converges, and we conclude by Theorem 20.21 that A is compact. □

If X and Y are metric spaces with metrics d_X and d_Y, resp., then $f : X \to Y$ is *uniformly continuous* if given ε, there exists δ such that $d_Y(f(x), f(y)) < \varepsilon$ whenever $d_X(x, y) < \delta$.

Proposition 20.24 *If X is a compact metric space, Y is a metric space, and $f : X \to Y$ is continuous, then f is uniformly continuous.*

Proof. Let $\varepsilon > 0$. For each $x \in X$, there exists r_x such that if $y \in B(x, 2r_x)$, then $d_Y(f(x), f(y)) < \varepsilon/2$. Choose a finite subcover

$$\{B(x_1, r_{x_1}) \ldots, B(x_n, r_{x_n})\}$$

of X. Let $\delta = \min(r_{x_1}, \ldots, r_{x_n})$. If $z, z' \in X$ with $d_X(z, z') < \delta$, then $z \in B(x_i, r_{x_i})$ for some i. By our choice of δ, $z' \in B(x_i, 2r_{x_i})$.

Then

$$d_Y(f(z), f(z')) \le d_Y(f(z), f(x_i)) + d_Y(f(x_i), f(z')) < \varepsilon,$$

which is what we wanted. $\square$

The final topic we discuss in this section is the completion of a metric space. If X and Y are metric spaces, a map $\varphi : X \to Y$ is an *isometry* if $d_Y(\varphi(x), \varphi(y)) = d_X(x, y)$ for all $x, y \in X$, where d_X is the metric for X and d_Y the one for Y. A metric space X^* is the *completion* of a metric space X is there is an isometry φ of X into X^* such that $\varphi(X)$ is dense in X^* and X^* is complete.

Theorem 20.25 *If X is a metric space, then it has a completion X^*.*

Of course, if X is already complete, its completion is X itself and φ is the identity map.

Proof. *Step 1.* We define X^* and a metric d^*. To do this, we introduce the set X' consisting of the set of Cauchy sequences in X. Thus $\{x_n\} \in X'$ if $\{x_n\}$ is a Cauchy sequence with respect to the metric d of X. Let us say $\{x_n\} \sim \{y_n\}$ if $\lim_{n \to \infty} d(x_n, y_n) = 0$. It is routine to check that this is an equivalence relation between elements of X'. We let X^* be the set of equivalence classes in X'. We denote the equivalence class containing $\{x_n\} \in X'$ by $\overline{x_n}$.

Let us define

$$d^*(\overline{x_n}, \overline{y_n}) = \lim_{n \to \infty} d(x_n, y_n).$$

This will be our metric for X^*, but before we prove that it is a metric, we first need to make sure that the limit in the definition exists. If $\{x_n\}$ and $\{y_n\}$ are Cauchy sequences in X, given ε there exists N such that $d(x_m, x_n) < \varepsilon$ and $d(y_m, y_n) < \varepsilon$ if $m, n \ge N$. For $m, n \ge N$,

$$d(x_n, y_n) \le d(x_n, x_N) + d(x_N, y_N) + d(y_N, y_n) \le 2\varepsilon + d(x_N, y_N)$$

and

$$d(x_N, y_N) \le d(x_N, x_n) + d(x_n, y_n) + d(y_n, y_N) \le 2\varepsilon + d(x_n, y_n).$$

Thus
$$|d(x_n, y_n) - d(x_N, y_N)| \leq 2\varepsilon,$$
and the same holds with n replaced by m. Therefore
$$|d(x_m, y_m) - d(x_n, y_n)| \leq |d(x_m, y_m) - d(x_N, y_N)|$$
$$+ |d(x_n, y_n) - d(x_N, y_N)|$$
$$\leq 2\varepsilon.$$

This proves that $\{d(x_n, y_n)\}$ is a Cauchy sequence of real numbers. Since $\mathbb{R}$ is complete, then $d(x_n, y_n)$ has a limit as $n \to \infty$.

Step 2. We prove that d^* is a metric. First of all, if $\{x_n\} \sim \{x_n'\}$, then
$$\lim |d(x_n, y_n) - d(x_n', y_n)| \leq \lim d(x_n, x_n') = 0,$$
and the definition of d^* does not depend on what representative of $\overline{x_n}$ we choose.

It is routine to check that $d^*(\overline{x_n}, \overline{y_n}) \geq 0$, that $d^*(\overline{x_n}, \overline{y_n})$ equals $d^*(\overline{y_n}, \overline{x_n})$, and that
$$d^*(\overline{x_n}, \overline{z_n}) \leq d^*(\overline{x_n}, \overline{y_n}) + d^*(\overline{y_n}, \overline{z_n}).$$

If $d^*(\overline{x_n}, \overline{y_n}) = 0$, then $\lim d(x_n, y_n) = 0$, so $\{x_n\} \sim \{y_n\}$, and hence $\overline{x_n} = \overline{y_n}$. Therefore d^* is a metric.

Step 3. We define the isometry φ and show that $\varphi(X)$ is dense in X^*. If $x \in X$, let $\varphi(x)$ be the equivalence class containing the sequence $(x, x, \ldots)$, that is, the sequence where each element is x. It is clear that this is a map from X into X^* and that it is an isometry.

If $\overline{x_n}$ is an element of X^* and $\varepsilon > 0$, then $\{x_n\}$ is a Cauchy sequence in X and there exists N such that $d(x_n, x_{n'}) < \varepsilon/2$ if $n, n' \geq N$. We see that
$$d^*(\varphi(x_N), \overline{x_n}) = \lim_{n \to \infty} d(x_N, x_n) \leq \varepsilon/2.$$

Therefore the ball of radius ε about $\overline{x_n}$ contains a point of $\varphi(X)$. Since $\overline{x_n}$ and ε were arbitrary, $\varphi(X)$ is dense in X^*.

Step 4. It remains to show that X^* is complete. Let $\{z_n\}$ be a Cauchy sequence in X^*. By Step 3, for each n there exists $y_n \in \varphi(X)$ such that $d^*(z_n, y_n) < 1/n$. Since
$$d^*(y_m, y_n) \leq d^*(y_m, z_m) + d^*(z_m, z_n) + d^*(z_n, y_n)$$
$$\leq d^*(z_m, z_n) + \frac{1}{m} + \frac{1}{n},$$

we conclude that $\{y_n\}$ is also a Cauchy sequence in X^*.

Each element $y_n \in \varphi(X)$ is of the form $y_n = (x_n, x_n, \ldots)$ for some $x_n \in X$. Because φ is an isometry, then $\{x_n\}$ is a Cauchy sequence in X. Given ε there exists N such that $d(x_m, x_n) < \varepsilon$ if $m, n \geq N$. Let $x = \overline{x_n} \in X^*$. We have

$$d^*(y_m, x) = \lim_{n \to \infty} d(x_m, x_n) \leq \varepsilon$$

if $m \geq N$. Thus y_m converges to x with respect to the metric d^*. Finally,

$$\limsup_{n \to \infty} d^*(z_n, x) \leq \limsup_{n \to \infty} d^*(z_n, y_n) + \limsup_{n \to \infty} d^*(y_n, x) = 0,$$

and we conclude that z_n converges to x with respect to the metric d^*. Therefore X^* is complete. $\qquad\square$

20.5   Separation properties

We define some types of topological spaces. Each successive definition implies the existence of a larger class of open sets and consequently are spaces that can be the domain of more continuous functions.

A topological space X is a T_1 *space* if whenever $x \neq y$, there exists an open set G such that $x \in G$ and $y \notin G$. X is a *Hausdorff space* if whenever $x \neq y$, there exist open sets G and H such that $x \in G$, $y \in H$, and $G \cap H = \emptyset$. We say that x and y are *separated* by the open sets G and H.

A space X is a *completely regular space* if X is a T_1 space and whenever F is a closed subset of X and $x \notin F$, there exists a continuous real-valued function f taking values in $[0, 1]$ such that $f(x) = 0$ and $f(y) = 1$ for all $y \in F$. Finally, a space X is a *normal space* if X is a T_1 space and whenever E and F are disjoint closed sets in X, there exist disjoint open sets G and H such that $E \subset G$ and $F \subset H$.

Clearly Hausdorff spaces are T_1 spaces. If X is completely regular and x and y are distinct points, let $F = \{y\}$. We will see in a moment that F is closed. Let f be a continuous function such that $f(x) = 0$ and $f = 1$ on F, i.e., $f(y) = 1$. If we let

$G = f^{-1}((-\infty, 1/2))$ and $H = f^{-1}((1/2, \infty))$, then G and H are open sets separating x and y. Therefore, except for showing that F is closed, we have shown that completely regular spaces are Hausdorff spaces. Finally, as a consequence of Urysohn's lemma in Section 20.6, we will see that normal spaces are completely regular spaces.

Let us develop a few of the properties of these spaces.

Proposition 20.26 *If X is a T_1 space and $y \in X$, then $\{y\}$ is a closed set.*

Proof. Let $F = \{y\}$. If $x \in F^c$, there exists an open set G such that $x \in G$ and $y \notin G$. Therefore x is not a limit point of F. This proves F is closed. $\square$

If X is a metric space and $x, y \in X$, set $r = d(x, y)$ and then $G = B(x, r/2)$ and $H = B(y, r/2)$ are open sets separating x and y. Therefore metric spaces are Hausdorff spaces.

Proposition 20.27 *The product of a non-empty class of Hausdorff spaces is a Hausdorff space.*

Proof. Let $\{X_\alpha\}$, $\alpha \in I$, be a non-empty collection of Hausdorff spaces and let $X = \prod_{\alpha \in I} X_\alpha$. If $x, y \in X$ are distinct points, then $\pi_\alpha(x) \neq \pi_\alpha(y)$ for at least one index α, where we recall that π_α is the projection of X onto X_α. Then there exist open sets G_0, H_0 in X_α that separate $\pi_\alpha(x)$ and $\pi_\alpha(y)$. The sets $\pi_\alpha^{-1}(G_0)$ and $\pi_\alpha^{-1}(H_0)$ are open sets in X, in fact they are subbasic open sets, which separate x and y. $\square$

Proposition 20.28 *Let X be a Hausdorff space, F a compact subset of X, and $x \notin F$. There exist disjoint open sets G and H such that $x \in G$ and $F \subset H$.*

Proof. For each $y \in F$ choose disjoint open sets G_y and H_y such that $x \in G_y$ and $y \in H_y$. The collection $\{H_y\}$, $y \in F$, is an open cover for F. Let $\{H_{y_1}, \ldots, H_{y_n}\}$ be a finite subcover. Then $F \subset H = \cup_{i=1}^n H_{y_i}$ and H is open, and $x \in G = \cap_{i=1}^n G_{y_i}$ and G is open. Moreover G and H are disjoint. $\square$

Corollary 20.29 *Compact subsets of a Hausdorff space are closed.*

Proof. If F is compact and $x \notin F$, construct G and H as in Proposition 20.28. Then G is an open set containing x disjoint from F, hence x is not a limit point of F. It follows that F is closed. □

Finally we prove

Theorem 20.30 *If X is a compact Hausdorff space, then X is a normal space.*

Proof. Let E and F be disjoint closed subsets of X. Since X is compact, then E and F are compact. Using Proposition 20.28, if $x \in E$, find disjoint open sets G_x and H_x such that $x \in G_x$ and $F \subset H_x$. Then $\{G_x\}$, $x \in E$, is an open cover for E. Let $\{G_{x_1}, \ldots, G_{x_n}\}$ be a finite subcover. Then $G = \cup_{i=1}^{n} G_{x_i}$ is an open set containing E that is disjoint from the open set $H = \cap_{i=1}^{n} H_{x_i}$ which contains F. □

Compact Hausdorff spaces, and their close cousins locally compact Hausdorff spaces, share many of the same properties as metric spaces and are often as useful.

20.6 Urysohn's lemma

We prove *Urysohn's lemma*, which shows that normal spaces have a plentiful supply of continuous functions. In particular, disjoint closed subsets of compact Hausdorff spaces can be separated by continuous functions. Another consequence is that normal spaces are completely regular spaces.

Lemma 20.31 *Let E and F be disjoint closed subsets of a normal space X. There exists a continuous real-valued function taking values in $[0, 1]$ such that $f = 0$ on E and $f = 1$ on F.*

Proof. By the definition of normal space, there exist disjoint open sets G and H such that $E \subset G$ and $F \subset H$. Let $N_{1/2} = G$. Notice that no point of F is a limit point of $N_{1/2}$, so $\overline{N_{1/2}} \subset F^c$. We thus have

$$E \subset N_{1/2} \subset \overline{N_{1/2}} \subset F^c.$$

Now similarly, E and $N^c_{1/2}$ are disjoint closed sets, so there exists an open set, call it $N_{1/4}$, such that

$$E \subset N_{1/4} \subset \overline{N_{1/4}} \subset N_{1/2}.$$

In the same way there exists an open set called $N_{3/4}$ such that

$$\overline{N_{1/2}} \subset N_{3/4} \subset \overline{N_{3/4}} \subset F^c.$$

We continue in this way, finding N_t for each dyadic rational t (that is, each rational of the form $m/2^n$ for some $n \geq 1$ and $1 \leq m \leq 2^n - 1$) such that if $s < t$ are dyadic rationals, then

$$E \subset N_s \subset \overline{N_s} \subset N_t \subset \overline{N_t} \subset F^c.$$

Define $f(x) = 0$ if x is in every N_t and

$$f(x) = \sup\{t \leq 1 : t \text{ a dyadic rational}, x \notin N_t\}$$

otherwise. Clearly f takes values in $[0,1]$, f is 0 on E and f is 1 on F. We need to show that f is continuous.

We show that $\{x : f(x) < a\} = \cup_{t<a} N_t$, where the union is over the dyadic rationals. First, suppose $f(x) < a$. If there does not exist a dyadic rational t less than a with $x \in N_t$, then $x \notin N_t$ for all $t < a$ and then $f(x) \geq a$, a contradiction. Thus $\{x : f(x) < a\}$ is contained in $\cup_{t<a} N_t$. On the other hand, if $x \in N_t$ for some $t < a$, then $f(x) \leq t < a$. We then have

$$f^{-1}((-\infty, a)) = \{x : f(x) < a\} = \cup_{t<a} N_t,$$

which is the union of open sets, and therefore $f^{-1}((-\infty, a))$ is an open set.

If $f(x) > a$, then $x \notin N_t$ for some $t > a$, hence $x \notin \overline{N_s}$ for some $s > a$. If $x \notin \overline{N_t}$ for some $t > a$, then $x \notin N_t$, and so $f(x) > a$. Therefore

$$f^{-1}((a, \infty)) = \{x : f(x) > a\} = \cup_{t>a} (\overline{N_t})^c,$$

which again is open.

The collection of sets of the form $(-\infty, a)$ and (a, ∞) form a subbase for the topology of the real line. That f is continuous follows by Proposition 20.10. $\square$

One may think of ∂N_t as being the contour line of level t for the graph of f.

Corollary 20.32 *If X is a compact Hausdorff space, K is a compact subset of X, and G is an open subset of X containing K, then there exists a continuous function f that is 1 on K and such that the support of f is contained in G.*

Recall that the support of a function f is the closure of the set $\{y : f(y) \neq 0\}$.

Proof. K and G^c are disjoint compact subsets of X, and by Urysohn's lemma there exists a continuous function f_0 that is 0 on G^c and 1 on K. If we let $f = 2(f_0 - \frac{1}{2})^+$, then f is 1 on K.

If x is in the support of f, then every neighborhood of x intersects the set $\{y : f(y) > 0\} = \{y : f_0(y) > \frac{1}{2}\}$. Since f_0 is continuous, then $f_0(x) \geq \frac{1}{2}$, which implies x is not in G^c. Thus, if x is in the support of f, then $x \in G$. $\square$

Remark 20.33 If we want our function to take values in $[a, b]$ and be equal to a on E and b on F, we just let f be the function given by Urysohn's lemma and then use

$$g(x) = (b - a)f(x) + a.$$

Remark 20.34 In Chapter 17 we proved the Riesz representation theorem, which identifies the positive linear functionals on $\mathcal{C}(X)$. We proved the theorem there under the assumption that X was a compact metric space. In fact, the theorem still holds if X is a compact Hausdorff space. We use Corollary 20.32 to guarantee that given a compact set K contained in an open set G, there exists a continuous function f with support in G that is equal to 1 on K. Once we have the existence of such functions, the rest of the proof in Chapter 17 goes through without change.

20.7   Tietze extension theorem

Let C be the Cantor set and let $A = C \times C$. If f is a continuous function mapping A to $[0, 1]$, can we extend f to a continuous function mapping $[0, 1]^2$ to $[0, 1]$? In this case, one can construct an extension by hand, but what about similar extensions in more abstract settings? This question arises frequently in analysis, and the Tietze extension theorem is a result that allows one to do this extension in many cases.

Theorem 20.35 *Let X be a normal space, F a closed subspace, and $f : F \to [a, b]$ a continuous function. There exists a continuous function $\overline{f} : X \to [a, b]$ which is an extension of f, that is, $\overline{f}|_F = f$.*

Proof. The proof is trivial if $a = b$, and we may therefore suppose $a < b$ and also that $[a, b]$ is the smallest closed interval containing the range of f. By considering the function

$$2\frac{f(x) - a}{b - a} - 1,$$

we may without loss of generality assume $a = -1$ and $b = 1$.

We will define $\overline{f}$ as the limit of an infinite sum of functions g_i. To define the sum, let $f_0 = f$ and let $A_0 = \{x \in F : f(x) \leq -1/3\}$, $B_0 = \{x \in F : f(x) \geq 1/3\}$. Since F is closed and f is a continuous function on F, then A_0 and B_0 are disjoint closed subsets of X. By Urysohn's lemma and Remark 20.33, there exists a continuous function $g_0 : X \to [-1/3, 1/3]$ such that g_0 is equal to $-1/3$ on A_0 and g_0 is equal to $1/3$ on B_0.

Define $f_1 = f_0 - g_0$ on F. Then f_1 is continuous on F and observe that $f_1 : F \to [-2/3, 2/3]$. Let A_1 be the set $\{x \in F : f_1(x) \leq -2/9\}$, let $B_1 = \{x \in F : f_1(x) \geq 2/9\}$, and use Remark 20.33 to find $g_1 : X \to [-2/9, 2/9]$ that is continuous and such that g_1 equals $-2/9$ on A_1 and g_1 equals $2/9$ on B_1.

Let $f_2 = f_1 - g_1 = f_0 - (g_0 + g_1)$ on F. Note $f_2 : F \to [-4/9, 4/9]$. We define A_2, B_2 and continue. We observe that the function g_2 we obtain from Remark 20.33 will take values in $[-4/27, 4/27]$. We set $f_3 = f_2 - g_2 = f_0 - (g_0 + g_1 + g_2)$.

We obtain a sequence $g_0, g_1, g_2, \ldots$ of continuous functions on X such that $|g_i(x)| \leq (1/3)(2/3)^i$. Therefore $\sum_{i=0}^{\infty} g_i$ converges uniformly on X. The uniform limit of continuous functions is continuous, so the sum, which we call $\overline{f}$, is continuous on X. Since $\sum_{i=0}^{\infty} \frac{1}{3}\left(\frac{2}{3}\right)^i = 1$, $\overline{f}$ takes values in $[-1, 1]$.

It remains to prove that $\overline{f}$ is an extension of f. We had $|f_1(x)| \leq 2/3$, $|f_2(x)| \leq 4/9$, and in general $|f_i(x)| \leq (2/3)^i$ for each i and each $x \in F$. Since

$$f_i = f_0 - (g_0 + g_1 + \cdots + g_{i-1})$$

and $|f_i(x)| \to 0$ uniformly over $x \in F$, then we conclude $\overline{f} = \sum_{i=0}^{\infty} g_i = f_0 = f$ on F. □

The condition that F be closed cannot be omitted. For example, if $X = [0, 1]$, $F = (0, 1]$, and $f(x) = \sin(1/x)$, there is no way to extend f continuously to X.

20.8   Urysohn embedding theorem

We alluded earlier to the fact that compact Hausdorff spaces can sometimes substitute for metric spaces. In fact, a second countable compact Hausdorff space can be made into a metric space. That is the essential content of the *Urysohn embedding theorem*, also known as the *Urysohn metrization theorem*.

Let $(X, \mathcal{T})$ be a topological space. X is *metrizable* if there exists a metric d such that a set is open with respect to the metric d if and only if it is in $\mathcal{T}$. To be a bit more precise, if $x \in G \in \mathcal{T}$, there exists r such that $B(x, r) \subset G$ and also $B(x, r) \in \mathcal{T}$ for each x and $r > 0$.

We will embed second countable normal spaces into $[0, 1]^{\mathbb{N}}$, where $\mathbb{N} = \{1, 2, \ldots\}$. We define a metric on $[0, 1]^{\mathbb{N}}$ by

$$\rho(x, y) = \sum_{i=1}^{\infty} 2^{-i} |x_i - y_i| \tag{20.3}$$

if $x = (x_1, x_2, \ldots)$ and $y = (y_1, y_2, \ldots)$.

Theorem 20.36 *Let X be a second countable normal space. There exists a homeomorphism φ of X onto a subset of $[0, 1]^{\mathbb{N}}$. In particular, X is metrizable.*

Proof. The proof is almost trivial if X is finite; just take a finite subset of $[0,1]^N$. Therefore we assume X is infinite. Let $\{G_1, G, \ldots\}$ be a countable base. We may assume none of these is equal to $\emptyset$ or X. The set of pairs (G_i, G_j) such that $\overline{G}_i \subset G_j$ is countably infinite, and we label these pairs $P_1, P_2, \ldots$. For each pair P_n there exists a continuous function f_n taking values in $[0,1]$ such that f_n is 0 on $\overline{G}_i$ and 1 on G_j^c; this follows by Urysohn's lemma. If $x \in X$, define $\varphi(x) = (f_1(x), f_2(x), \ldots)$. Clearly φ maps X into $[0,1]^N$.

If $x \neq y$, there exists an open set G in the countable base such that $x \in G$ and $y \notin G$. Since X is normal, there exist disjoint open sets H_1 and H_2 such that $\{x\} \subset H_1$ and $G^c \subset H_2$. Then $x \notin \overline{H}_2$ and there exists K in the countable base such that $x \in K$ and $K \cap \overline{H}_2 = \emptyset$. If follows that $K \cap H_2 = \emptyset$, or $K \subset H_2^c$. If $z \in \overline{K}$, then $z \in \overline{H_2^c} = H_2^c \subset G$, using the fact that H_2^c is closed. We then have $\overline{K} \subset G$, and the pair (K, G) will be a P_n for some n. Since $f_n(x) = 0$ and $f_n(y) = 1$, then $\varphi(x) \neq \varphi(y)$, or φ is one-to-one.

We next prove that φ is continuous. Let $\varepsilon > 0$ and let $x \in X$. By Exercise 20.9 it suffices to prove that there exists an open set G containing x such that if $y \in G$, then $\rho(\varphi(x), \varphi(y)) < \varepsilon$. Choose M large enough so that $\sum_{n=M+1}^{\infty} 2^{-n} < \varepsilon/2$. It therefore is enough to find G such that $|f_n(x) - f_n(y)| < \varepsilon/2$ if $n \leq M$. Each f_n is continuous, so there exists G_n open and containing x such that $|f_n(x) - f_n(y)| < \varepsilon/2$ if $y \in G_n$. We then let $G = \cap_{n=1}^{M} G_n$.

Finally we need to show that φ^{-1} is continuous on $\varphi(X)$. It suffices to show that if $x \in X$ and G is an open set containing x, there exists δ such that if $\rho(\varphi(x), \varphi(y)) < \delta$, then $y \in G$. We may suppose that there is a pair $P_n = (G_i, G_j)$ such that $x \in G_i \subset \overline{G}_i \subset G_j \subset G$. If we choose δ small enough so that $2^n \delta < 1/2$, then $|f_n(x) - f_n(y)| < 1/2$. Since $x \in G_i$, then $f_n(x) = 0$. Since $f_n = 1$ on G_j^c, then we cannot have $y \in G_n^c$, or else $|f_n(x) - f_n(y)| \geq 1/2$. Therefore $y \in G_j \subset G$. $\qquad\square$

Remark 20.37 Exercise 20.17 asks you to prove that the topology on $[0,1]^N$ arising from the metric ρ is the same as the product topology. By the Tychonoff theorem, $[0,1]^N$ is compact. Therefore our proof additionally shows that we can embed X as a subset of $[0,1]^N$, a compact set. This is often useful. For example, every metric space is normal (Exercise 20.28) and every separable metric space is second countable, so every separable metric space can be embedded in a compact metric space.

20.9 Locally compact Hausdorff spaces

A topological space is *locally compact* if each point has a neighborhood with compact closure. In this section we consider *locally compact Hausdorff spaces*, often abbreviated as *LCH* . We will show that one can add a point at infinity to make these into compact Hausdorff spaces. This allows one to provide more general versions of the Riesz representation theorem, the Ascoli-Arzelà theorem, and the Stone-Weierstrass theorem; see [4] for details as to these applications.

Let $(X, \mathcal{T})$ be a locally compact Hausdorff space. Let ∞ denote a point not in X and let $X^* = X \cup \{\infty\}$. Define $\mathcal{T}^*$ to consist of X^*, all elements of $\mathcal{T}$, and all sets $G \subset X^*$ such that G^c is compact in $(X, \mathcal{T})$. We can easily check that $\mathcal{T}^*$ is a topology.

Theorem 20.38 $(X^*, \mathcal{T}^*)$ *is a compact Hausdorff space.*

The space X^* is known as the *Alexandroff one-point compactification* of X. Sometimes it is called simply the *one-point compactification* of X. The point ∞ is referred to as the *point at infinity*.

Proof. First we show X^* is compact. We make the observation that if G is open in $(X^*, \mathcal{T}^*)$, then $G \cap X$ is open in $(X, \mathcal{T})$. If $\{G_\alpha\}$ is an open cover for X^*, there will be at least one β such that $\infty \in G_\beta$. Then G_β^c will be compact with respect to $(X, \mathcal{T})$. The collection $\{G_\alpha \cap X\}$ will be an open cover (with respect to $(X, \mathcal{T})$) of G_β^c, so there is a finite subcover $\{G_{\alpha_1} \cap X, \ldots, G_{\alpha_n} \cap X\}$ of X. Then $\{G_\beta, G_{\alpha_1}, \ldots, G_{\alpha_n}\}$ is an open cover for X^*.

Secondly we show X^* is Hausdorff. Any two points in X can be separated by open sets in $\mathcal{T}$, which are open in $\mathcal{T}^*$. Thus we need only to show that we can separate any point $x \in X$ and the point ∞. If $x \in X$, then x has a neighborhood A whose closure is compact. Then $(\overline{A})^c$ will be an open set in $(X^*, \mathcal{T}^*)$ containing ∞ which is disjoint from A. Since A is a neighborhood of x, there exists a set G that is open in $\mathcal{T}$ such that $x \in G \subset A$. Then $(\overline{A})^c$ and G are open sets in $\mathcal{T}^*$ separating x and ∞. $\qquad\square$

20.10   Stone-Čech compactification

In this section we give another, more complicated, compactification. Given a completely regular space X, we find a compact Hausdorff space $\overline{X}$ such that X is dense in $\overline{X}$ and every bounded continuous function on X can be extended to a bounded continuous function on $\overline{X}$. The space $\overline{X}$ is called the *Stone-Čech compactification* of X, and is traditionally written $\beta(X)$.

This is an amazing theorem. Suppose $X = (0, 1]$. This is a metric space, hence a normal space, hence a completely regular space. The function $f(x) = \sin(1/x)$ cannot be extended to have domain $[0, 1]$, so evidently $\beta(X) \neq [0, 1]$. Yet there is a compactification of $(0, 1]$ for which $\sin(1/x)$ does have a continuous extension.

Here is the theorem.

Theorem 20.39 *Let X be a completely regular space. There exists a compact Hausdorff space $\beta(X)$ and a homeomorphism φ mapping X into a dense subset of $\beta(X)$ such that if f is a bounded continuous function from X to $\mathbb{R}$, then $f \circ \varphi^{-1}$ has a bounded continuous extension to $\beta(X)$.*

Before proving this theorem, let us sort out what the theorem says. Let $Y = \varphi(X) \subset \beta(X)$. Since φ is a homeomorphism, every bounded continuous function f on X corresponds to a bounded continuous function $\widetilde{f}$ on Y. The relationship is given by $\widetilde{f}(y) = f \circ \varphi^{-1}(y)$. The assertion is that $\widetilde{f}$ has a bounded continuous extension to $\beta(X)$.

Proof. Let I be the collection of bounded continuous functions on X and if $f \in I$, let $J_f = [-\inf_{x \in X} f(x), \sup_{x \in X} f(x)]$, the range of f. Each J_f is a finite closed interval. Let $X^* = \prod_{f \in I} J_f$. Then X^* will be a compact Hausdorff space. Define $\varphi : X \to X^*$ by $\pi_f(\varphi(x)) = f(x)$; in other words, $\varphi(x)$ is in the product space, and its fth coordinate is $f(x)$. Thinking through the definitions, note that we have

$$\pi_f \circ \varphi = f.$$

Finally, let $\beta(X)$ be the closure of $\varphi(X)$ in X^*. Since X^* is compact and $\beta(X)$ is closed, then $\beta(X)$ is compact. Subspaces of Hausdorff spaces are Hausdorff.

Now that we have defined $\beta(X)$, the rest of the proof will follow fairly easily from the definitions. First we prove that φ is one-to-one. If $x \neq y$, then since X is completely regular, there exists $f \in I$ such that $f(x) \neq f(y)$. This means $\pi_f(\varphi(x)) \neq \pi_f(\varphi(y))$, which proves $\varphi(x) \neq \varphi(y)$.

Next we show φ is continuous. We let G be a subbasic open set in X^* and prove that $\varphi^{-1}(G \cap \varphi(X))$ is open in X. The continuity will then follow by Proposition 20.10. If G is a subbasic open set in X^*, then $G = \pi_f^{-1}(H)$ for some open set in $\mathbb{R}$. Note

$$\varphi^{-1}(G \cap \varphi(X)) = \{x \in X : \varphi(x) \in G\} = \{x \in X : \pi_f \circ \varphi(x) \in H\}$$
$$= \{x \in X : f(x) \in H\} = f^{-1}(H).$$

However, $f^{-1}(H)$ is open because H is open in $\mathbb{R}$ and f is continuous.

We show φ^{-1} is continuous on $\varphi(X)$. Let $\varphi(x) \in \varphi(X)$ and let H be an open set in X containing x. To prove continuity, we show there exists an open set G in X^* such that if $\varphi(y) \in G$, then $y \in H$. We will then apply Exercise 20.9.

Let x be a point in X and H^c a closed set in X not containing x. Since X is completely regular, there exists $f \in I$ such that $f(x) = 0$ and $f = 1$ on H^c. Let $G = \{z \in X^* : \pi_f(x) < 1/2\}$. Since the projection π_f is continuous and $(-\infty, 1/2)$ is an open set in $\mathbb{R}$, then G is open in X^*. If $\varphi(y) \in G$, then $f(y) = \pi_f \circ \varphi(y) < 1/2$. This implies that $y \notin H^c$, hence $y \in H$, which is what we wanted.

It remains to prove the assertion about the extension. We have $f = \pi_f \circ \varphi$. Therefore on $\varphi(X)$, we have $f \circ \varphi^{-1} = \pi_f \circ \varphi \circ \varphi^{-1} = \pi_f$. Clearly π_f has a bounded continuous extension to X^*, hence to $\beta(X)$. $\qquad \square$

We remark that there can only be one bounded continuous extension of each function in I because X (or more precisely, its image $\varphi(X)$) is dense in $\beta(X)$.

20.11 Ascoli-Arzelà theorem

Let X be a compact Hausdorff space and let $\mathcal{C}(X)$ be the set of continuous functions on X. Since X is compact, $f(X)$ is compact and hence bounded if $f \in \mathcal{C}(X)$.

We make $\mathcal{C}(X)$ into a metric space by setting

$$d(f, g) = \sup_{x \in X} |f(x) - g(x)|,$$

the usual supremum norm. In this section we characterize the compact subsets of $\mathcal{C}(X)$.

A subset $\mathcal{F}$ of $\mathcal{C}(X)$ is *equicontinuous* if given ε and x there is an open set G containing x such that if $y \in G$ and $f \in \mathcal{F}$, then $|f(y) - f(x)| < \varepsilon$. What makes equicontinuity stronger than continuity is that the same ε works for every $f \in \mathcal{F}$.

Here is the Ascoli-Arzelà theorem. You may have seen this in an undergraduate analysis; the novelty here is that we allow X to be a compact Hausdorff space and the proof avoids the use of the "diagonalization procedure."

Theorem 20.40 *Let X be a compact Hausdorff space and let $\mathcal{C}(X)$ be the set of continuous functions on X. A subset $\mathcal{F}$ of $\mathcal{C}(X)$ is compact if and only if the following three conditions hold:*
(1) $\mathcal{F}$ is closed;
(2) $\sup_{f \in \mathcal{F}} |f(x)| < \infty$ for each $x \in X$;
(3) $\mathcal{F}$ is equicontinuous.

In (2), we require $\sup_{f \in \mathcal{F}} |f(x)|$ to be finite, but the size can depend on x.

Proof. First we show that if (1), (2), and (3) hold, then $\mathcal{F}$ is compact. Since $\mathcal{C}(X)$ is a metric space, which is complete by Exercise 20.23, and $\mathcal{F}$ is a closed subset of $\mathcal{C}(X)$, then $\mathcal{F}$ is complete. We will show $\mathcal{F}$ is compact by showing it is totally bounded and then appealing to Theorem 20.23.

Let $\varepsilon > 0$. For each $x \in X$ there is an open set G_x such that if $y \in G_x$ and $f \in \mathcal{F}$, then $|f(y) - f(x)| < \varepsilon/3$. Since $\{G_x\}$, $x \in X$, is an open cover of X and X is compact, we can cover X by a finite subcover $\{G_{x_1}, \ldots, G_{x_n}\}$. Since $\sup_{f \in \mathcal{F}} |f(x_i)|$ is bounded for each i, we can find M such that

$$\sup_{f \in \mathcal{F}, 1 \leq i \leq n} |f(x_i)| \leq M.$$

Let $a_1, \ldots, a_r$ be real numbers such that every point in $[-M, M]$ is within $\varepsilon/3$ of one of the a_j. If $\{a_{j_1}, \ldots, a_{j_n}\}$ is a subset of

$\{a_1, \ldots, a_r\}$, let

$$H(a_{j_1}, \ldots, a_{j_n}) = \{f \in \mathcal{F} : |f(x_i) - a_{j_i}| < \varepsilon/3, 1 \le i \le n\}.$$

There are at most r^n sets of this form. For each one that is non-empty, select an element $g_{a_{j_1}, \ldots, a_{j_n}} \in H(a_{j_1}, \ldots, a_{j_n})$. We claim this collection of functions is an ε-net for $\mathcal{F}$.

If $f \in \mathcal{F}$, choose $a_{j_1}, \ldots, a_{j_n}$ such that $|a_{j_i} - f(x_i)| < \varepsilon/3$ for each $1 \le i \le n$. The set $H(a_{j_1}, \ldots, a_{j_n})$ is non-empty because it contains f. If $y \in X$, then $y \in G_{x_{i_0}}$ for some i_0. Then

$$|f(y) - g_{a_{j_1}, \ldots, a_{j_n}}(y)| \le |f(y) - f(x_{i_0})| + |f(x_{i_0}) - g_{a_{j_1}, \ldots, a_{j_n}}(x_{i_0})|$$
$$+ |g_{a_{j_1}, \ldots, a_{j_n}}(x_{i_0}) - g_{a_{j_1}, \ldots, a_{j_n}}(y)|.$$

The first and third terms on the right hand side of the inequality are less than $\varepsilon/3$ because f and $g_{a_{j_1}, \ldots, a_{j_n}}$ are in $\mathcal{F}$ and $y \in G_{x_{i_0}}$. The second term is less than $\varepsilon/3$ because $g_{a_{j_1}, \ldots, a_{j_n}}(x_{i_0}) = a_{j_{i_0}}$ and we chose $a_{j_{i_0}}$ to be within $\varepsilon/3$ of $f(x_{i_0})$. This proves that

$$\sup_{y \in X} |f(y) - g_{a_{j_1}, \ldots, a_{j_n}}(y)| < \varepsilon.$$

Therefore $\{g_{a_{j_1}, \ldots, a_{j_n}}\}$ is a finite ε-net, and hence $\mathcal{F}$ is totally bounded.

Now suppose $\mathcal{F}$ is compact. (1) follows because $\mathcal{C}(X)$ is a metric space and compact subsets of a metric space are closed. Fix x. The map $\tau_x : \mathcal{C}(X) \to \mathbb{R}$ given by $\tau_x f = f(x)$ is continuous, so $\{f(x)\}$, $f \in \mathcal{F}$, is the image under τ_x of $\mathcal{F}$. Since $\mathcal{F}$ is compact and τ_x is continuous, then $\tau_x(\mathcal{F})$ is compact and hence bounded, which proves (2).

Finally, let $\varepsilon > 0$. Since $\mathcal{F}$ is a compact subset of a metric space, there exists a finite $\varepsilon/3$-net: $\{f_1, \ldots, f_n\}$. If $x \in X$, for some i between 1 and n there exists an open set G_i such that $|f_i(y) - f_i(x)| < \varepsilon/3$ if $y \in G_i$. Let $G = \cap_{i=1}^n G_i$. If $y \in G$ and $g \in \mathcal{F}$, there exists i such that $d(g, f_i) < \varepsilon/3$, and so

$$|g(y) - g(x)| \le |g(y) - f_i(y)| + |f_i(y) - f_i(x)| + |f_i(x) - g(x)| < \varepsilon.$$

This proves that $\mathcal{F}$ is equicontinuous. $\square$

The most useful consequence of the Ascoli-Arzelà theorem is that if (2) and (3) hold for a family $\mathcal{F}$, then any sequence in $\mathcal{F}$ has a subsequence which converges uniformly (the limit is not necessarily in $\mathcal{F}$ unless we also assume $\mathcal{F}$ is closed).

20.12   Stone-Weierstrass theorems

The Stone-Weierstrass theorems are a pair of theorems, one for real-valued functions and one for complex-valued functions, that allow one to approximate continuous functions. They are very useful; we have seen one application already in Section 19.4.

First we prove the *Weierstrass approximation theorem*, which allows one to approximate real-valued functions on a compact interval.

There are a number of different proofs. We give one that uses some ideas from Section 15.3, although we prove what we need here from scratch.

Theorem 20.41 *Let $[a, b]$ be a finite subinterval of $\mathbb{R}$, g a continuous function on $[a, b]$, and $\varepsilon > 0$. Then there exists a polynomial $P(x)$ such that*

$$\sup_{x \in [a,b]} |g(x) - P(x)| < \varepsilon.$$

Proof. Let

$$\varphi_\beta(x) = \frac{1}{\sqrt{2\pi}\beta} e^{-x^2/2\beta^2}.$$

We saw in Exercise 11.18 that $\int_{\mathbb{R}} \varphi_1(x)\,dx = 1$, and by a change of variables, $\int_{\mathbb{R}} \varphi_\beta(x)\,dx = 1$ for every $\beta > 0$. Also, again by a change of variables and the dominated convergence theorem, if $\delta > 0$,

$$\int_{[-\delta,\delta]^c} \varphi_\beta(x)\,dx = \int_{|x|>\delta/\beta} \varphi_1(x)\,dx \to 0$$

as $\beta \to 0$.

Without loss of generality we may assume that g is not identically zero. Extend g to all of $\mathbb{R}$ by setting $g(x) = 0$ if $x < a - 1$ or $x > b + 1$ and letting g be linear on $[a - 1, a]$ and on $[b, b + 1]$. Then g has compact support and is continuous on $\mathbb{R}$.

Step 1. We prove that

$$g * \varphi_\beta(x) = \int g(x - y)\varphi_\beta(y)\,dy$$

will be close to $g(x)$, uniformly over $x \in \mathbb{R}$, if β is small enough. Let $\varepsilon > 0$ and choose δ such that $|g(z) - g(z')| < \varepsilon/4$ if $|z - z'| < \delta$.

Since the integral of φ_β is 1, we have

$$|g * \varphi_\beta(x) - g(x)| = \left| \int [g(x-y) - g(x)]\varphi_\beta(y)\, dy \right| \qquad (20.4)$$

$$\leq \int_{|y|>\delta} |g(x-y) - g(x)|\varphi_\beta(y)\, dy$$

$$+ \int_{|y|\leq\delta} |g(x-y) - g(x)|\varphi_\beta(y)\, dy.$$

The term on the second line of (20.4) is less than or equal to

$$\|g\|_\infty \int_{|y|>\delta} \varphi_\beta(y)\, dy,$$

which will be less than $\varepsilon/4$ if we take β small enough. The term on the last line of (20.4) is less than or equal to

$$(\varepsilon/4) \int_{|y|\leq\delta} \varphi_\beta(y)\, dy \leq (\varepsilon/4) \int_{\mathbb{R}} \varphi_\beta(y)\, dy = \varepsilon/4$$

by our choice of δ. Therefore

$$|g * \varphi_\beta(x) - g(x)| \leq \varepsilon/2$$

uniformly over $x \in \mathbb{R}$ if we take β small enough.

Step 2. Next we approximate $\varphi_\beta(x)$ by a polynomial. Let $N = 2[|a| + |b| + 1]$. For every $M > 0$, the Taylor series for e^x about 0 converges uniformly on $[-M, M]$. To see this, we see that the remainder term satisfies

$$\left| \sum_{k=n+1}^{\infty} \frac{x^k}{k!} \right| \leq \sum_{k=n+1}^{\infty} \frac{M^k}{k!} \to 0$$

as $n \to \infty$. Thus, by replacing x by $-x^2/2\beta^2$ in the Taylor series for e^x and taking n large enough, there is a polynomial Q such that

$$\sup_{-N \leq x \leq N} |Q(x) - \varphi_\beta(x)| < \frac{\varepsilon}{4N\|g\|_\infty}.$$

Now

$$|g * \varphi_\beta(x) - g * Q(x)| = \left| \int g(x-y)[\varphi_\beta(y) - Q(y)]\, dy \right| \qquad (20.5)$$

$$\leq \int |g(x-y)|\, |\varphi_\beta(y) - Q(y)|\, dy.$$

If $x \in [a, b]$, then $g(x-y)$ will be non-zero only if $x-y \in [a-1, b+1]$, which happens only if $|y| \leq 2[|a| + |b| + 1]$. Thus the last line of (20.5) is bounded by

$$\|g\|_\infty \int_{-N}^{N} |\varphi_\beta(y) - Q(y)| \, dy \leq \varepsilon/2.$$

Step 3. Finally, by a change of variables,

$$g * Q(x) = \int g(y)Q(x - y) \, dy.$$

Since Q is a polynomial, then $Q(x - y)$ is a polynomial in x and y, and we can write

$$Q(x - y) = \sum_{j=0}^{n} \sum_{k=0}^{n} c_{jk} x^j y^k$$

for some constants c_{jk}. Then

$$g * Q(x) = \sum_{j=0}^{n} \left(\int \sum_{k=0}^{n} c_{jk} y^k g(y) \, dy \right) x^j,$$

which is a polynomial. Therefore we have approximated g on $[a, b]$ by a polynomial $g * Q$ to within ε, uniformly on the interval $[a, b]$. $\square$

For an alternate proof, see Theorem 21.13.

Let X be a topological space and let $\mathcal{C}(X)$ be the set of real-valued continuous functions on X. Let $\mathcal{A}$ be a subset of $\mathcal{C}(X)$. We say $\mathcal{A}$ is an *algebra of functions* if $f + g$, cf, and fg are in $\mathcal{A}$ whenever $f, g \in \mathcal{A}$ and c is a real number. We say $\mathcal{A}$ is a *lattice of functions* if $f \wedge g$ and $f \vee g$ are in $\mathcal{A}$ whenever $f, g \in \mathcal{A}$. Recall $f \wedge g(x) = \min(f(x), g(x))$ and $f \vee g(x) = \max(f(x), g(x))$.

We say $\mathcal{A}$ *separates points* if whenever $x \neq y$ are two points in X, there exists $f \in \mathcal{A}$ (depending on x and y) such that $f(x) \neq f(y)$. We say $\mathcal{A}$ *vanishes at no point* of X if whenever $x \in X$, there exists $g \in \mathcal{A}$ (depending on x) such that $g(x) \neq 0$.

Lemma 20.42 *Suppose $\mathcal{A}$ is a subset of $\mathcal{C}(X)$, not necessarily an algebra of functions or a lattice of functions, such that $\mathcal{A}$ separates*

points and vanishes at no point. Suppose $f + g$ *and* cf *are in* $\mathcal{A}$
whenever $f, g \in \mathcal{A}$ *and* c *is a real number.*
(1) If x *and* y *are two distinct points in* X *and* a, b *are two real
numbers, there exists a function* $f \in \mathcal{A}$ *(depending on* x, y, a, b*)
such that* $f(x) = a$ *and* $f(y) = b$.
(2) If x *is a point in* X *and* b *a real number, there exists a function*
$g \in \mathcal{A}$ *(depending on* x, b*) such that* $g(x) = b$.

Proof. (1) Let h be a function such that $h(x) \neq h(y)$. Then let

$$f(z) = a\frac{h(z) - h(y)}{h(x) - h(y)} + b\frac{h(z) - h(x)}{h(y) - h(x)}.$$

(2) Let k be a function such that $k(x) \neq 0$. Then $g(z) = bk(z)/k(x)$ will be the desired function. $\qquad\square$

Theorem 20.43 *Let X be a compact Hausdorff space and let $\mathcal{A}$
be a collection of continuous functions that separate points, that
vanish at no point, that is a lattice of functions, and such that
$f + g$ and cf are in $\mathcal{A}$ whenever $f, g \in \mathcal{A}$ and c is a real number.
Then $\mathcal{A}$ is dense in $\mathcal{C}(X)$.*

Saying $\mathcal{A}$ is dense in $\mathcal{C}(X)$ is equivalent to saying that if $\varepsilon > 0$
and $f \in \mathcal{C}(X)$, there exists $g \in \mathcal{A}$ such that $\sup_{x \in X} |f(x) - g(x)| < \varepsilon$. Thus we can approximate any continuous function in $\mathcal{C}(X)$ by
an element of $\mathcal{A}$.

Proof. Let $\varepsilon > 0$ and let $f \in \mathcal{C}(X)$. Fix $x \in X$ for the moment.
If $y \neq x$, let h_y be an element of $\mathcal{A}$ such that $h_y(x) = f(x)$ and
$h_y(y) = f(y)$. Choose an open set G_y containing y such that
$h_y(z) < f(z) + \varepsilon$ for all $z \in G_y$. This is possible because f and
h_y are continuous; we use Exercise 20.9. The collection $\{G_y\}$,
$y \in X$, is an open cover for X, hence there is a finite subcover
$\{G_{y_1}, \ldots, G_{y_n}\}$. Define

$$k_x(z) = h_{y_1}(z) \wedge \cdots \wedge h_{y_n}(z).$$

Note that $k_x \in \mathcal{A}$, $k_x(x) = f(x)$, and $k_x(z) < f(z) + \varepsilon$ for every
$z \in X$. We used the fact that $\mathcal{A}$ is a lattice of functions.

We construct such a function k_x for each $x \in X$. Let H_x be an
open set containing x such that $k_x(z) > f(z) - \varepsilon$ for each $z \in H_x$.

This is possible because k_x and f are continuous, $k_x(x) = f(x)$, and Exercise 20.9. The collection $\{H_x\}$, $x \in X$, is an open cover of X. Let $\{H_{x_1}, \ldots, H_{x_m}\}$ be a finite subcover. Let

$$g(z) = k_{x_1}(z) \vee \cdots \vee k_{x_m}(z).$$

Then $g \in \mathcal{A}$ and $g(z) > f(z) - \varepsilon$ for each z. Moreover, since $k_{x_i} < f + \varepsilon$ for each i, then $g(z) < f(z) + \varepsilon$ for each z. Therefore $\sup_{z \in X} |g(z) - f(z)| < \varepsilon$. This proves $\mathcal{A}$ is dense in $\mathcal{C}(X)$. $\square$

Here is the version of the Stone-Weierstrass theorem for real-valued functions.

Theorem 20.44 *Suppose X is a compact Hausdorff space and $\mathcal{A}$ is an algebra of continuous functions that separates points and vanishes at no point. Then $\mathcal{A}$ is dense in $\mathcal{C}(X)$.*

Proof. We make the observation that if $\mathcal{A}$ is an algebra, then $\overline{\mathcal{A}}$, the closure of $\mathcal{A}$, is also an algebra. In view of Theorem 20.43, we need only show $\overline{\mathcal{A}}$ is also a lattice of functions. Since $\overline{\mathcal{A}}$ is closed, if it is dense in $\mathcal{C}(X)$, it must equal $\mathcal{C}(X)$.

Thus we need to show that if $f_1, f_2 \in \overline{\mathcal{A}}$, then $f_1 \wedge f_2$ and $f_1 \vee f_2$ are also in $\mathcal{A}$. Since

$$f_1 \wedge f_2 = \tfrac{1}{2}(f_1 + f_2 - |f_1 - f_2|), \qquad f_1 \vee f_2 = \tfrac{1}{2}(f_1 + f_2 + |f_1 - f_2|),$$

it is enough to show that if $f \in \overline{\mathcal{A}}$, then $|f| \in \overline{\mathcal{A}}$.

Let $\varepsilon > 0$ and suppose $f \in \overline{\mathcal{A}}$. Then there exists $g \in \mathcal{A}$ such that $\sup_{x \in X} |f(x) - g(x)| < \varepsilon/4$. Let $M = \|g\|_\infty$. Since the function $x \to |x|$ is continuous, by the Weierstrass approximation theorem (Theorem 20.41), there exists a polynomial P such that $\sup_{y \in [-M,M]} |P(y) - |y|| < \varepsilon/4$. In particular, $|P(0)| < \varepsilon/4$. If we let $R(y) = P(y) - P(0)$, then R is a polynomial with zero constant term such that

$$\sup_{y \in [-M,M]} |R(y) - |y|| < \varepsilon/2.$$

Since $\mathcal{A}$ is an algebra, then g, g^2, g^3, etc. are in $\mathcal{A}$, and hence $R(g) \in \mathcal{A}$. Here $R(g)$ is the function defined by $R(g)(x) = R(g(x))$. We have

$$\sup_{x \in X} |R(g)(x) - |g(x)|| < \varepsilon/2.$$

We know that $|f(x) - g(x)| < \varepsilon/4$ for all $x \in X$, hence

$$|\, |f(x)| - |g(x)|\, | < \varepsilon/4.$$

We conclude that

$$\sup_{x \in X} |R(g)(x) - |f(x)|\, | < \varepsilon.$$

We have thus found an element of $\mathcal{A}$, namely, $R(g)$, that is within ε of $|f|$, uniformly over $x \in X$. Since ε is arbitrary, we conclude that $|f| \in \overline{\mathcal{A}}$, and the proof is complete. □

The above theorem has an extension to complex-valued functions. Let $\mathcal{C}(X, \mathbb{C})$ be the set of complex-valued continuous functions on a topological space X. As usual we use $\overline{f}$ for the complex conjugate of f. When we say that $\mathcal{A}$ is an algebra of complex-valued functions, we require that $f + g$, fg, and cf be in $\mathcal{A}$ when $f, g \in \mathcal{A}$ and c is a complex number.

We now present the version of the Stone-Weierstrass theorem for complex-valued functions.

Theorem 20.45 *Suppose X is a compact Hausdorff space and $\mathcal{C}(X, \mathbb{C})$ is the set of complex-valued continuous functions on X. Let $\mathcal{A}$ be an algebra of continuous complex-valued functions that separates points and vanishes at no point. Suppose in addition that $\overline{f}$ is in $\mathcal{A}$ whenever f is in $\mathcal{A}$. Then the closure of $\mathcal{A}$ is $\mathcal{C}(X, \mathbb{C})$.*

Proof. Let $\mathcal{R}$ be the set of real-valued functions in $\mathcal{A}$. Clearly $\mathcal{R}$ is an algebra of continuous functions, that is, $f + g$, fg, and cf are in $\mathcal{R}$ whenever $f, g \in \mathcal{R}$ and c is a real number. If $f \in \mathcal{A}$, then $\overline{f} \in \mathcal{A}$, and therefore $\operatorname{Re} f = (f + \overline{f})/2$ and $\operatorname{Im} f = (f - \overline{f})/2i$ are in $\mathcal{A}$. Hence if $f \in \mathcal{A}$, then the real part and imaginary parts of f are in $\mathcal{R}$.

If $x \in X$, there exists $f \in \mathcal{A}$ such that $f(x) \neq 0$. This means that either $\operatorname{Re} f$ or $\operatorname{Im} f$ (or both) are non-zero, and hence $\mathcal{R}$ vanishes at no point of X. If $x \neq y$ are two points in X, there exists a function g such that $g(x) \neq g(y)$. Therefore at least one $\operatorname{Re} g(x) \neq \operatorname{Re} g(y)$ or $\operatorname{Im} g(x) \neq \operatorname{Im} g(y)$ (or perhaps both hold). This implies that $\mathcal{R}$ separates points.

By the real-valued version of the Stone-Weierstrass theorem, Theorem 20.44, $\mathcal{R}$ is dense in the collection of real-valued continuous functions on X. If $f \in \mathcal{C}(X, \mathbb{C})$, we can approximate $\operatorname{Re} f$ to within $\varepsilon/\sqrt{2}$ by a function $k_1 \in \mathcal{R}$ and also $\operatorname{Im} f$ to within $\varepsilon/\sqrt{2}$ by a function $k_2 \in \mathcal{R}$. Then $k_1 + ik_2$ will be in $\mathcal{A}$ and approximates f to within ε. $\qquad\qquad\qquad\qquad\qquad\qquad\qquad\qquad\qquad\square$

Example 20.46 The assumption that the complex conjugate of every function in $\mathcal{A}$ be in $\mathcal{A}$ cannot be eliminated. To see this, you need to know the following fact from complex analysis. This can be proved in a number of ways. Using Morera's theorem (see, e.g., [8]) gives a quick proof.

If f_n is a sequence of complex-valued functions that are analytic in the open unit disk in the complex plane and continuous on the closed unit disk and f_n converges uniformly to a function f on the closed unit disk, then f is analytic in the open unit disk as well.

To see why this example shows that the inclusion of complex conjugates is necessary, let $\mathcal{A}$ be the collection of complex-valued functions that are analytic in the open unit disk in the plane and continuous on the closed unit disk. Since the function 1 and the function z are in $\mathcal{A}$, then $\mathcal{A}$ vanishes at no point and separates points. Clearly $\mathcal{A}$ is an algebra of functions. The function $\overline{z}$ is not in the closure of $\mathcal{A}$ because it is not analytic in the open unit disk (or anywhere else); it doesn't satisfy the Cauchy-Riemann equations.

20.13   Connected sets

If X is a topological space, X is *disconnected* if there exist two disjoint non-empty open sets G and H such that $X = G \cup H$. The space X is *connected* if X is not disconnected. A subset A of X is *connected* if there do not exist two disjoint open sets G and H such that $A \subset G \cup H$, $A \cap G \neq \emptyset$, and $A \cap H \neq \emptyset$.

A subset A is connected if A, viewed as a topological space with the relative topology, is connected.

The most obvious example of connected sets are intervals in $\mathbb{R}$, but there is something to prove. Note that J being an interval can be characterized by the fact that if $x \leq z$ are in J and $x \leq y \leq z$, then $y \in J$.

Proposition 20.47 *A subset A of the real line with the usual topology is connected if and only if A is an interval.*

Proof. First we prove that if A is not an interval, then A is not connected. If A is not an interval, there exists $x < z \in A$ and $y \notin A$ such that $x < y < z$. If we let $G = (-\infty, y)$ and $H = (y, \infty)$, the sets G and H are our two disjoint open sets that show A is not connected.

Now we show that if A is an interval, then A is connected. Suppose not. Then there exist disjoint open sets G and H whose union contains A and which each intersect A. Pick a point $x \in A \cap G$ and a point $y \in A \cap H$. We may assume $x < y$, for if not, we reverse the labels of G and H. Let $t = \sup\{s : s \in [x, y] \cap G\}$. Since $x \leq t \leq y$, then $t \in A$. If $t \in G$, then since G is open, there exists a point $u \in (t, y)$ that is in G. Since A is an interval and $x \leq t < u < y$, then $u \in A$. This contradicts t being an upper bound. If $t \in H$, then since H is open, there exists $\varepsilon < t - x$ such that $(t - \varepsilon, t] \subset H$. This contradicts t being the least upper bound. Therefore we have found a point $t \in A$ that is not in $G \cup H$, a contradiction, and therefore A is connected. $\square$

Analogously to the situation with compact spaces, continuous functions map connected spaces to connected spaces.

Theorem 20.48 *Suppose f is a continuous function from a connected topological space X onto a topological space Y. Then Y is connected.*

Proof. If Y is not connected, there exist disjoint open sets G and H whose union is Y. Then since f is continuous, $f^{-1}(G)$ and $f^{-1}(H)$ are disjoint open sets whose union is X, contradicting that X is connected. Therefore Y must be connected. $\square$

A corollary of Theorem 20.48 and Proposition 20.47 is the *intermediate value theorem*.

Corollary 20.49 *Suppose f is a continuous function from a connected topological space X into the reals, $a < b \in \mathbb{R}$, and there exist points x and y in X such that $f(x) = a$ and $f(y) = b$. If $a < c < b$, there exists a point $z \in X$ such that $f(z) = c$.*

Proof. By Theorem 20.48, $f(X)$ is connected. By Proposition 20.47, $f(X)$ must be an interval. Since $a, b \in f(X)$ and $a < c < b$, then $c \in f(X)$. $\square$

We now prove that the product of connected topological spaces is connected. We begin with a lemma.

Lemma 20.50 *Suppose $\{A_\alpha\}$, $\alpha \in I$, is a non-empty collection of connected subsets of X such that $\cap_{\alpha \in I} A_\alpha \neq \emptyset$. Then $A = \cup_{\alpha \in I} A_\alpha$ is connected.*

Proof. If A is not connected, there exist disjoint open sets G and H which both intersect A and whose union contains A. Let x be any point of $\cap_{\alpha \in I} A_\alpha$. Suppose $x \in G$, the other case being similar. If $\alpha \in I$, then A_α is connected and $x \in A_\alpha \cap G$. Since $A_\alpha \subset A \subset G \cup H$, we must have $A_\alpha \subset G$, or else we get a contradiction to A_α being connected. This is true for each $\alpha \in I$, so $A = \cup_{\alpha \in I} A_\alpha \subset G$. This contradicts A having a non-empty intersection with H, and we conclude that A is connected. $\square$

We do the case of finite products of connected topological spaces separately; we will need this result for the general proof.

Lemma 20.51 *Suppose $X_1, \ldots, X_n$ are finitely many connected topological spaces. Then $X = \prod_{j=1}^{n} X_j$ is connected.*

Proof. Suppose X is not connected. Then there exist disjoint non-empty open sets G and H whose union is X. Pick $x \in G$ and $y \in H$. For $k = 0, \ldots, n$, let z^k be the point whose first k coordinates are the same as those of x and whose remaining coordinates are the same as y. Thus if $z^k = (z_1^k, \ldots, z_n^k)$, $x = (x_1, \ldots, x_n)$, and $y = (y_1, \ldots, y_n)$, then $z_i^k = x_i$ if $i \leq k$ and $z_i^k = y_i$ if $i > k$.

Since $z^0 = x \in G$ and $z^n = y \in H$, there exists $k < n$ such that $z^k \in G$ and $z^{k+1} \in H$. Let

$$F = \{x_1\} \times \cdots \times \{x_k\} \times X_{k+1} \times \{y_{k+2}\} \times \cdots \times \{y_n\}.$$

It is routine to check that F is homeomorphic to X_{k+1}, so there exists a continuous function f such that $F = f(X_{k+1})$, and by Theorem 20.48, F is connected. But $z^k \in F \cap G$ and $z^{k+1} \in F \cap H$, so $F \cap G$ and $F \cap H$ are two non-empty disjoint relatively open sets whose union is F, contradicting that F is connected. $\qquad\square$

Theorem 20.52 *Suppose $\{X_\alpha\}$, $\alpha \in I$, is a non-empty collection of connected topological spaces. Then $X = \prod_{\alpha \in I} X_\alpha$ is connected.*

Proof. The idea of the proof is to find a set E that is a connected subset of X and such that $\overline{E}$ is equal to X.

Fix $x \in X$. Define $D(x; \alpha_{i_1}, \ldots \alpha_{i_n})$ to be the set of points y of X all of whose coordinates agree with the corresponding coordinates of x except for the $\alpha_{i_1}, \ldots, \alpha_{i_n}$ ones. That is, if $\alpha \notin \{\alpha_{i_1}, \ldots, \alpha_{i_n}\}$, then $\pi_\alpha(x) = \pi_\alpha(y)$. Note that $D(x; \alpha_{i_1}, \ldots, \alpha_{i_n})$ is homeomorphic to $(\prod_{j=1}^n X_{\alpha_j}) \times (\prod_{\alpha \neq \alpha_{i_1}, \ldots, \alpha_{i_n}} X_\alpha)$. As in the proof of Lemma 20.51, $D(x, \alpha_{i_1}, \ldots, \alpha_{i_n})$ is connected.

Let

$$E_n(x) = \cup_{\alpha_{i_1}, \ldots, \alpha_{i_n} \in I} D(x; \alpha_{i_1}, \ldots, \alpha_{i_n}).$$

We see that $E_n(x)$ is the set of points y in X such that at most n coordinates of y differ from the corresponding coordinates of x. Since x is in each $D(x; \alpha_{i_1}, \ldots, \alpha_{i_n})$, then by Lemma 20.50, $E_n(x)$ is connected. Let $E = \cup_{n \geq 1} E_n(x)$. By Lemma 20.50 again, since $x \in E_n(x)$ for each n, then E is connected.

We now show that E is dense in X. Let $z \in X$ and let G be an open set containing x. We must prove that G contains a point of E. By the definition of the product topology, there exists a basic open set H such that $x \in H \subset G$, where $H = \pi_{\alpha_1}^{-1}(K_1) \cap \cdots \cap \pi_{\alpha_n}^{-1}(K_n)$, $n \geq 1$, each $\alpha_i \in I$, and each K_i is open in X_{α_i}. Pick $y_{\alpha_i} \in K_i$ for each i, and let y be the point whose α_i^{th} coordinate is y_{α_i} for $i \leq n$ and all its other coordinates agree with the corresponding coordinates of x. Thus $\pi_{\alpha_i}(y) = y_{\alpha_i}$ for $i \leq n$ and if $\alpha \neq \alpha_1, \ldots, \alpha_n$, then $\pi_\alpha(y) = \pi_\alpha(x)$. The point y is thus in $E_n(x)$ and also in H, therefore $y \in H \subset G$ and $y \in E_n(x) \subset E$.

We finish the proof by showing X is connected. If not, $X = G \cup H$, where G and H are disjoint non-empty open subsets of X. Since E is connected, either $E \cap G$ or $E \cap H$ is empty; let us assume the latter. Choose $x \in H$. Then H is an open set containing x that does not intersect G, hence $x \notin \overline{E}$. However E is dense in X, a contradiction. We conclude X is connected. $\square$

20.14   Exercises

Exercise 20.1 If X is a non-empty set and $\mathcal{T}_1$ and $\mathcal{T}_2$ are two topologies on X, prove that $\mathcal{T}_1 \cap \mathcal{T}_2$ is also a topology on X.

Exercise 20.2 If G is open in a topological space X and A is dense in X, show that $\overline{G} = \overline{G \cap A}$.

Exercise 20.3 Let X be $\mathbb{R}^2$ with the usual topology and say that $x \sim y$ if $x = Ay$ for some matrix A of the form

$$A = \begin{pmatrix} \cos\theta & -\sin\theta \\ \sin\theta & \cos\theta \end{pmatrix},$$

with $\theta \in \mathbb{R}$. Geometrically, $x \sim y$ if x can be obtained from y by a rotation of $\mathbb{R}^2$ about the origin.
(1) Show that $\sim$ is an equivalence relationship.
(2) Show that the quotient space is homeomorphic to $[0, \infty)$ with the usual topology.

Exercise 20.4 Prove that every metric space is first countable.

Exercise 20.5 Let X be an uncountable set of points and let $\mathcal{T}$ consist of all subsets A of X such that A^c is finite and let $\mathcal{T}$ also contain the empty set. Prove that X is a topological space that is not first countable.

Exercise 20.6 Give an example of a metric space which is not second countable.

Exercise 20.7 Prove that a subset A of X is dense if and only if A intersects every open set.

Exercise 20.8 Let $X = \prod_{\alpha \in I} X_\alpha$, where I is a non-empty index set. Prove that a net $\langle x_\beta \rangle$ in X converges to x if and only if the net $\langle \pi_\alpha(x_\beta) \rangle$ converges to $\pi_\alpha(x)$ for each $\alpha \in I$.

Exercise 20.9 Let f map a topological space X into a topological space Y. Prove that f is continuous if and only if whenever $x \in X$ and G is an open set in Y containing $f(x)$, there exists an open set H is X such that $f(y) \in G$ whenever $y \in H$.

Exercise 20.10 Let X and Y be topological spaces and $y_0 \in Y$. Prove that $X \times \{y_0\}$, with the relative topology derived from $X \times Y$, is homeomorphic to X.

Exercise 20.11 Let X, Y, and Z be topological spaces. Suppose $f : X \to Y$ and $g : Y \to Z$ are continuous functions. Prove that $g \circ f$ is a continuous function from X to Z.

Exercise 20.12 Suppose that X and Y are topological spaces and $f : X \to Y$ such that $f(x_n)$ converges to $f(x)$ whenever x_n converges to x. Is f necessarily continuous? If not, give a counterexample.

Exercise 20.13 Prove that $f : X \to Y$ is continuous if and only if the net $\langle f(x_\alpha) \rangle$ converges to $f(x)$ whenever the net $\langle x_\alpha \rangle$ converges to x.

Exercise 20.14 Let X be the collection of Lebesgue measurable functions on $[0, 1]$ furnished with the topology of pointwise convergence. Say that $f \sim g$ for $f, g \in X$ if $f = g$ a.e. Describe the quotient topology.

Exercise 20.15 A set A has the *Lindelöf property* if every open cover of A has a countable subcover. Prove that a metric space X has the Lindelöf property if and only if X is separable.

Exercise 20.16 Find an example of a compact set that is not closed.

Exercise 20.17 Show that the product topology on $[0, 1]^{\mathbb{N}}$ and the topology generated by the metric ρ of (20.3) in Section 20.8 are the same.

Exercise 20.18 Let $\{X_\alpha\}$, $\alpha \in I$ be a non-empty collection of topological spaces and let $X = \prod_{\alpha \in I} X_\alpha$. A sequence $\{x_n\}$ in X *converges pointwise* to x if $\pi_\alpha(x_n) \to \pi_\alpha(x)$ for each $\alpha \in I$. Prove that x_n converges to x pointwise if and only if x_n converges to x with respect to the product topology of X.

Exercise 20.19 This exercise illustrates why the notion of sequences is not that useful for general topological spaces. Let X be the space of real-valued bounded functions on $[0,1]$. X can be identified with $\mathbb{R}^{[0,1]}$ and we furnish X with the product topology. Let E be the set of Borel measurable functions on $[0,1]$.
(1) Show that E is dense in X.
(2) Let N be a set in $[0,1]$ that is not Borel measurable. Let $f = \chi_N$. Prove that there does not exist a sequence in E that converges to f, but that every neighborhood of f contains points of E.

Exercise 20.20 Prove that if I is a non-empty countable set and each X_α, $\alpha \in I$, is second countable, then $\prod_{\alpha \in I} X_\alpha$ is second countable.

Exercise 20.21 If X is a metric space, define

$$A^\delta = \{x \in X, d(x, A) < \delta\},$$

where $d(x, A) = \inf_{y \in A} d(x, y)$. For closed subsets of X, define

$$d_H(E, F) = \inf\{\delta : E \subset F^\delta \text{ and } F \subset E^\delta\}.$$

(1) Prove that d_H is a metric. (This is called the *Hausdorff metric.*)

(2) Suppose X is compact. Is the set of closed subsets with metric d_H necessarily compact? Prove, or else give a counterexample.

Exercise 20.22 Prove that if $\{x_n\}$ is a Cauchy sequence in a metric space X and a subsequence of $\{x_n\}$ converges to a point x, then the full sequence converges to x.

Exercise 20.23 Prove that if X is a topological space, then $\mathcal{C}(X)$ is a complete metric space.

Exercise 20.24 Prove that a sequence $\{x_n\}$ converges to a point x if and only if every subsequence $\{x_{n_j}\}$ has a further subsequence which converges to x.

Exercise 20.25 Let A be a subset of a metric space X. Prove that if A is totally bounded, then $\overline{A}$ is also totally bounded.

Exercise 20.26 Let X be a topological set such that the set $\{y\}$ is closed for each $y \in X$. Prove that X is a T_1 space.

Exercise 20.27 Find two disjoint closed subsets E and F of $\mathbb{R}$ such that $\inf_{x \in E, y \in F} |x - y| = 0$.

Exercise 20.28 Prove that every metric space is a normal space.

Exercise 20.29 Prove that a space X is a Hausdorff space if and only if every net converges to at most one point.

Exercise 20.30 Show that a closed subspace of a normal space is normal.

Exercise 20.31 Prove that $[0, 1]^{[0,1]}$ with the product topology is not metrizable.

Exercise 20.32 Prove that if X is metrizable and I is countable and non-empty, then X^I is metrizable.

Exercise 20.33 Let X be a locally compact Hausdorff space and X^* its one point compactification. A continuous function f mapping X to $\mathbb{R}$ is said to *vanish at infinity* if given $\varepsilon > 0$ there exists a compact set K such that $|f(x)| < \varepsilon$ for $x \notin K$. Prove that f vanishes at infinity if and only if f is the restriction of a continuous function $\overline{f} : X^* \to \mathbb{R}$ with $\overline{f}(\infty) = 0$.

Exercise 20.34 Prove that the one-point compactification of $\mathbb{R}^n$ is homeomorphic to the n-sphere $\{x \in \mathbb{R}^{n+1} : \|x\| = 1\}$.

Exercise 20.35 A sequence $\{f_n\}$ in $\mathcal{C}(X)$ is said to *converge uniformly on compact sets* to a function $f \in \mathcal{C}(X)$ if $\{f_n\}$ converges

to f uniformly on K whenever K is a compact subset of X.
(1) Give an example of a sequence $\{f_n\}$ in $\mathcal{C}(\mathbb{R})$ that converges uniformly to 0 on compact sets but such that $\{f_n\}$ does not converge uniformly to 0 on $\mathbb{R}$.
(2) Let X be a σ-compact locally compact Hausdorff space, $M > 0$, and $\{f_n\}$ an equicontinuous sequence in $\mathcal{C}(X)$ such that $|f_n(x)| \leq M$ for all $x \in X$ and all n. Prove there exists a subsequence that converges uniformly on compact sets.

Exercise 20.36 Show that $\mathbb{R}^{\mathbb{N}}$ is not locally compact.

Exercise 20.37 Show that $C([0,1])$ is not locally compact.

Exercise 20.38 Prove that if $\{X_\alpha\}$, $\alpha \in I$, is a non-empty collection of Hausdorff spaces such that $\prod_{\alpha \in I} X_\alpha$ is locally compact, then each X_α is also locally compact.

Exercise 20.39 A real-valued function f on a subset X of $\mathbb{R}$ is *Hölder continuous of order* α if there exists M such that

$$|f(x) - f(y)| \leq M|x - y|^\alpha$$

for each $x, y \in X$. Suppose $0 < \alpha \leq 1$ and let $X = [0,1]$. Prove that

$$\left\{ f \in C([0,1]) : \sup_{x \in [0,1]} |f(x)| \leq 1, \quad \sup_{x,y \in [0,1], x \neq y} \frac{|f(x) - f(y)|}{|x - y|^\alpha} \leq 1 \right\}$$

is compact in $C([0,1])$.

Exercise 20.40 Let $K : [0,1]^2 \to \mathbb{R}$ be continuous and let L be the set of Lebesgue measurable functions f on $[0,1]$ such that $\|f\|_\infty \leq 1$. For $f \in L$, define $Tf(x) = \int_0^1 K(x,y)f(y)\,dy$. Prove that $\{Tf; f \in L\}$ is an equicontinuous family in $C([0,1])$.

Exercise 20.41 Prove that if X is a compact metric space, then $\mathcal{C}(X)$ is separable.

Exercise 20.42 Let $X = [0, \infty]$ be the one point compactification of $[0, \infty)$, the non-negative real numbers with the usual metric. Let $\mathcal{A}$ be the collection of all finite linear combinations

$$\sum_{j=1}^n a_j e^{-\lambda_j x},$$

where the a_j are real and each $\lambda_j \geq 0$.

(1) Prove that $\mathcal{A}$ is a dense subset of $\mathcal{C}(X)$.

(2) Prove that if f_1 and f_2 are two continuous integrable functions from $[0, \infty)$ to $\mathbb{R}$ that vanish at infinity and which have the same *Laplace transform*, that is, $\int_0^\infty e^{-\lambda x} f_1(x)\, dx = \int_0^\infty e^{-\lambda x} f_2(x)\, dx$ for all $\lambda \geq 0$, then $f_1(x) = f_2(x)$ for all x.

Exercise 20.43 Suppose X and Y are compact Hausdorff spaces. Let $\mathcal{A}$ be the collection of real-valued functions in $\mathcal{C}(X \times Y)$ of the form

$$\sum_{i=1}^n a_i g_i(x) h_i(y),$$

where $n \geq 1$, each $a_i \in \mathbb{R}$, each $g_i \in \mathcal{C}(X)$, and each $h_i \in \mathcal{C}(Y)$. Prove that $\mathcal{A}$ is dense in $\mathcal{C}(X \times Y)$.

Exercise 20.44 Let X be a compact Hausdorff space and suppose $\mathcal{A}$ is an algebra of continuous functions that separates points. Prove that either $\mathcal{A}$ is dense in $\mathcal{C}(X)$ or else there exists a point $x \in X$ such that $\overline{\mathcal{A}} = \{f \in \mathcal{C}(X) : f(x) = 0\}$.

Exercise 20.45 Prove that if $f : [0,1] \to \mathbb{R}$ and $g : [0,1] \to \mathbb{R}$ are continuous functions such that

$$\int_0^1 f(x) x^n\, dx = \int_0^1 g(x) x^n\, dx$$

for $n = 0, 1, 2, \ldots$, then $f = g$.

Exercise 20.46 Let X be the closed unit disk in the complex plane. A polynomial in z and $\overline{z}$ is a function of the form

$$P(z) = \sum_{j=1}^n \sum_{k=1}^n a_{jk} z^j \overline{z}^k,$$

where a_{jk} are complex numbers. Prove that if f is a function in $\mathcal{C}(X, \mathbb{C})$, then f can be uniformly approximated by polynomials in z and $\overline{z}$.

Exercise 20.47 Prove that if B is a Banach space, then B is connected.

Exercise 20.48 Prove that if A is a convex subset of a Banach space, then A is connected.

Exercise 20.49 A topological space X is *arcwise connected* if whenever $x, y \in X$, there exists a continuous function f from $[0, 1]$ into X such that $f(0) = x$ and $f(1) = y$.
(1) Prove that if X is arcwise connected, then X is connected.
(2) Let $A_1 = \{(x, y) \in \mathbb{R}^2 : y = \sin(1/x), 0 < x \leq 1\}$ and $A_2 = \{(x, y) \in \mathbb{R}^2 : x = 0, -1 \leq y \leq 1\}$. Let $X = A_1 \cup A_2$ with the relative topology derived from $\mathbb{R}^2$. Prove that X is connected but not arcwise connected.

Exercise 20.50 If X is a topological space, a *component* of X is a connected subset of X that is not properly contained in any other connected subset of X. Prove that each $x \in X$ is contained in a unique component of X.

Exercise 20.51 A topological space X is *totally disconnected* if the components are all single points.
(1) Prove that the Cantor set with the relative topology derived from the real line is totally disconnected.
(2) Prove that if $\{X_\alpha\}$, $\alpha \in I$, is a non-empty collection of totally disconnected spaces, then $X = \prod_{\alpha \in I} X_\alpha$ is totally disconnected.

Exercise 20.52 Prove that a topological space X is connected if and only if for each pair $x, y \in X$ there is a connected subspace of X containing both x and y.

Exercise 20.53 Let X be a connected space. Suppose $f : X \to \mathbb{R}$ is continuous and non-constant. Prove that X is uncountable.

Exercise 20.54 Suppose $\{A_\alpha\}$, $\alpha \in I$, is a non-empty collection of connected subsets of a topological space X with the property that $A_\alpha \cap A_\beta \neq \emptyset$ for each $\alpha, \beta \in I$. Prove that $\cup_{\alpha \in I} A_\alpha$ is connected.

Chapter 21

Probability

Although some of the terminology and concepts of probability theory derive from its origins in gambling theory and statistics, the mathematical foundations of probability are based in real analysis. For example, a probability is just a measure with total mass one, and one of the main theorems, the strong law of large numbers, is an assertion about almost everywhere convergence.

In this chapter we introduce some of the major concepts of probability theory, including independence, the laws of large numbers, conditional expectation, martingales, weak convergence, characteristic functions, and the central limit theorem. We finish by constructing two different types of probabilities on infinite dimensional spaces.

21.1 Definitions

A *probability space* is a triple $(\Omega, \mathcal{F}, \mathbb{P})$, where Ω is an arbitrary set, $\mathcal{F}$ is a σ-field of subsets of Ω, and $\mathbb{P}$ is a probability on $(\Omega, \mathbb{P})$. A *σ-field* is exactly the same thing as a σ-algebra. A *probability* or *probability measure* is a positive measure whose total mass is 1, so that $\mathbb{P}(\Omega) = 1$. Elements of $\mathcal{F}$ are called *events*. Elements of Ω are often denoted ω.

Instead of saying a property occurs almost everywhere, we talk about properties occurring *almost surely*, written *a.s.* Real-valued measurable functions from Ω to $\mathbb{R}$ are called *random variables* and are usually denoted by X or Y or other capital letters. Often one sees "random variable" abbreviated by "*r.v.*"

The Lebesgue integral of a random variable X with respect to a probability measure $\mathbb{P}$ is called the *expectation* or the *expected value* of X, and we write $\mathbb{E}\,X$ for $\int X\,d\mathbb{P}$. The notation $\mathbb{E}\,[X; A]$ is used for $\int_A X\,d\mathbb{P}$.

The random variable 1_A is the function that is one if $\omega \in A$ and zero otherwise. It is called the *indicator* of A since the term characteristic function in probability refers to the Fourier transform. Events such as $\{\omega : X(\omega) > a\}$ are almost always abbreviated by $(X > a)$. An expression such as

$$(X > a, Y > b)$$

means $\{\omega : X(\omega) > a \text{ and } Y(\omega) > b\}$; the comma means "and."

Given a random variable X, the σ-field generated by X, denoted $\sigma(X)$ is the collection of events $(X \in A)$, A a Borel subset of $\mathbb{R}$. If we have several random variables: $X_1, X_2, \ldots, X_n$, we write $\sigma(X_1, \ldots, X_n)$ for the σ-field generated by the collection of events

$$\{(X_i \in A) : A \text{ a Borel subset of } \mathbb{R}, i = 1, \ldots, n.\}$$

This definition is extended in the obvious way when there are infinitely many random variables X_i.

Given a random variable X, we can define a probability on $(\mathbb{R}, \mathcal{B})$ where $\mathcal{B}$ is the Borel σ-field on $\mathbb{R}$, by

$$\mathbb{P}_X(A) = \mathbb{P}(X \in A), \qquad A \in \mathcal{B}. \tag{21.1}$$

The probability $\mathbb{P}_X$ is called the *law* of X or the *distribution* of X. We define $F_X : \mathbb{R} \to [0, 1]$ by

$$F_X(x) = \mathbb{P}_X((-\infty, x]) = \mathbb{P}(X \le x). \tag{21.2}$$

The function F_X is called the *distribution function* of X. Note that F_X is an increasing function whose corresponding Lebesgue-Stieltjes measure is $\mathbb{P}_X$.

Proposition 21.1 *The distribution function F_X of a random variable X satisfies:*
(1) F_X is increasing;
(2) F_X is right continuous with limits from the left existing;
(3) $\lim_{x \to \infty} F_X(x) = 1$ and $\lim_{x \to -\infty} F_X(x) = 0$.

Proof. These follow directly from elementary properties of measures. For example, if $x \leq y$, then $(X \leq x) \subset (X \leq y)$, and (1) follows. If $x_n \downarrow x$, then $(X \leq x_n) \downarrow (X \leq x)$, and so $F_X(x_n) = \mathbb{P}(X \leq x_n) \to \mathbb{P}(X \leq x) = F_X(x)$, since $\mathbb{P}(X \leq x_1) \leq 1$. This proves that F_X is right continuous. Since F_X is increasing, $\lim_{y \to x-} F_X(y)$ exists. This proves (2). The proof of (3) is left to the reader. □

Any function $F : \mathbb{R} \to [0,1]$ satisfying (1)-(3) of Proposition 21.1 is called a distribution function, whether or not it comes from a random variable.

Proposition 21.2 *Suppose F is a distribution function. There exists a random variable X such that $F = F_X$.*

Proof. Let $\Omega = [0,1]$, $\mathcal{F}$ the Borel σ-field, and $\mathbb{P}$ Lebesgue measure. Define $X(\omega) = \sup\{y : F(y) < \omega\}$. If $X(\omega) \leq x$, then $F(y) > \omega$ for all $y > x$. By right continuity, $F(x) \geq \omega$. On the other hand, if $\omega \leq F(x)$, then $x \notin \{y : F(y) < \omega\}$, so $X(\omega) \leq x$. Therefore $\{\omega : X(\omega) \leq x\} = \{\omega : 0 \leq \omega \leq F(x)\}$, and we conclude $F_X(x) = F(x)$. □

In the above proof, essentially $X = F^{-1}$. However F may have jumps or be constant over some intervals, so some care is needed in defining X.

Certain distributions or laws appear very often. We list some of them.

(1) *Bernoulli.* A random variable X is a Bernoulli random variable with parameter p if $\mathbb{P}(X = 1) = p$, $\mathbb{P}(X = 0) = 1 - p$ for some $p \in [0,1]$.

(2) *Binomial.* A random variable X is a binomial random variable with parameters n and p if $\mathbb{P}(X = k) = \binom{n}{k} p^k (1-p)^{n-k}$, where n is a positive integer, $0 \leq k \leq n$, and $p \in [0,1]$. Here $\binom{n}{k} = n!/k!(n-k)!$.

(3) *Geometric.* A random variable X is a geometric random variable with parameter p if $\mathbb{P}(X = k) = (1-p)p^k$, where $p \in [0,1]$ and k is a non-negative integer.

(4) *Poisson.* If $\mathbb{P}(X = k) = e^{-\lambda}\lambda^k/k!$ for k a non-negative integer and $\lambda > 0$, then X is called a Poisson random variable with parameter λ.

If F is absolutely continuous, we call $f = F'$ the *density* of F. Some examples of distributions characterized by densities are the following.

(5) *Exponential.* Let $\lambda > 0$. For $x > 0$ let $f(x) = \lambda e^{-\lambda x}$. If X has a distribution function whose density is equal to f, then X is said to be an exponential random variable with parameter λ.

(6) *Standard normal.* Define $f(x) = \frac{1}{\sqrt{2\pi}}e^{-x^2/2}$. If the distribution function of X has f as its density, then X is a standard normal random variable. Thus

$$\mathbb{P}(X \in A) = \mathbb{P}_X(A) = \frac{1}{\sqrt{2\pi}} \int_A e^{-x^2/2}\, dx.$$

Exercise 11.18 shows that $\mathbb{P}_X$ has total mass 1 and so is a probability measure.

We can use the law of a random variable to calculate expectations.

Proposition 21.3 *Suppose g is Borel measurable and suppose g is either bounded or non-negative. Then*

$$\mathbb{E}\, g(X) = \int g(x)\, \mathbb{P}_X(dx).$$

Proof. If g is the indicator of an event A, this is just the definition of $\mathbb{P}_X$. By linearity, the result holds for simple functions g. By approximating a non-negative measurable function from below by simple functions and using the monotone convergence theorem, the result holds for non-negative functions g, and by linearity again, it holds for bounded and measurable g.   □

If F_X has a density f_X, then $\mathbb{P}_X(dx) = f_X(x)\, dx$. If X is integrable, that is, $\mathbb{E}\,|X| < \infty$, we have

$$\mathbb{E}\, X = \int x f_X(x)\, dx$$

and in any case we have

$$\mathbb{E} X^2 = \int x^2 f_X(x) \, dx,$$

although both sides of the equality might be infinite.

We define the *mean* of a random variable to be its expectation, and the *variance* of a random variable is defined by

$$\operatorname{Var} X = \mathbb{E} (X - \mathbb{E} X)^2.$$

The square root of the variance of X is called the *standard deviation* of X. From the definition of variance, it is clear that $\operatorname{Var}(X + c) = \operatorname{Var} X$ for any constant c.

Note

$$\operatorname{Var} X = \mathbb{E} [X^2 - 2X \cdot \mathbb{E} X + (\mathbb{E} X)^2] = \mathbb{E} X^2 - (\mathbb{E} X)^2. \quad (21.3)$$

Immediate consequences of this are that $\operatorname{Var} X \leq \mathbb{E} X^2$ and that $\operatorname{Var}(cX) = c^2 \operatorname{Var} X$ for any constant c.

It is an exercise in calculus to see that the mean of a standard normal random variable is zero and its variance is one; use the fact that $xe^{-x^2/2}$ is an odd function to see that the mean is zero and use integration by parts to calculate the variance.

An equality that is useful is the following.

Proposition 21.4 *If $X \geq 0$ a.s. and $p > 0$, then*

$$\mathbb{E} X^p = \int_0^\infty p\lambda^{p-1} \mathbb{P}(X > \lambda) \, d\lambda.$$

The proof will show that this equality is also valid if we replace $\mathbb{P}(X > \lambda)$ by $\mathbb{P}(X \geq \lambda)$.

Proof. Use the Fubini theorem and write

$$\int_0^\infty p\lambda^{p-1} \mathbb{P}(X > \lambda) \, d\lambda = \mathbb{E} \int_0^\infty p\lambda^{p-1} 1_{(\lambda,\infty)}(X) \, d\lambda$$

$$= \mathbb{E} \int_0^X p\lambda^{p-1} \, d\lambda = \mathbb{E} X^p.$$

This completes the proof. $\qquad\qquad\qquad\qquad\qquad\qquad\qquad\qquad \square$

We have already seen the *Chebyshev inequality* in Lemma 10.4. In probability notation, the Chebyshev inequality says

$$\mathbb{P}(X \geq a) \leq \frac{\mathbb{E}\,X}{a}$$

if $X \geq 0$.

If we apply this to $X = (Y - \mathbb{E}\,Y)^2$, we obtain

$$\mathbb{P}(|Y - \mathbb{E}\,Y| \geq a) = \mathbb{P}((Y - \mathbb{E}\,Y)^2 \geq a^2) \leq \text{Var}\,Y/a^2. \qquad (21.4)$$

Remark 21.5 Observe that if g is a convex function and and x_0 is in the domain of g, then there is a line through $(x_0, g(x_0))$ such that the graph of g lies above this line. When g is differentiable at x_0, the tangent line is the one we want, but such a line exists even at points where g is not differentiable.

This remark allows us to prove *Jensen's inequality*, not to be confused with the Jensen formula of complex analysis.

Proposition 21.6 *Suppose g is convex and X and $g(X)$ are both integrable. Then*
$$g(\mathbb{E}\,X) \leq \mathbb{E}\,g(X).$$

Proof. If $x_0 \in \mathbb{R}$, we have

$$g(x) \geq g(x_0) + c(x - x_0)$$

for some constant c by Remark 21.5. Set $x = X(\omega)$ and take expectations to obtain

$$\mathbb{E}\,g(X) \geq g(x_0) + c(\mathbb{E}\,X - x_0).$$

Now choose x_0 equal to $\mathbb{E}\,X$. $\qquad\qquad\qquad\qquad\square$

21.2 Independence

Let us say two events A and B are *independent* if $\mathbb{P}(A \cap B) = \mathbb{P}(A)\mathbb{P}(B)$. The events $A_1, \ldots, A_n$ are independent if

$$\mathbb{P}(A_{i_1} \cap A_{i_2} \cap \cdots \cap A_{i_j}) = \mathbb{P}(A_{i_1})\mathbb{P}(A_{i_2}) \cdots \mathbb{P}(A_{i_j})$$

whenever $1 \leq i_1 < \ldots < i_j \leq n$. When n is three, for example, not only must $\mathbb{P}(A_1 \cap A_2 \cap A_3)$ factor properly, but so must $\mathbb{P}(A_1 \cap A_2)$, $\mathbb{P}(A_1 \cap A_3)$, and $\mathbb{P}(A_2 \cap A_3)$.

Proposition 21.7 *If A and B are independent, then A^c and B are independent.*

Proof. We write

$$\mathbb{P}(A^c \cap B) = \mathbb{P}(B) - \mathbb{P}(A \cap B) = \mathbb{P}(B) - \mathbb{P}(A)\mathbb{P}(B)$$
$$= \mathbb{P}(B)(1 - \mathbb{P}(A)) = \mathbb{P}(B)\mathbb{P}(A^c),$$

which proves the proposition. □

We say two σ-fields $\mathcal{F}$ and $\mathcal{G}$ are independent if A and B are independent whenever $A \in \mathcal{F}$ and $B \in \mathcal{G}$. Two random variables X and Y are independent if the σ-field generated by X and the σ-field generated by Y are independent. We define the independence of n σ-fields or n random variables in the obvious way.

If we have an infinite sequence of events $\{A_n\}$, we say they are independent if every finite subset of them is independent. We define independence for an infinite sequence of random variables similarly.

Remark 21.8 If f and g are Borel functions and X and Y are independent, then $f(X)$ and $g(Y)$ are independent. This follows because the σ-field generated by $f(X)$ is a sub-σ-field of the one generated by X, and similarly for $g(Y)$.

If $\{A_n\}$ is a sequence of events, define $(A_n \text{ i.o.})$, read "A_n *infinitely often*," by

$$(A_n \text{ i.o.}) = \cap_{n=1}^\infty \cup_{i=n}^\infty A_i.$$

This set consists of those ω that are in infinitely many of the A_n.

We now state one of the most useful tools in probability theory, the *Borel-Cantelli lemma*. Note for the first part of the lemma that no assumption of independence is made. Also note that we have used the proof of the first part of the Borel-Cantelli lemma several times already without calling it by that name; see, e.g., the proofs of Proposition 13.2 and Theorem 15.4.

Lemma 21.9 (Borel-Cantelli lemma) *Let $\{A_n\}$ be a sequence of events.*
(1) If $\sum_n \mathbb{P}(A_n) < \infty$, then $\mathbb{P}(A_n \text{ i.o.}) = 0$.
(2) Suppose in addition that the A_n are independent events. If $\sum_n \mathbb{P}(A_n) = \infty$, then $\mathbb{P}(A_n \text{ i.o.}) = 1$.

Proof. (1) We have

$$\mathbb{P}(A_n \text{ i.o.}) = \lim_{n \to \infty} \mathbb{P}(\cup_{i=n}^{\infty} A_i).$$

However,

$$\mathbb{P}(\cup_{i=n}^{\infty} A_i) \le \sum_{i=n}^{\infty} \mathbb{P}(A_i),$$

and the right hand side tends to zero as $n \to \infty$.

(2) Write

$$\mathbb{P}(\cup_{i=n}^{N} A_i) = 1 - \mathbb{P}(\cap_{i=n}^{N} A_i^c) = 1 - \prod_{i=n}^{N} \mathbb{P}(A_i^c) = 1 - \prod_{i=n}^{N}(1 - \mathbb{P}(A_i)).$$

By the mean value theorem, $1 - e^{-x} \le x$, or $1 - x \le e^{-x}$, so we have that the right hand side is greater than or equal to

$$1 - \exp\left(-\sum_{i=n}^{N} \mathbb{P}(A_i)\right).$$

As $N \to \infty$, this tends to 1, so $\mathbb{P}(\cup_{i=n}^{\infty} A_i) = 1$. This holds for all n, which proves the result.   □

The following is known as the multiplication theorem.

Theorem 21.10 *If X, Y, and XY are integrable and X and Y are independent, then $\mathbb{E}[XY] = (\mathbb{E}X)(\mathbb{E}Y)$.*

Proof. First suppose that X and Y are both non-negative and bounded by a positive integer M. Let

$$X_n = \sum_{k=0}^{M2^n} \frac{k}{2^n} 1_{[k/2^n,(k+1)/2^n)}(X),$$

and define Y_n similarly. We see that $1_{[k/2^n,(k+1)/2^n)}(X)$ is independent of $1_{[j/2^n,(j+1)/2^n)}(Y)$ for each j and k by Remark 21.8. Then

$$\mathbb{E}\left[X_n Y_n\right]$$

$$= \sum_{k=0}^{M2^n} \sum_{j=0}^{M2^n} \frac{k}{2^n} \cdot \frac{j}{2^n} \mathbb{E}\left[1_{[k/2^n,(k+1)/2^n)}(X)1_{[j/2^n,(j+1)/2^n)}(Y)\right]$$

$$= \sum_{k=0}^{M2^n} \sum_{j=0}^{M2^n} \frac{k}{2^n} \cdot \frac{j}{2^n} \mathbb{P}(X \in [k/2^n,(k+1)/2^n),$$

$$Y \in [j/2^n,(j+1)/2^n))$$

$$= \sum_{k=0}^{M2^n} \sum_{j=0}^{M2^n} \frac{k}{2^n} \cdot \frac{j}{2^n} \mathbb{P}(X \in [k/2^n,(k+1)/2^n))$$

$$\times \mathbb{P}(Y \in [j/2^n,(j+1)/2^n))$$

$$= \left(\sum_{k=0}^{M2^n} \frac{k}{2^n} \mathbb{E}\left[1_{[k/2^n,(k+1)/2^n)}(X)\right] \right)$$

$$\times \left(\sum_{j=0}^{M2^n} \frac{j}{2^n} \mathbb{E}\left[1_{[j/2^n,(j+1)/2^n)}(Y)\right] \right)$$

$$= (\mathbb{E}\,X_n)(\mathbb{E}\,Y_n).$$

If we let $n \to \infty$, by the dominated convergence theorem, we obtain our theorem in the case when X and Y are non-negative and bounded by M.

If X and Y are non-negative but not necessarily bounded, use Remark 21.8 to see that $X \wedge M$ and $Y \wedge M$ are independent, so

$$\mathbb{E}\left[(X \wedge M)(Y \wedge M)\right] = \mathbb{E}\left[X \wedge M\right] \mathbb{E}\left[Y \wedge M\right].$$

Letting $M \to \infty$ and using the monotone convergence theorem, we have $\mathbb{E}\left[XY\right] = (\mathbb{E}\,X)(\mathbb{E}\,Y)$ when X and Y are non-negative. Finally, writing $X = X^+ - X^-$ and $Y = Y^+ - Y^-$, we obtain the multiplication theorem for the general case by linearity. $\square$

Remark 21.11 If $X_1,\ldots,X_n$ are independent, then so are the random variables $X_1 - \mathbb{E}\,X_1,\ldots,X_n - \mathbb{E}\,X_n$. Assuming all the random variables are integrable,

$$\mathbb{E}\left[(X_1 - \mathbb{E}\,X_1) + \cdots (X_n - \mathbb{E}\,X_n)\right]^2$$
$$= \mathbb{E}\,(X_1 - \mathbb{E}\,X_1)^2 + \cdots + \mathbb{E}\,(X_n - \mathbb{E}\,X_n)^2,$$

using the multiplication theorem to show that the expectations of the cross product terms are zero. We have thus shown

$$\mathrm{Var}\,(X_1 + \cdots + X_n) = \mathrm{Var}\,X_1 + \cdots + \mathrm{Var}\,X_n. \qquad (21.5)$$

In words, if the X_i are independent, then the variance of the sum is equal to the sum of the variances.

21.3   Weak law of large numbers

Suppose X_n is a sequence of independent random variables. Suppose also that they all have the same distribution, that is, $\mathbb{P}_{X_n} = \mathbb{P}_{X_1}$ for all n. This situation comes up so often it has a name, *independent and identically distributed*, which is abbreviated *i.i.d.* In this case, $\mathbb{P}(X_n \in A) = \mathbb{P}(X_1 \in A)$ for all n and all Borel sets A. We also see that $\mathbb{E}\,X_n = \mathbb{E}\,X_1$, $\mathrm{Var}\,X_n = \mathrm{Var}\,X_1$, and so on.

Define $S_n = \sum_{i=1}^{n} X_i$. S_n is called a *partial sum process*. S_n/n is the average value of the first n of the X_i's. We say a sequence of random variables $\{Y_n\}$ *converges in probability* to a random variable Y if it converges in measure with respect to the measure $\mathbb{P}$. Recall that this means that for each $\varepsilon > 0$,

$$\mathbb{P}(|Y_n - Y| > \varepsilon) \to 0$$

as $n \to \infty$.

The weak law of large numbers (we will do the strong law of large numbers in Section 21.4) is a version of the law of averages.

Theorem 21.12 *Suppose the X_i are i.i.d. and $\mathbb{E}\,X_1^2 < \infty$. Then $S_n/n \to \mathbb{E}\,X_1$ in probability.*

Proof. Since the X_i are i.i.d., they all have the same expectation, and so $\mathbb{E}\,S_n = n\mathbb{E}\,X_1$. Hence $\mathbb{E}\,(S_n/n - \mathbb{E}\,X_1)^2$ is the variance of

S_n/n. If $\varepsilon > 0$, by (21.4),

$$\mathbb{P}(|S_n/n - \mathbb{E}\,X_1| > \varepsilon) = \mathbb{P}((S_n/n - \mathbb{E}\,X_1)^2 > \varepsilon^2) \qquad (21.6)$$
$$\leq \frac{\text{Var}\,(S_n/n)}{\varepsilon^2}.$$

Using Remark 21.11, the right hand side is equal to

$$\frac{\sum_{i=1}^n \text{Var}\,X_i}{n^2\varepsilon^2} = \frac{n\,\text{Var}\,X_1}{n^2\varepsilon^2}. \qquad (21.7)$$

Since $\mathbb{E}\,X_1^2 < \infty$, then $\text{Var}\,X_1 < \infty$, and the result follows by letting $n \to \infty$. $\qquad\square$

A nice application of the weak law of large numbers is a proof of the Weierstrass approximation theorem. Recall from undergraduate probability that the sum of n i.i.d. Bernoulli random variables with parameter p is a binomial random variable with parameters n and p. To see this, if S_n is the sum of i.i.d. Bernoulli random variables $X_1, \ldots, X_n$, then S_n is the number of the X_i that are equal to 1. The probability that the first k of the X_i's are 1 and the rest 0 is $p^k(1-p)^{n-k}$, using independence. The probability that the last k of the X_i's are 1 and the rest 0 is the same, and we get the same probability for any configuration of k ones and $n-k$ zeroes. There are $\binom{n}{k}$ such configurations, so $\mathbb{P}(S_n = k) = \binom{n}{k} p^k(1-p)^{n-k}$. For another proof of this fact, see Remark 21.42.

An easy computation shows that $\text{Var}\,X_1 = p(1-p)$, and so, using Remark 21.11, $\text{Var}\,S_n = np(1-p)$.

Theorem 21.13 *Suppose f is a continuous function on $[0,1]$ and $\varepsilon > 0$. There exists a polynomial P such that*

$$\sup_{x \in [0,1]} |f(x) - P(x)| < \varepsilon.$$

Proof. Let

$$P_n(x) = \sum_{k=0}^n f(k/n) \binom{n}{k} x^k(1-x)^{n-k}.$$

Clearly P is a polynomial. Since f is continuous, there exists M such that $|f(x)| \leq M$ for all $x \in [0,1]$ and there exists δ such that $|f(x) - f(y)| < \varepsilon/2$ whenever $|x - y| \leq \delta$.

Let X_i be i.i.d. Bernoulli random variables with parameter x. Then S_n, the partial sum, is a binomial random variable, and hence $P_n(x) = \mathbb{E}\, f(S_n/n)$. The mean of S_n/n is x. We have

$$
\begin{aligned}
|P_n(x) - f(x)| &= |\mathbb{E}\, f(S_n/n) - f(\mathbb{E}\, X_1)| \\
&\leq \mathbb{E}\, |f(S_n/n) - f(\mathbb{E}\, X_1)| \\
&= \mathbb{E}\, [\, |f(S_n/n) - f(\mathbb{E}\, X_1)|;\, |S_n/n - \mathbb{E}\, X_1| \leq \delta\,] \\
&\quad + \mathbb{E}\, [\, |f(S_n/n) - f(\mathbb{E}\, X_1)|;\, |S_n/n - \mathbb{E}\, X_1| > \delta\,] \\
&\leq \varepsilon/2 + 2M\mathbb{P}(|S_n/n - x| > \delta).
\end{aligned}
$$

By (21.6) and (21.7) the second term on the last line will be less than or equal to

$$
2M\mathrm{Var}\, X_1/n\delta^2 \leq 2Mx(1-x)/n\delta^2 \leq 2Mn\delta^2,
$$

which will be less than $\varepsilon/2$ if n is large enough, uniformly in x. $\square$

In the next section we prove the strong law of large numbers. There we get a stronger result than Theorem 21.12 with weaker hypotheses. There are, however, versions of the weak law of large numbers that have weaker hypotheses than Theorem 21.16.

21.4   Strong law of large numbers

The strong law of large numbers is the mathematical formulation of the law of averages. If one tosses a fair coin over and over, the proportion of heads should converge to $1/2$. Mathematically, if X_i is 1 if the ith toss turns up heads and 0 otherwise, then we want S_n/n to converge with probability one to $1/2$, where $S_n = X_1 + \cdots + X_n$.

Before stating and proving the strong law of large numbers, we need three facts from calculus. First, recall that if $b_n \to b$ are real numbers, then

$$
\frac{b_1 + \cdots + b_n}{n} \to b. \tag{21.8}
$$

Second, there exists a constant c_1 such that

$$
\sum_{k=n}^{\infty} \frac{1}{k^2} \leq \frac{c_1}{n}. \tag{21.9}
$$

(To prove this, recall the proof of the integral test and compare the sum to $\int_{n-1}^{\infty} x^{-2}\, dx$ when $n \geq 2$.) Third, suppose $a > 1$ and k_n is the largest integer less than or equal to a^n. Note $k_n \geq a^n/2$. Then

$$\sum_{\{n:k_n \geq j\}} \frac{1}{k_n^2} \leq \sum_{\{n:a^n \geq j\}} \frac{4}{a^{2n}} \leq \frac{4}{j^2} \cdot \frac{1}{1 - a^{-2}} \qquad (21.10)$$

by the formula for the sum of a geometric series.

We also need two probability estimates.

Lemma 21.14 *If $X \geq 0$ a.s. and $\mathbb{E}\, X < \infty$, then*

$$\sum_{n=1}^{\infty} \mathbb{P}(X \geq n) < \infty.$$

Proof. Since $\mathbb{P}(X \geq x)$ increases as x decreases,

$$\sum_{n=1}^{\infty} \mathbb{P}(X \geq n) \leq \sum_{n=1}^{\infty} \int_{n-1}^{n} \mathbb{P}(X \geq x)\, dx$$

$$= \int_{0}^{\infty} \mathbb{P}(X \geq x)\, dx = \mathbb{E}\, X,$$

which is finite. $\qquad \square$

Lemma 21.15 *Let $\{X_n\}$ be an i.i.d. sequence with each $X_n \geq 0$ a.s. and $\mathbb{E}\, X_1 < \infty$. Define*

$$Y_n = X_n 1_{(X_n \leq n)}.$$

Then

$$\sum_{k=1}^{\infty} \frac{\operatorname{Var} Y_k}{k^2} < \infty.$$

Proof. Since $\operatorname{Var} Y_k \leq \mathbb{E}\, Y_k^2$,

$$\sum_{k=1}^{\infty} \frac{\operatorname{Var} Y_k}{k^2} \leq \sum_{k=1}^{\infty} \frac{\mathbb{E}\, Y_k^2}{k^2} = \sum_{k=1}^{\infty} \frac{1}{k^2} \mathbb{E}\left[X_k^2; X_k \leq k\right]$$

$$= \sum_{k=1}^{\infty} \frac{1}{k^2} \int_0^{\infty} 1_{(x \leq k)} 2x \mathbb{P}(X_k > x) \, dx$$

$$= \int_0^{\infty} \sum_{k=1}^{\infty} \frac{1}{k^2} 1_{(x \leq k)} 2x \mathbb{P}(X_1 > x) \, dx$$

$$\leq c_1 \int_0^{\infty} \frac{1}{x} \cdot 2x \mathbb{P}(X_1 > x) \, dx$$

$$= 2c_1 \int_0^{\infty} \mathbb{P}(X_1 > x) \, dx = 2c_1 \mathbb{E} \, X_1 < \infty.$$

We used the fact that the X_k are i.i.d., the Fubini theorem, Proposition 21.4, and (21.9). $\qquad\square$

We now state and prove the *strong law of large numbers*.

Theorem 21.16 *Suppose* $\{X_i\}$ *is an i.i.d. sequence with* $\mathbb{E} \, X_1 < \infty$. *Let* $S_n = \sum_{i=1}^{n} X_i$. *Then*

$$\frac{S_n}{n} \to \mathbb{E} \, X_1, \qquad \text{a.s.}$$

Proof. By writing each X_n as $X_n^+ - X_n^-$ and considering the positive and negative parts separately, it suffices to suppose each $X_n \geq 0$. Define $Y_k = X_k 1_{(X_k \leq k)}$ and let $T_n = \sum_{i=1}^{n} Y_i$. The main part of the argument is to prove that $T_n/n \to \mathbb{E} \, X_1$ a.s.

Step 1. Let $a > 1$ and let k_n be the largest integer less than or equal to a^n. Let $\varepsilon > 0$ and let

$$A_n = \left(\frac{|T_{k_n} - \mathbb{E} \, T_{k_n}|}{k_n} > \varepsilon \right).$$

By (21.4)

$$\mathbb{P}(A_n) \leq \frac{\text{Var} \, (T_{k_n}/k_n)}{\varepsilon^2} = \frac{\text{Var} \, T_{k_n}}{k_n^2 \varepsilon^2} = \frac{\sum_{j=1}^{k_n} \text{Var} \, Y_j}{k_n^2 \varepsilon^2}.$$

Then

$$\sum_{n=1}^{\infty} \mathbb{P}(A_n) \leq \sum_{n=1}^{\infty} \sum_{j=1}^{k_n} \frac{\text{Var} \, Y_j}{k_n^2 \varepsilon^2}$$

$$= \frac{1}{\varepsilon^2} \sum_{j=1}^{\infty} \sum_{\{n: k_n \geq j\}} \frac{1}{k_n^2} \operatorname{Var} Y_j$$

$$\leq \frac{4(1 - a^{-2})^{-1}}{\varepsilon^2} \sum_{j=1}^{\infty} \frac{\operatorname{Var} Y_j}{j^2}$$

by (21.10). By Lemma 21.15, $\sum_{n=1}^{\infty} \mathbb{P}(A_n) < \infty$, and by the Borel-Cantelli lemma, $\mathbb{P}(A_n \text{ i.o.}) = 0$. This means that for each ω except for those in a null set, there exists $N(\omega)$ such that if $n \geq N(\omega)$, then $|T_{k_n}(\omega) - ET_{k_n}|/k_n < \varepsilon$. Applying this with $\varepsilon = 1/m$, $m = 1, 2, \ldots$, we conclude

$$\frac{T_{k_n} - \mathbb{E} T_{k_n}}{k_n} \to 0, \qquad \text{a.s.}$$

Step 2. Since

$$\mathbb{E} Y_j = \mathbb{E}[X_j; X_j \leq j] = \mathbb{E}[X_1; X_1 \leq j] \to \mathbb{E} X_1$$

by the dominated convergence theorem as $j \to \infty$, then by (21.8)

$$\frac{\mathbb{E} T_{k_n}}{k_n} = \frac{\sum_{j=1}^{k_n} \mathbb{E} Y_j}{k_n} \to \mathbb{E} X_1.$$

Therefore $T_{k_n}/k_n \to \mathbb{E} X_1$ a.s.

Step 3. If $k_n \leq k \leq k_{n+1}$, then

$$\frac{T_k}{k} \leq \frac{T_{k_{n+1}}}{k_{n+1}} \cdot \frac{k_{n+1}}{k_n}$$

since we are assuming that the X_k are non-negative. Therefore

$$\limsup_{k \to \infty} \frac{T_k}{k} \leq a \mathbb{E} X_1, \qquad \text{a.s.}$$

Similarly, $\liminf_{k \to \infty} T_k/k \geq (1/a)\mathbb{E} X_1$ a.s. Since $a > 1$ is arbitrary,

$$\frac{T_k}{k} \to \mathbb{E} X_1, \qquad \text{a.s.}$$

Step 4. Finally,

$$\sum_{n=1}^{\infty} \mathbb{P}(Y_n \neq X_n) = \sum_{n=1}^{\infty} \mathbb{P}(X_n > n) = \sum_{n=1}^{\infty} \mathbb{P}(X_1 > n) < \infty$$

by Lemma 21.14. By the Borel-Cantelli lemma, $\mathbb{P}(Y_n \neq X_n \text{ i.o.}) = 0$. In particular, $Y_n - X_n \to 0$ a.s. By (21.8) we have

$$\frac{T_n - S_n}{n} = \sum_{i=1}^{n} \frac{(Y_i - X_i)}{n} \to 0, \qquad \text{a.s.,}$$

hence $S_n/n \to \mathbb{E}\,X_1$ a.s.   □

21.5 Conditional expectation

It is fairly common in probability theory for there to be more than one σ-field present. For example, if $X_1, X_2, \ldots$ is a sequence of random variables, one might let $\mathcal{F}_n = \sigma(X_1, \ldots, X_n)$, which means that $\mathcal{F}_n$ is the σ-field generated by the collection of sets $(X_i \in A)$ for $i = 1, 2, \ldots, n$ and A a Borel subset of $\mathbb{R}$.

If $\mathcal{F} \subseteq \mathcal{G}$ are two σ-fields and X is an integrable $\mathcal{G}$ measurable random variable, the *conditional expectation* of X given $\mathcal{F}$, written $\mathbb{E}[X \mid \mathcal{F}]$ and read as "the expectation (or expected value) of X given $\mathcal{F}$," is any $\mathcal{F}$ measurable random variable Y such that $\mathbb{E}[Y; A] = \mathbb{E}[X; A]$ for every $A \in \mathcal{F}$. The *conditional probability* of $A \in \mathcal{G}$ given $\mathcal{F}$ is defined by $\mathbb{P}(A \mid \mathcal{F}) = \mathbb{E}[1_A \mid \mathcal{F}]$. When $\mathcal{F} = \sigma(Y)$, one usually writes $\mathbb{E}[X \mid Y]$ for $\mathbb{E}[X \mid \mathcal{F}]$.

If Y_1, Y_2 are two $\mathcal{F}$ measurable random variables such that $\mathbb{E}[Y_1; A] = \mathbb{E}[Y_2; A]$ for all $A \in \mathcal{F}$, then $Y_1 = Y_2$ a.s. by Proposition 8.1. In other words, conditional expectation is unique up to a.s. equivalence.

In the case X is already $\mathcal{F}$ measurable, $\mathbb{E}[X \mid \mathcal{F}] = X$. This follows from the definition.

If X is independent of $\mathcal{F}$, $\mathbb{E}[X \mid \mathcal{F}] = \mathbb{E}\,X$. To see this, if $A \in \mathcal{F}$, then 1_A and X are independent, and by the multiplication theorem

$$\mathbb{E}[X; A] = \mathbb{E}[X 1_A] = (\mathbb{E}\,X)(\mathbb{E}\,1_A) = \mathbb{E}[\mathbb{E}\,X; A].$$

For another example which ties this definition with the one used in elementary probability courses, suppose $\{A_i\}$ is a finite collection of disjoint sets whose union is Ω, $\mathbb{P}(A_i) > 0$ for all i, and $\mathcal{F}$ is the

σ-field generated by the A_i's. Then

$$\mathbb{P}(A \mid \mathcal{F}) = \sum_i \frac{\mathbb{P}(A \cap A_i)}{\mathbb{P}(A_i)} 1_{A_i}.$$

This follows since the right-hand side is $\mathcal{F}$ measurable and its expectation over any set A_j is $\mathbb{P}(A \cap A_j)$ because

$$\mathbb{E}\left[\sum_i \frac{\mathbb{P}(A \cap A_i)}{\mathbb{P}(A_i)} 1_{A_i}; A_j\right] = \frac{\mathbb{P}(A \cap A_j)}{\mathbb{P}(A_j)} \mathbb{E}\left[1_{A_j}; A_j\right] = \mathbb{P}(A \cap A_j).$$

For a very concrete example, suppose we toss a fair coin independently 5 times and let X_i be 1 or 0 depending whether the ith toss was a heads or tails. Let A be the event that there were 5 heads and let $\mathcal{F}_i = \sigma(X_1, \ldots, X_i)$. Then $\mathbb{P}(A) = 1/32$ while $\mathbb{P}(A \mid \mathcal{F}_1)$ is equal to $1/16$ on the event $(X_1 = 1)$ and 0 on the event $(X_1 = 0)$. $\mathbb{P}(A \mid \mathcal{F}_2)$ is equal to $1/8$ on the event $(X_1 = 1, X_2 = 1)$ and 0 otherwise.

Proposition 21.17 *If $\mathcal{F} \subset \mathcal{G}$ and X is integrable and $\mathcal{G}$ measurable, then*

$$\mathbb{E}\left[\mathbb{E}\left[X \mid \mathcal{F}\right]\right] = \mathbb{E}\, X.$$

Proof. We write

$$\mathbb{E}\left[\mathbb{E}\left[X \mid \mathcal{F}\right]\right] = \mathbb{E}\left[\mathbb{E}\left[X \mid \mathcal{F}\right]; \Omega\right] = \mathbb{E}\left[X; \Omega\right] = \mathbb{E}\, X,$$

using the definition of conditional expectation. □

The following is easy to establish and is left to the reader.

Proposition 21.18 *(1) If $X \geq Y$ are both integrable, then*

$$\mathbb{E}\left[X \mid \mathcal{F}\right] \geq \mathbb{E}\left[Y \mid \mathcal{F}\right], \qquad \text{a.s.}$$

(2) If X and Y are integrable and $a \in \mathbb{R}$, then

$$\mathbb{E}\left[aX + Y \mid \mathcal{F}\right] = a\mathbb{E}\left[X \mid \mathcal{F}\right] + \mathbb{E}\left[Y \mid \mathcal{F}\right], \qquad \text{a.s.}$$

It is easy to check that limit theorems such as the monotone convergence and dominated convergence theorems have conditional expectation versions, as do inequalities like Jensen's and Chebyshev's inequalities. Thus, for example, we have the following.

Proposition 21.19 (Jensen's inequality for conditional expectations) *If g is convex and X and $g(X)$ are integrable,*

$$\mathbb{E}\left[g(X) \mid \mathcal{F}\right] \geq g(\mathbb{E}\left[X \mid \mathcal{F}\right]), \quad \text{a.s.}$$

A key fact is the following.

Proposition 21.20 *If X and XY are integrable and Y is measurable with respect to $\mathcal{F}$, then*

$$\mathbb{E}\left[XY \mid \mathcal{F}\right] = Y\mathbb{E}\left[X \mid \mathcal{F}\right]. \tag{21.11}$$

Proof. If $A \in \mathcal{F}$, then for any $B \in \mathcal{F}$,

$$\mathbb{E}\left[1_A \mathbb{E}\left[X \mid \mathcal{F}\right]; B\right] = \mathbb{E}\left[\mathbb{E}\left[X \mid \mathcal{F}\right]; A \cap B\right] = \mathbb{E}\left[X; A \cap B\right]$$
$$= \mathbb{E}\left[1_A X; B\right].$$

Since $1_A \mathbb{E}\left[X \mid \mathcal{F}\right]$ is $\mathcal{F}$ measurable, this shows that (21.11) holds when $Y = 1_A$ and $A \in \mathcal{F}$.

Using linearity shows that (21.11) holds whenever Y is $\mathcal{F}$ measurable and is a simple random variable. Taking limits and using the dominated convergence theorem, the equality holds when Y is non-negative, $\mathcal{F}$ measurable, and X and XY are integrable. Finally, using linearity again, we have (21.11) when Y is $\mathcal{F}$ measurable and X and XY are integrable. $\qquad\square$

We have two other equalities.

Proposition 21.21 *If $\mathcal{E} \subset \mathcal{F} \subset \mathcal{G}$, then*

$$\mathbb{E}\left[\mathbb{E}\left[X \mid \mathcal{F}\right] \mid \mathcal{E}\right] = \mathbb{E}\left[X \mid \mathcal{E}\right] = \mathbb{E}\left[\mathbb{E}\left[X \mid \mathcal{E}\right] \mid \mathcal{F}\right].$$

Proof. The second equality holds because $\mathbb{E}\left[X \mid \mathcal{E}\right]$ is $\mathcal{E}$ measurable, hence $\mathcal{F}$ measurable. To show the first equality, let $A \in \mathcal{E}$. Then since A is also in $\mathcal{F}$,

$$\mathbb{E}\left[\mathbb{E}\left[\mathbb{E}\left[X \mid \mathcal{F}\right] \mid \mathcal{E}\right]; A\right] = \mathbb{E}\left[\mathbb{E}\left[X \mid \mathcal{F}\right]; A\right] = \mathbb{E}\left[X; A\right]$$
$$= \mathbb{E}\left[\mathbb{E}\left[X \mid \mathcal{E}\right]; A\right].$$

Since both sides are $\mathcal{E}$ measurable, the equality follows. $\qquad\square$

To show the existence of $\mathbb{E}\left[X \mid \mathcal{F}\right]$, we proceed as follows.

Proposition 21.22 *If X is integrable, then $\mathbb{E}\left[X \mid \mathcal{F}\right]$ exists.*

Proof. Using linearity, we need only consider $X \geq 0$. Define a measure $\mathbb{Q}$ on $\mathcal{F}$ by $\mathbb{Q}(A) = \mathbb{E}\left[X; A\right]$ for $A \in \mathcal{F}$. This is clearly absolutely continuous with respect to $\mathbb{P}|_{\mathcal{F}}$, the restriction of $\mathbb{P}$ to $\mathcal{F}$. Let $Y = \mathbb{E}\left[X \mid \mathcal{F}\right]$ be the Radon-Nikodym derivative of $\mathbb{Q}$ with respect to $\mathbb{P}|_{\mathcal{F}}$. Recalling the statement of the Radon-Nikodym theorem (Theorem 13.4), we see that the Radon-Nikodym derivative of $\mathbb{Q}$ with respect to $\mathbb{P}|_{\mathcal{F}}$ is $\mathcal{F}$ measurable. Then if $A \in \mathcal{F}$,

$$\mathbb{E}\left[X; A\right] = \mathbb{Q}(A) = \int_A Y \, d\mathbb{P}|_{\mathcal{F}} = \int_A Y \, d\mathbb{P} = \mathbb{E}\left[Y; A\right].$$

The third equality holds because both Y and A are $\mathcal{F}$ measurable. Thus Y is the desired random variable. $\qquad\square$

21.6 Martingales

In this section we consider martingales. These are a very useful tool in probability. They also have applications to real analysis, and they are fundamental to the theory of financial mathematics.

Let $\mathcal{F}$ be a σ-field and let $\{\mathcal{F}_n\}$ be an increasing sequence of σ-fields, each of which is contained in $\mathcal{F}$. That is, $\mathcal{F}_1 \subset \mathcal{F}_2 \subset \cdots$ and $\mathcal{F}_n \subset \mathcal{F}$ for each n. A sequence of random variables M_n is *adapted* to $\{\mathcal{F}_n\}$ if for each n, M_n is $\mathcal{F}_n$ measurable.

M_n is a *martingale* with respect to an increasing family of σ-fields $\{\mathcal{F}_n\}$ if M_n is adapted to $\mathcal{F}_n$, M_n is integrable for each n, and

$$\mathbb{E}\left[M_{n+1} \mid \mathcal{F}_n\right] = M_n, \qquad \text{a.s.}, \qquad n = 1, 2, \ldots. \qquad (21.12)$$

When the σ-fields are not specified and we talk about M_n being a martingale, it is understood that $\mathcal{F}_n = \sigma(M_1, \ldots, M_n)$.

If X_n is a sequence of adapted integrable random variables with

$$\mathbb{E}\left[X_{n+1} \mid \mathcal{F}_n\right] \geq X_n, \qquad \text{a.s.}, \qquad n = 1, 2, \ldots, \qquad (21.13)$$

we call X_n a *submartingale*. If instead we have

$$\mathbb{E}\left[X_{n+1} \mid \mathcal{F}_n\right] \leq X_n, \qquad \text{a.s.}, \qquad n = 1, 2, \ldots,$$

we call X_n a *supermartingale*.

Let us look at some examples. If X_i is a sequence of mean zero integrable i.i.d. random variables and S_n is the partial sum process, then $M_n = S_n$ is a martingale, since

$$\mathbb{E}\left[M_{n+1} \mid \mathcal{F}_n\right] = M_n + \mathbb{E}\left[M_{n+1} - M_n \mid \mathcal{F}_n\right]$$
$$= M_n + \mathbb{E}\left[M_{n+1} - M_n\right] = M_n,$$

using independence and the fact that S_n is measurable with respect to $\mathcal{F}_n$.

If the X_i's have variance one and $M_n = S_n^2 - n$, then

$$\mathbb{E}\left[S_{n+1}^2 \mid \mathcal{F}_n\right] = \mathbb{E}\left[(S_{n+1} - S_n)^2 \mid \mathcal{F}_n\right] + 2S_n\mathbb{E}\left[S_{n+1} \mid \mathcal{F}_n\right]$$
$$+ S_n^2$$
$$= 1 + S_n^2,$$

using independence. It follows that M_n is a martingale.

Another example is the following: suppose $X \in L^1$ and define $M_n = \mathbb{E}\left[X \mid \mathcal{F}_n\right]$. Then M_n is a martingale.

If M_n is a martingale, g is a convex function, and $g(M_n)$ is integrable for each n, then by Jensen's inequality for conditional expectations,

$$\mathbb{E}\left[g(M_{n+1}) \mid \mathcal{F}_n\right] \geq g(\mathbb{E}\left[M_{n+1} \mid \mathcal{F}_n\right]) = g(M_n),$$

or $g(M_n)$ is a submartingale. Similarly if g is convex and increasing on $[0, \infty)$ and M_n is a positive submartingale, then $g(M_n)$ is a submartingale because

$$\mathbb{E}\left[g(M_{n+1}) \mid \mathcal{F}_n\right] \geq g(\mathbb{E}\left[M_{n+1} \mid F_n\right]) \geq g(M_n).$$

We next want to talk about stopping times. Suppose we have an increasing sequence of σ-fields $\{\mathcal{F}_n\}$ contained in a σ-field $\mathcal{F}$. Let $\mathcal{F}_\infty = \sigma(\cup_{n=1}^\infty \mathcal{F}_n)$. A random variable N (which is $\mathcal{F}$ measurable) from Ω to $\{0, 1, 2, \ldots\} \cup \{\infty\}$ is called a *stopping time* if for each finite n, $(N \leq n) \in \mathcal{F}_n$. A stopping time is also called an *optional time* .

The intuition is that if $\mathcal{F}_n$ is what you know at time n, then at each n you know whether to stop or not. For example, if $X_1, X_2, \ldots$

is a sequence of random variables adapted to the increasing family of σ-fields $\mathcal{F}_n$ and A is a Borel subset of $\mathbb{R}$, then

$$N = \min\{k \geq 0 : X_k \in A\}$$

is a stopping time. In words, N is the first time that one of the X_n is in the set A. To show that N is a stopping time, we write

$$(N \leq n) = \cup_{k=1}^{n}(X_k \in A).$$

On the other hand, if $L = \max\{k \leq 9 : X_k \in A\} \wedge 9$, the last time X_k is in A up to time 9, and $\mathcal{F}_n = \sigma(X_1, \ldots, X_n)$, it can be shown that L is not a stopping time. The intuition here is that one cannot know whether $(L \leq 2)$ without looking into the future at $X_3, \ldots, X_9$.

Proposition 21.23 *(1) Fixed times n are stopping times.*
(2) If N_1 and N_2 are stopping times, then so are $N_1 \wedge N_2$ and $N_1 \vee N_2$.
(3) If N_n is an increasing sequence of stopping times, then so is $N = \sup_n N_n$.
(4) If N_n is a decreasing sequence of stopping times, then so is $N = \inf_n N_n$.
(5) If N is a stopping time, then so is $N + n$.

Proof. We prove (2) and (3) and leave the remaining assertions to the reader. Since

$$(N_1 \wedge N_2 \leq n) = (N_1 \leq n) \cup (N_2 \leq n)$$

and

$$(N_1 \vee N_2 \leq n) = (N_1 \leq n) \cap (N_2 \leq n),$$

then $(N_1 \wedge N_2 \leq n)$ and $(N_1 \vee N_2 \leq n)$ are in $\mathcal{F}_n$ for each n, and we obtain (2). We see (3) holds because

$$(\sup_i N_i \leq n) = \cap_i(N_i \leq n) \in \mathcal{F}_n$$

for each n. $\qquad\square$

Note that if one takes expectations in (21.12), one has $\mathbb{E}\, M_n = \mathbb{E}\, M_{n-1}$, and by induction $\mathbb{E}\, M_n = \mathbb{E}\, M_0$. The *optional stopping theorem* of Doob says that the same is true if we replace n by a stopping time N. When we write M_N, we first evaluate $N(\omega)$, and then we look at $M_n(\omega)$ if $n = N(\omega)$.

Theorem 21.24 *Let $\{\mathcal{F}_n\}$ be an increasing family of σ-fields, each contained in a σ-field $\mathcal{F}$. Let M_n be a martingale with respect to $\{\mathcal{F}_n\}$ and let N be a stopping time bounded by a positive integer K. Then $\mathbb{E}\, M_N = \mathbb{E}\, M_K$.*

Proof. We write

$$\mathbb{E}\, M_N = \sum_{k=0}^{K} \mathbb{E}\,[M_N; N = k] = \sum_{k=0}^{K} \mathbb{E}\,[M_k; N = k].$$

Note $(N = k) = (N \leq k) - (N \leq k - 1)$ is $\mathcal{F}_j$ measurable if $j \geq k$, so

$$\begin{aligned}
\mathbb{E}\,[M_k; N = k] &= \mathbb{E}\,[M_{k+1}; N = k] \\
&= \mathbb{E}\,[M_{k+2}; N = k] \\
&= \ldots = \mathbb{E}\,[M_K; N = k].
\end{aligned}$$

Hence

$$\mathbb{E}\, M_N = \sum_{k=0}^{K} \mathbb{E}\,[M_K; N = k] = \mathbb{E}\, M_K = \mathbb{E}\, M_0.$$

This completes the proof. $\square$

The assumption that N be bounded cannot be entirely dispensed with. For example, let M_n be the partial sums of a sequence of i.i.d. random variable that take the values ± 1, each with probability $\frac{1}{2}$. If $N = \min\{i : M_i = 1\}$, we will see in Remark 21.30 later on that $N < \infty$ a.s., but $\mathbb{E}\, M_N = 1 \neq 0 = \mathbb{E}\, M_0$.

The same proof as that in Theorem 21.24 gives the following corollary.

Corollary 21.25 *If N is bounded by K and M_n is a submartingale, then $\mathbb{E}\, M_N \leq \mathbb{E}\, M_K$.*

The first interesting consequence of the optional stopping theorems is *Doob's inequality*. If M_n is a martingale, denote $M_n^* = \max_{i \leq n} |M_i|$.

Theorem 21.26 *If M_n is a martingale or a positive submartingale,*

$$\mathbb{P}(M_n^* \geq a) \leq \mathbb{E}\,|M_n|/a.$$

Proof. Let $N = \min\{j : |M_j| \geq a\} \wedge n$. Since $g(x) = x$ is convex, $|M_n|$ is a submartingale. If $A = (M_n^* \geq a)$, then by the optional stopping theorem,

$$\mathbb{E}\,|M_n| \geq \mathbb{E}\,|M_N| \geq \mathbb{E}\,[\,|M_N|; A] \geq a\mathbb{P}(A).$$

Dividing both sides by a gives the desired inequality. $\square$

Note that if $|M_n|$ is bounded by a real number K, then

$$|M_{n-1}| \leq \mathbb{E}\,[\,|M_n|\,\mid\,\mathcal{F}_{n-1}] \leq K,$$

and by induction $|M_j| \leq K$ for each j. Hence $\|M_n^*\|_\infty \leq \|M_n\|_\infty$. By the Marcinkiewicz interpolation theorem (Theorem 24.1, which we will prove in Chapter 24) and Theorem 21.26, we see that for each $p \in (1, \infty)$ there exists a constant c_p such that

$$\|M_n^*\|_p \leq c_p\|M_n\|_p. \tag{21.14}$$

This can also be proved by a variation of the proof of Theorem 21.26. The inequalities (21.14) are also referred to as *Doob's inequalities*.

The martingale convergence theorem is another important consequence of optional stopping. The main step is the *upcrossing lemma*. The number of upcrossings of an interval $[a, b]$ is the number of times a sequence of random variables crosses from below a to above b.

To be more exact, let

$$S_1 = \min\{k : X_k \leq a\}, \quad T_1 = \min\{k > S_1 : X_k \geq b\},$$

and

$$S_{i+1} = \min\{k > T_i : X_k \leq a\}, \quad T_{i+1} = \min\{k > S_{i+1} : X_k \geq b\}.$$

The number of upcrossings U_n before time n is

$$U_n = \max\{j : T_j \leq n\}.$$

Lemma 21.27 (Upcrossing lemma) *If X_k is a submartingale,*

$$\mathbb{E}\,U_n \leq \frac{\mathbb{E}\,[(X_n - a)^+]}{b - a}.$$

Proof. The number of upcrossings of $[a, b]$ by X_k is the same as the number of upcrossings of $[0, b-a]$ by $Y_k = (X_k - a)^+$. Moreover Y_k is still a submartingale. If we obtain the inequality for the the the number of upcrossings of the interval $[0, b - a]$ by the process Y_k, we will have the desired inequality for upcrossings of X.

Thus we may assume $a = 0$. Fix n and define $Y_{n+1} = Y_n$. This will still be a submartingale. Define the S_i, T_i as above, and let $S_i' = S_i \wedge (n + 1)$, $T_i' = T_i \wedge (n + 1)$. Since $T_{i+1} > S_{i+1} > T_i$, then $T_{n+1}' = n + 1$.

We write

$$\mathbb{E}\, Y_{n+1} = \mathbb{E}\, Y_{S_1'} + \sum_{i=0}^{n+1} \mathbb{E}\, [Y_{T_i'} - Y_{S_i'}] + \sum_{i=0}^{n+1} \mathbb{E}\, [Y_{S_{i+1}'} - Y_{T_i'}].$$

All the summands in the third term on the right are non-negative since Y_k is a submartingale. For the jth upcrossing, $Y_{T_j'} - Y_{S_j'} \geq b - a$, while $Y_{T_j'} - Y_{S_j'}$ is always greater than or equal to 0. Therefore

$$\sum_{i=0}^{\infty} (Y_{T_i'} - Y_{S_i'}) \geq (b - a)U_n,$$

and then

$$\mathbb{E}\, U_n \leq \mathbb{E}\, Y_{n+1}/(b - a) \tag{21.15}$$

as desired. □

This leads to the *martingale convergence theorem*.

Theorem 21.28 *If X_n is a submartingale such that $\sup_n \mathbb{E}\, X_n^+ < \infty$, then X_n converges a.s. as $n \to \infty$.*

Proof. Let $U(a, b) = \lim_{n \to \infty} U_n$. For each a, b rational, by the monotone convergence theorem,

$$\mathbb{E}\, U(a, b) \leq \sup_n \mathbb{E}\, (X_n - a)^+/(b - a) < \infty.$$

Thus $U(a, b) < \infty$ a.s. If $N_{a,b}$ is the set of ω such that $U(a, b) = \infty$, then $\mathbb{P}(N_{a,b}) = 0$. Let $N = \cup_{a,b \in \mathbb{Q}, a < b} N_{a,b}$. If $\omega \notin N$, then the sequence $X_n(\omega)$ cannot have $\limsup X_n(\omega) > \liminf X_n(\omega)$; if this held, we could find rationals a and b such that $\liminf X_n(\omega) < a <$

$b < \limsup X_n(\omega)$, and then the number of upcrossings of $[a, b]$ would be infinite. Therefore X_n converges a.s., although we still have to rule out the possibility of the limit being infinite. Since X_n is a submartingale, $\mathbb{E}\, X_n \geq \mathbb{E}\, X_0$, and thus

$$\mathbb{E}\,|X_n| = \mathbb{E}\, X_n^+ + \mathbb{E}\, X_n^- = 2\mathbb{E}\, X_n^+ - \mathbb{E}\, X_n \leq 2\mathbb{E}\, X_n^+ - \mathbb{E}\, X_0.$$

By Fatou's lemma, $\mathbb{E}\, \lim_n |X_n| \leq \sup_n \mathbb{E}\,|X_n| < \infty$, or X_n converges a.s. to a finite limit. $\qquad\square$

We show how one can use martingales to find certain hitting probabilities. If one is gambling and wins \$1 with probability $1/2$ and loses \$1 with probability $1/2$ on each play, what are the chances that one will reach \$1,000 before going broke if one starts with \$10?

Proposition 21.29 *Suppose the $Y_1, Y_2, \ldots$ are i.i.d. with*

$$\mathbb{P}(Y_1 = 1) = 1/2, \qquad \mathbb{P}(Y_1 = -1) = 1/2,$$

and $S_n = \sum_{i=1}^n Y_i$. Suppose a and b are positive integers. Let

$$N_{-a} = \min\{k : S_k = -a\}, \qquad N_b = \min\{k : S_k = b\},$$

and $N = N_{-a} \wedge N_b$. Then

$$\mathbb{P}(N_{-a} < N_b) = \frac{b}{a+b}, \qquad \mathbb{P}(N_b < N_{-a}) = \frac{a}{a+b}.$$

In addition, $\mathbb{E}\, N = ab$.

Proof. $S_n^2 - n$ is a martingale, so

$$\mathbb{E}\, S_{n \wedge N}^2 = \mathbb{E}\,(n \wedge N)$$

by the optional stopping theorem. Let $n \to \infty$. The right hand side converges to $\mathbb{E}\, N$ by the monotone convergence theorem. Since $S_{n \wedge N}$ is bounded in absolute value by $a + b$, the left hand side converges by the dominated convergence theorem to $\mathbb{E}\, S_N^2$, which is finite. It follows that $\mathbb{E}\, N$ is finite, hence N is finite almost surely.

S_n is a martingale, so $\mathbb{E}\, S_{n \wedge N} = \mathbb{E}\, S_0 = 0$. By the dominated convergence theorem and the fact that $N < \infty$ a.s., we have $S_{n \wedge N} \to S_N$, and so $\mathbb{E}\, S_N = 0$, or

$$-a\mathbb{P}(S_N = -a) + b\mathbb{P}(S_N = b) = 0.$$

We also have
$$\mathbb{P}(S_N = -a) + \mathbb{P}(S_N = b) = 1.$$

Solving these two equations for $\mathbb{P}(S_N = -a)$ and $\mathbb{P}(S_N = b)$ yields our first result. Since

$$\mathbb{E}\, N = \mathbb{E}\, S_N^2 = a^2 \mathbb{P}(S_N = -a) + b^2 \mathbb{P}(S_N = b),$$

substituting gives the second result. $\square$

Remark 21.30 Based on this proposition, if we let $a \to \infty$, we see that $\mathbb{P}(N_b < \infty) = 1$ and $\mathbb{E}\, N_b = \infty$.

21.7   Weak convergence

We will see in Section 21.9 that if the X_i are i.i.d. random variables that are mean zero and have variance one and $S_n = \sum_{i=1}^{n} X_i$, then $S_n/\sqrt{n}$ converges in the sense that

$$\mathbb{P}(S_n/\sqrt{n} \in [a, b]) \to \mathbb{P}(Z \in [a, b]),$$

where Z is a standard normal random variable. We want to set up the framework for this type of convergence.

We say distribution functions F_n *converges weakly* to a distribution function F if $F_n(x) \to F(x)$ for all x at which F is continuous. We say X_n *converges weakly* to X if F_{X_n} converges weakly to F_X. We sometimes say X_n *converges in distribution* or *converges in law* to X. Probabilities μ_n on $\mathbb{R}$ with the Borel σ-field converge weakly if their corresponding distribution functions converges, that is, if $F_{\mu_n}(x) = \mu_n(-\infty, x]$ converges weakly. If x is a point at which F is continuous, then x is called a *continuity point* of F. A warning: weak convergence in probability is not the same as weak convergence in functional analysis; see Exercise 21.28.

An example that illustrates why we restrict the convergence to continuity points of F is the following. Let $X_n = 1/n$ with probability one, and $X = 0$ with probability one. Then $F_{X_n}(x)$ is 0 if $x < 1/n$ and 1 otherwise. $F_{X_n}(x)$ converges to $F_X(x)$ for all x except $x = 0$.

Proposition 21.31 *The random variables X_n converge weakly to X if and only if $\mathbb{E}\,g(X_n) \to \mathbb{E}\,g(X)$ for all g bounded and continuous.*

The idea that $\mathbb{E}\,g(X_n)$ converges to $\mathbb{E}\,g(X)$ for all g bounded and continuous makes sense for any metric space and is used as a definition of weak convergence for X_n taking values in general metric spaces.

Proof. First suppose $\mathbb{E}\,g(X_n)$ converges to $\mathbb{E}\,g(X)$ for all bounded and continuous g. Let x be a point where F_X is continuous, let $\varepsilon > 0$, and choose δ such that $|F(y) - F(x)| < \varepsilon$ if $|y - x| < \delta$. Choose g continuous such that g is one on $(-\infty, x]$, takes values between 0 and 1, and is 0 on $[x + \delta, \infty)$. Then

$$
\limsup_{n \to \infty} F_{X_n}(x) \leq \limsup_{n \to \infty} \mathbb{E}\,g(X_n)
$$
$$
= \mathbb{E}\,g(X) \leq F_X(x + \delta)
$$
$$
\leq F(x) + \varepsilon.
$$

Similarly, if h is a continuous function taking values between 0 and 1 that is 1 on $(-\infty, x - \delta]$ and 0 on $[x, \infty)$,

$$
\liminf_{n \to \infty} F_{X_n}(x) \geq \liminf_{n \to \infty} \mathbb{E}\,h(X_n) = \mathbb{E}\,h(X) \geq F_X(x - \delta) \geq F(x) - \varepsilon.
$$

Since ε is arbitrary, $F_{X_n}(x) \to F_X(x)$.

Now suppose X_n converges weakly to X. If a and b are points at which F and also each of the F_{X_n} are continuous, then

$$
\mathbb{E}\,1_{(a,b]}(X_n) = \mathbb{P}(a < X_n \leq b) = \mathbb{P}(X_n \leq b) - \mathbb{P}(X_n \leq a)
$$
$$
= F_{X_n}(b) - F_{X_n}(a) \to F(b) - F(a)
$$
$$
= \mathbb{P}(X \leq b) - \mathbb{P}(X \leq a)
$$
$$
= \mathbb{P}(a < X \leq b) = \mathbb{E}\,1_{(a,b]}(X).
$$

By taking linear combinations, we have $\mathbb{E}\,g(X_n) \to \mathbb{E}\,g(X)$ for every g which is a step function where the end points of the intervals are continuity points for all the F_{X_n} and for F_X. The set of points that are not a continuity point for some F_{X_n} or for F_X is countable. Since we can approximate any continuous function uniformly on an interval by step functions which jump only at points that are continuity points for all the F_n and for F, we have $\mathbb{E}\,g(X_n) \to$

$\mathbb{E}\, g(X)$ for all g such that the support of g is a closed interval whose endpoints are continuity points of F_X and g is continuous on its support.

Let $\varepsilon > 0$ and choose M large such that $F_X(M) > 1 - \varepsilon$ and $F_X(-M) < \varepsilon$ and so that M and $-M$ are continuity points of F_X and of the F_{X_n}. By the above argument,

$$\mathbb{E}\,(g1_{[-M,M]})(X_n) \to \mathbb{E}\,(g1_{[-M,M]})(X),$$

where g is a bounded continuous function. The difference between $\mathbb{E}\,(g1_{[-M,M]})(X)$ and $\mathbb{E}\, g(X)$ is bounded by

$$\|g\|_\infty \mathbb{P}(X \notin [-M, M]) \leq 2\varepsilon \|g\|_\infty.$$

Similarly, when X is replaced by X_n, the difference is bounded by

$$\|g\|_\infty \mathbb{P}(X_n \notin [-M, M]) \to \|g\|_\infty \mathbb{P}(X \notin [-M, M]).$$

So for n large, the difference between $\mathbb{E}\, g(X_n)$ and $\mathbb{E}\, g(X)$ is less than

$$3\varepsilon \|g\|_\infty + \varepsilon.$$

Since ε is arbitrary, $\mathbb{E}\, g(X_n) \to \mathbb{E}\, g(X)$ whenever g is bounded and continuous. $\qquad\square$

Let us examine the relationship between weak convergence and convergence in probability.

Proposition 21.32 *(1) If X_n converges to X in probability, then X_n converges weakly to X.*
(2) If X_n converges weakly to a constant, then X_n converges in probability.
(3) If X_n converges weakly to X and Y_n converges weakly to a constant c, then $X_n + Y_n$ converges weakly to $X + c$ and $X_n Y_n$ converges weakly to cX.

Part (3) is known as *Slutsky's theorem*.

Proof. To prove (1), let g be a bounded and continuous function. If n_j is any subsequence, then there exists a further subsequence such that $X(n_{j_k})$ converges almost surely to X. (We sometimes write $X(n)$ for X_n here.) Then by the dominated convergence theorem, $\mathbb{E}\, g(X(n_{j_k})) \to \mathbb{E}\, g(X)$. That suffices to show $\mathbb{E}\, g(X_n)$ converges to $\mathbb{E}\, g(X)$. Hence by Proposition 21.31 we see that X_n converges weakly to X.

For (2), if X_n converges weakly to c,

$$\mathbb{P}(X_n - c > \varepsilon) = \mathbb{P}(X_n > c + \varepsilon) = 1 - \mathbb{P}(X_n \leq c + \varepsilon)$$
$$\to 1 - \mathbb{P}(c \leq c + \varepsilon) = 0.$$

We use the fact that if Y is identically equal to c, then $c + \varepsilon$ is a point of continuity for F_Y. Similarly $\mathbb{P}(X_n - c \leq -\varepsilon) \to 0$, so $\mathbb{P}(|X_n - c| > \varepsilon) \to 0$.

We now prove the first part of (3), leaving the second part for the reader. Let x be a point such that $x - c$ is a continuity point of F_X. Choose ε so that $x - c + \varepsilon$ is again a continuity point. Then

$$\mathbb{P}(X_n + Y_n \leq x) \leq \mathbb{P}(X_n + c \leq x + \varepsilon) + \mathbb{P}(|Y_n - c| > \varepsilon)$$
$$\to \mathbb{P}(X \leq x - c + \varepsilon).$$

Thus $\limsup \mathbb{P}(X_n + Y_n \leq x) \leq \mathbb{P}(X + c \leq x + \varepsilon)$. Since ε can be as small as we like and $x - c$ is a continuity point of F_X, then

$$\limsup \mathbb{P}(X_n + Y_n \leq x) \leq \mathbb{P}(X + c \leq x).$$

The lim inf is done similarly. $\qquad \square$

Example 21.33 We give an example where X_n converges weakly but does not converge in probability. Let $\{X_n\}$ be an i.i.d. sequence of Bernoulli random variable with parameter $1/2$. Clearly X_n converges weakly to a Bernoulli random variable with parameter $1/2$ since F_{X_n} is constant in n. If X_n converges in probability, then there exists a subsequence $\{X_{n_j}\}$ that converges a.s by Proposition 10.2. But if $A_j = (X_{n_{2j}} = 0, X_{n_{2j+1}} = 1)$, the independence of the X_n's tells us that $\mathbb{P}(A_j) = 1/4$ for each j. By the definition of A_j, we see that the A_j are independent, so by the Borel-Cantelli lemma, $\mathbb{P}(A_j \text{ i.o.}) = 1$. This contradicts the assertion that X_{n_j} converges a.s.

We say a sequence of distribution functions $\{F_n\}$ is *tight* if for each $\varepsilon > 0$ there exists M such that $F_n(M) \geq 1 - \varepsilon$ and $F_n(-M) \leq \varepsilon$ for all n. A sequence of random variables is tight if the corresponding distribution functions are tight; this is equivalent to $\mathbb{P}(|X_n| \geq M) \leq \varepsilon$. The following theorem is known as *Helly's theorem*.

Theorem 21.34 *Let F_n be a sequence of distribution functions that is tight. There exists a subsequence n_j and a distribution function F such that F_{n_j} converges weakly to F.*

If X_n is identically equal to n, then $F_{X_n} \to 0$. However, the constant function 0 is not a distribution function. This does not contradict Helly's theorem since the X_n are not tight.

Proof. Let q_k be an enumeration of the rationals. Since $F_n(q_k) \in [0, 1]$, any subsequence has a further subsequence that converges. Use Cantor's diagonalization method (see Remark 21.35) so that $F_{n_j}(q_k)$ converges for each q_k and call the limit $F(q_k)$. F is increasing, and define $F(x) = \inf_{\{k: q_k \geq x\}} F(q_k)$. We see that F is right continuous and increasing.

If x is a point of continuity of F and $\varepsilon > 0$, then there exist r and s rational such that $r < x < s$ and $F(s) - \varepsilon < F(x) < F(r) + \varepsilon$. Then

$$\liminf_{j \to \infty} F_{n_j}(x) \geq \liminf_{j \to \infty} F_{n_j}(r) = F(r) > F(x) - \varepsilon$$

and

$$\limsup_{j \to \infty} F_{n_j}(x) \leq \limsup_{j \to \infty} F_{n_j}(s) = F(s) < F(x) + \varepsilon.$$

Since ε is arbitrary, $F_{n_j}(x) \to F(x)$.

Since the F_n are tight, there exists M such that $F_n(-M) < \varepsilon$. Then $F(-M) \leq \varepsilon$, which implies $\lim_{x \to -\infty} F(x) = 0$. Showing $\lim_{x \to \infty} F(x) = 1$ is similar. Therefore F is in fact a distribution function. $\qquad \square$

Remark 21.35 Cantor's diagonalization method may be familiar to you from the proof of the Ascoli-Arzelà theorem from undergraduate analysis. In our context it works as follows. The sequence $\{F_n(q_1)\}$ is a sequence of real numbers bounded between 0 and 1, and so $\{F_n\}$ has a subsequence, which we label as $\{F_{1,j}\}$, $j = 1, 2, \ldots$, such that $F_{1,j}(q_1)$ converges as $j \to \infty$. Next the subsequence $\{F_{1,j}(q_2)\}$ is a sequence of real numbers bounded between 0 and 1, so there exists a further subsequence $\{F_{2,j}\}$, $j = 1, 2, \ldots$ such that $F_{2,j}(q_2)$ converges. Since $\{F_{2,j}\}$ is a subsequence of $\{F_{1,j}\}$, then $F_{1,j}(q_1)$ still converges. Take a further subsequence of $\{F_{2,j}\}$, which we will call $\{F_{3,j}\}$, such that $F_{3,j}(q_3)$ converges. Continue.

We note that $F_{m,j}(q_i)$ converges whenever $i \leq m$. We now consider the subsequence $\{F_{m,m}\}$. This is a subsequence of our original sequence $\{F_n\}$. Furthermore, for each k, $\{F_{m,m}\}$, $m = k, k+1, \ldots$, is a subsequence of $\{F_{k,j}\}$. (The first $k-1$ elements of $\{F_{m,m}\}$ might not be elements of $\{F_{k,j}\}$.) Therefore $F_{m,m}(q_k)$ converges. We have thus found a subsequence of our original sequence $\{F_n\}$ that converges at each q_k.

We conclude this section by giving an easily checked criterion for tightness.

Proposition 21.36 *Suppose there exists* $\varphi : [0,\infty) \to [0,\infty)$ *that is increasing and* $\varphi(x) \to \infty$ *as* $x \to \infty$. *If* $\sup_n \mathbb{E}\,\varphi(|X_n|) < \infty$, *then the* X_n *are tight.*

Proof. Let $\varepsilon > 0$ and let $c = \sup_n \mathbb{E}\,\varphi(|X_n|)$. Choose M such that $\varphi(x) \geq c/\varepsilon$ if $x > M$. Then

$$\mathbb{P}(|X_n| > M) \leq \int \frac{\varphi(|X_n|)}{c/\varepsilon} 1_{(|X_n|>M)}\, d\mathbb{P} \leq \frac{\varepsilon}{c} \mathbb{E}\,\varphi(|X_n|) \leq \varepsilon.$$

Thus the X_n are tight. $\qquad\square$

21.8 Characteristic functions

We define the *characteristic function* of a random variable X by $\varphi_X(u) = \mathbb{E}\,e^{iux}$ for $u \in \mathbb{R}$.

Note that $\varphi_X(u) = \int e^{iux}\,\mathbb{P}_X(dx)$. Therefore if X and Y have the same law, they have the same characteristic function. Also, if the law of X has a density, that is, $\mathbb{P}_X(dx) = f_X(x)\,dx$, then $\varphi_X(u) = \int e^{iux} f_X(x)\,dx$, so in this case the characteristic function is the same as the Fourier transform of f_X.

Proposition 21.37 $\varphi(0) = 1$, $|\varphi(u)| \leq 1$, $\varphi(-u) = \overline{\varphi(u)}$, *and* φ *is uniformly continuous.*

Proof. Since $|e^{iux}| \leq 1$, everything follows immediately from the definitions except the uniform continuity. For that we write

$$|\varphi(u+h) - \varphi(u)| = |\mathbb{E}\, e^{i(u+h)X} - \mathbb{E}\, e^{iuX}|$$
$$\leq \mathbb{E}\,|e^{iuX}(e^{ihX} - 1)| = \mathbb{E}\,|e^{ihX} - 1|.$$

Observe that $|e^{ihX} - 1|$ tends to 0 almost surely as $h \to 0$, so the right hand side tends to 0 by the dominated convergence theorem. Note that the right hand side is independent of u.   □

The definitions also imply

$$\varphi_{aX}(u) = \varphi_X(au)$$

and

$$\varphi_{X+b}(u) = e^{iub}\varphi_X(u).$$

Proposition 21.38 *If X and Y are independent, then*

$$\varphi_{X+Y}(u) = \varphi_X(u)\varphi_Y(u).$$

Proof. We have

$$\mathbb{E}\, e^{iu(X+Y)} = \mathbb{E}\,[e^{iuX}e^{iuY}] = (\mathbb{E}\, e^{iuX})(\mathbb{E}\, e^{iuY})$$

by the multiplication theorem.   □

Let us look at some examples of characteristic functions.

(1) *Bernoulli*: By direct computation, this is $pe^{iu} + (1 - p) = 1 - p(1 - e^{iu})$.

(2) *Poisson*: Here we have

$$\mathbb{E}\, e^{iuX} = \sum_{k=0}^{\infty} e^{iuk}e^{-\lambda}\frac{\lambda^k}{k!}$$
$$= e^{-\lambda}\sum\frac{(\lambda e^{iu})^k}{k!} = e^{-\lambda}e^{\lambda e^{iu}}$$
$$= e^{\lambda(e^{iu}-1)}.$$

(3) *Binomial*:

$$\mathbb{E}\, e^{iuX} = \sum_{k=0}^{n} e^{iuk} \binom{n}{k} p^k (1-p)^{n-k}$$

$$= \sum_{k=0}^{n} \binom{n}{k} (pe^{iu})^k (1-p)^{n-k} = (pe^{iu} + 1 - p)^n$$

by the binomial theorem.

(4) *Exponential*:

$$\int_0^\infty \lambda e^{iux} e^{-\lambda x}\, dx = \lambda \int_0^\infty e^{(iu-\lambda)x} dx = \frac{\lambda}{\lambda - iu}.$$

(5) *Standard normal*: We evaluated the Fourier transform of $\frac{1}{\sqrt{2\pi}} e^{-x^2/2}$ in Proposition 16.5, and obtained

$$\varphi(u) = \frac{1}{\sqrt{2\pi}} \int_{-\infty}^\infty e^{iux} e^{-x^2/2}\, dx = e^{-u^2/2}.$$

We proceed to the *inversion formula*, which gives a formula for the distribution function in terms of the characteristic function.

Theorem 21.39 *Let μ be a probability measure and let $\varphi(u) = \int e^{iux} \mu(dx)$. If $a < b$, then*

$$\lim_{T\to\infty} \frac{1}{2\pi} \int_{-T}^T \frac{e^{-iua} - e^{-iub}}{iu} \varphi(u)\, du \qquad (21.16)$$

$$= \mu(a,b) + \tfrac{1}{2}\mu(\{a\}) + \tfrac{1}{2}\mu(\{b\}).$$

If μ is point mass at 0, so $\varphi(u) = 1$, then the integrand in this case is $2\sin u/u$, which is not integrable. This shows that taking a limit cannot be avoided.

Proof. By the Fubini theorem,

$$\int_{-T}^T \frac{e^{-iua} - e^{-iub}}{iu} \varphi(u)\, du = \int_{-T}^T \int \frac{e^{-iua} - e^{-iub}}{iu} e^{iux} \mu(dx)\, du$$

$$= \int \int_{-T}^T \frac{e^{-iua} - e^{-iub}}{iu} e^{iux}\, du\, \mu(dx).$$

To justify this, we bound the integrand by $b - a$, using the mean value theorem.

Expanding e^{-iub} and e^{-iua} in terms of sines and cosines using Euler's formula, that is, $e^{i\theta} = \cos\theta + i\sin\theta$, and using the fact that cosine is an even function and sine an odd one, we are left with

$$\int 2\left[\int_0^T \frac{\sin(u(x-a))}{u}\, du - \int_0^T \frac{\sin(u(x-b))}{u}\, du\right] \mu(dx).$$

By Exercise 11.12 and the dominated convergence theorem, this tends to

$$\int \left[\pi \operatorname{sgn}(x-a) - \pi \operatorname{sgn}(x-b)\right] \mu(dx),$$

which is equal to the right hand side of (21.16). $\square$

A corollary to the inversion formula is the uniqueness theorem.

Corollary 21.40 *If $\varphi_X(u) = \varphi_Y(u)$ for all u, then $\mathbb{P}_X = \mathbb{P}_Y$.*

Proof. If a and b are points such that $\mathbb{P}_X(\{a\}) = 0$, $\mathbb{P}_X(\{b\}) = 0$, and the same for $\mathbb{P}_Y$, then the inversion formula shows that $\mathbb{P}_X((a,b]) = \mathbb{P}_Y((a,b])$, which is the same as saying

$$F_X(b) - F_X(a) = F_Y(b) - F_Y(a).$$

Taking a limit as $a \to -\infty$ but avoiding points that are not continuity points of both F_X and F_Y (there are only countably many of these), we have $F_X(b) = F_Y(b)$ if b is a continuity point of both F_X and F_Y. Since F_X and F_Y are right continuous, given x, we can take b decreasing to x but avoiding points that are not continuity points of both F_X and F_Y, and we obtain $F_X(x) = F_Y(x)$ for all x. Since $\mathbb{P}_X$ is the Lebesgue-Stieltjes measure associated with F_X and this is uniquely determined by F_X, we conclude $\mathbb{P}_X = \mathbb{P}_Y$. $\square$

A random variable X is a *normal random variable* with mean μ and variance σ^2 if it has the density

$$(2\pi\sigma^2)^{-1/2} e^{i\mu x - x^2/2\sigma^2}.$$

A normal random variable is also known as a *Gaussian random variable*. Some calculus shows if X is a normal random variable with mean μ and variance σ^2, then $(X - \mu)/\sigma$ is a standard normal random variable, and conversely, if Z is a standard normal random variable, then $\mu + \sigma Z$ is a normal random variable with mean μ and variance σ^2. If X is a normal random variable with mean μ and variance σ^2, then X has characteristic function

$$\mathbb{E}\, e^{iuX} = e^{iu\mu}\mathbb{E}\, e^{i(u\sigma)Z} = e^{iu\mu - \sigma^2 u^2/2},$$

where Z is a standard normal random variable.

The following proposition can be proved directly, but the proof using characteristic functions is much easier.

Proposition 21.41 *(1) If X and Y are independent, X is a normal random variable with mean a and variance b^2, and Y is a normal random variable with mean c and variance d^2, then $X + Y$ is normal random variable with mean $a + c$ and variance $b^2 + d^2$.*
(2) If X and Y are independent, X is a Poisson random variable with parameter λ_1, and Y is a Poisson random variable with parameter λ_2, then $X + Y$ is a Poisson random variable with parameter $\lambda_1 + \lambda_2$.
(3) If X and Y are independent, X is a binomial random variable with parameters m and p, and Y is a binomial random variable with parameters n and p, then $X + Y$ is a binomial random variable with parameters $m + n$ and p.

Proof. For (1), using Proposition 21.38,

$$\varphi_{X+Y}(u) = \varphi_X(u)\varphi_Y(u) = e^{iau - b^2 u^2/2}e^{icu - c^2 u^2/2}$$
$$= e^{i(a+c)u - (b^2 + d^2)u^2/2}.$$

Now use the uniqueness theorem.

Parts (2) and (3) are proved similarly.   □

Remark 21.42 Since a Bernoulli random variable with parameter p is the same as a binomial random variable with parameters 1 and p, then Proposition 21.41 and an induction argument shows that the sum of n independent Bernoulli random variables with parameter p is a binomial random variable with parameters n and p.

We will need the following result in our proof of the central limit theorem.

Proposition 21.43 *If* $\mathbb{E}\,|X|^k < \infty$ *for an integer* k, *then* φ_X *has a continuous derivative of order* k *and*

$$\varphi^{(k)}(u) = \int (ix)^k e^{iux}\, \mathbb{P}_X(dx).$$

In particular,

$$\varphi^{(k)}(0) = i^k \mathbb{E}\,X^k. \tag{21.17}$$

Proof. Write

$$\frac{\varphi(u+h) - \varphi(u)}{h} = \int \frac{e^{i(u+h)x} - e^{iux}}{h}\, \mathbb{P}(dx).$$

The integrand on the right is bounded by $|x|$. If $\int |x|\,\mathbb{P}_X(dx) < \infty$, we can use the dominated convergence theorem to obtain the desired formula for $\varphi'(u)$. As in the proof of Proposition 21.37, we see $\varphi'(u)$ is continuous. We do the case of general k by induction. Evaluating $\varphi^{(k)}$ at 0 gives (21.17).   □

We will use the following theorem in the proof of the central limit theorem.

Theorem 21.44 *Suppose* $\{X_n\}$ *is a tight sequence of random variables,* X *is another random variable, and* $\varphi_{X_n}(u) \to \varphi_X(u)$ *for each* $u \in \mathbb{R}$ *as* $n \to \infty$. *Then* X_n *converges weakly to* X.

Proof. If X_n does not converge weakly to X, there is a continuity point x for F_X such that $F_{X_n}(x)$ does not converge to $F_X(x)$. Helly's theorem, Theorem 21.34, shows there is subsequence $\{X_{n_j}\}$ which converges weakly, say to the random variable Y. By Proposition 21.31, $\varphi_{X_{n_j}}(u) = \mathbb{E}\,\exp(iuX_{n_j})$ converges to $\mathbb{E}\,\exp(iuY) = \varphi_Y(u)$ for each u. Therefore $\varphi_Y(u) = \varphi_X(u)$ for all u, and by the uniqueness theorem, $F_Y = F_X$. But then $F_{X_{n_j}}(x) \to F_X(x)$, a contradiction.   □

An n-dimensional vector $X = (X_1, \ldots, X_n)$ is a *random vector* if $X : \Omega \to \mathbb{R}^n$ is a measurable map. Here this means that

$X^{-1}(A) \in \mathcal{F}$ whenever $A \in \mathcal{B}_n$, where $\mathcal{B}_n$ is the Borel σ-field of $\mathbb{R}^n$. The *joint law* or *joint distribution* is the probability $\mathbb{P}_X$ on $(\mathbb{R}^n, \mathcal{B}_n)$ given by $\mathbb{P}_X(A) = \mathbb{P}(X \in A)$ for $A \in \mathcal{B}_n$. As in Proposition 21.31, we have

$$\mathbb{E}\, g(X) = \int_{\mathbb{R}^n} g(x)\, \mathbb{P}_X(dx)$$

whenever g is measurable and either non-negative or bounded.

If X is a random vector, $\varphi_X(u) = \mathbb{E}\, e^{iu\cdot X}$ for $u \in \mathbb{R}^n$ is called the *joint characteristic function* of X, where $x \cdot y$ denotes the usual inner product in $\mathbb{R}^n$.

If X^k is an n-dimensional random vector for each k and each coordinate X_i^k of X^k converges to Y_i in probability, then the dominated convergence theorem shows that

$$\varphi_{X^k}(u) = \mathbb{E}\, e^{iu\cdot X^k} \to \mathbb{E}\, e^{iu\cdot Y} = \varphi_Y(u) \qquad (21.18)$$

for each u, where $Y = (Y_1, \ldots, Y_n)$.

Proposition 21.45 *Suppose that $X = (X_1, \ldots, X_n)$ is an n-dimensional random vector. Then*

$$\varphi_X(u) = \prod_{j=1}^{n} \varphi_{X_j}(u_j), \qquad u = (u_1, \ldots, u_n), \qquad (21.19)$$

for all $u \in \mathbb{R}^n$ if and only if the X_i are independent.

Proof. If the X_i are independent, then we see that the characteristic function of X factors into the product of characteristic functions by using the multiplication theorem and writing

$$\varphi_X(u) = \mathbb{E}\, e^{iu\cdot X} = \mathbb{E}\, e^{i\sum_{j=1}^{n} u_j X_j} = \mathbb{E}\left[\prod_{j=1}^{n} e^{iu_j X_j}\right] \qquad (21.20)$$

$$= \prod_{j=1}^{n} \mathbb{E}\, e^{iu_j X_j} = \prod_{j=1}^{n} \varphi_{X_j}(u_j).$$

Suppose (21.19) holds for all u. Let $Y_1, \ldots, Y_n$ be independent random variables such that Y_i has the same law as X_i for each i; see

Exercise 21.39 for how to construct such Y_i. Let $Y = (Y_1, \ldots, Y_n)$. Then using the independence as in (21.20) we have

$$\varphi_Y(u) = \prod_{j=1}^{n} \varphi_{Y_j}(u_j).$$

Since Y_j has the same law as X_j, then this is equal to $\prod_{j=1}^{n} \varphi_{X_j}(u_j)$. By (21.19) this in turn is equal to $\varphi_X(u)$. Therefore $\varphi_Y(u) = \varphi_X(u)$ for all u.

If f is a C^∞ function with compact support, then $\widehat{f}$, the Fourier transform of f, will be integrable by Exercise 16.5. By the Fubini theorem,

$$\mathbb{E}\, f(X) = (2\pi)^{-n} \int_{\mathbb{R}^n} \widehat{f}(u) \mathbb{E}\, e^{-iu \cdot X}\, du$$

$$= (2\pi)^{-n} \int_{\mathbb{R}^n} \widehat{f}(u) \varphi_X(-u)\, du$$

$$= (2\pi)^{-n} \int_{\mathbb{R}^n} \widehat{f}(u) \varphi_Y(-u)\, du$$

$$= (2\pi)^{-n} \int_{\mathbb{R}^n} \widehat{f}(u) \mathbb{E}\, e^{-iu \cdot Y}\, du = \mathbb{E}\, f(Y).$$

By a limit argument, $\mathbb{P}(X \in A) = \mathbb{E}\, 1_A(X) = \mathbb{E}\, 1_A(Y) = \mathbb{P}(Y \in A)$ when A is a rectangle of the form $(a_1, b_1) \times \cdots \times (a_n, b_n)$. The collection of sets A for which $\mathbb{P}(X \in A) = \mathbb{P}(Y \in A)$ is easily seen to be a σ-field, and since it contains rectangles of the above form, it contains all Borel subsets of $\mathbb{R}^n$. Thus X and Y have the same law. In particular, the X_i are independent because the Y_i are. $\quad\square$

A collection $\{X_1, \ldots, X_n\}$ of random variables is called *jointly normal* or *jointly Gaussian* if there exist i.i.d. standard normal random variables $Z_1, \ldots, Z_m$ and real numbers b_{ij} and a_i such that

$$X_i = \sum_{j=1}^{m} b_{ij} Z_j + a_i, \qquad i = 1, 2, \ldots, n.$$

Given two random variables X and Y, the *covariance* of X and Y is defined by

$$\mathrm{Cov}\,(X, Y) = \mathbb{E}\,[(X - \mathbb{E}\, X)(Y - \mathbb{E}\, Y)].$$

In the case of a random vector X, we define the *covariance matrix* Σ by letting Σ be the matrix whose (i, j) entry is $\mathrm{Cov}\,(X_i, X_j)$. In the case of jointly normal random variables,

$$\mathrm{Cov}\,(X_i, X_j) = \mathbb{E}\left[\sum_{k=1}^{m}\sum_{\ell=1}^{m} b_{ik} b_{j\ell} Z_k Z_\ell\right] = \sum_{k=1}^{m} b_{ik} b_{jk},$$

using the fact that $\mathbb{E}\,Z_k Z_\ell$ is zero unless $k = \ell$, in which case it is one. If we let B be the matrix whose (i, j) entry is b_{ij} and use C^T to denote the transpose of a matrix C, we obtain

$$\Sigma = BB^T.$$

Let us compute the characteristic function of a jointly normal random vector when all the a_i are zero. If $Z = (Z_1, \ldots, Z_m)$ are i.i.d. standard normal random variables, then

$$\varphi_Z(u) = \prod_{j=1}^{m} \varphi_{Z_j}(u_j) = \prod_{j=1}^{m} e^{-u_j^2/2} = e^{-|u|^2/2},$$

where $u = (u_1, \ldots, u_m)$ and $|u| = (\sum_{j=1}^{m} u_j^2)^{1/2}$. When all the a_i are 0, we then have

$$\varphi_X(u) = \mathbb{E}\,e^{iu \cdot BZ} = \mathbb{E}\,e^{i\sum_{j=1}^{n}\sum_{k=1}^{m} u_j b_{jk} Z_k} \qquad (21.21)$$

$$= \mathbb{E}\,e^{i(uB) \cdot Z} = \varphi_Z(uB) = e^{-uBB^T u^T/2}.$$

21.9 Central limit theorem

The simplest case of the *central limit theorem* (CLT) is the case when the X_i are i.i.d. random variables with mean zero and variance one, and then the central limit theorem says that $S_n/\sqrt{n}$ converges weakly to a standard normal. We prove this case. The more complicated cases consider when the random variables are no longer identically distributed or independent.

We need the fact that if c_n are complex numbers converging to c, then $(1 + (c_n/n))^n \to e^c$. We leave the proof of this to the reader, with the warning that any proof using logarithms needs to be done with some care, since $\log z$ is a multi-valued function when z is complex.

We also use the fact that if $f : \mathbb{R} \to \mathbb{R}$ is twice continuously differentiable, then

$$\frac{f(x) - f(a) - f'(a)(x - a) - f''(a)(x - a)^2/2}{(x - a)^2} \to 0 \qquad (21.22)$$

as $x \to a$. This follows, for example, by applying l'Hôpital's rule twice.

Theorem 21.46 *Suppose the X_i are i.i.d. random variables with mean zero and variance one and $S_n = \sum_{i=1}^{n} X_i$. Then $S_n/\sqrt{n}$ converges weakly to a standard normal random variable.*

Proof. Since X_1 has finite second moment, then φ_{X_1} has a continuous second derivative by Proposition 21.43. By (21.22),

$$\varphi_{X_1}(u) = \varphi_{X_1}(0) + \varphi'_{X_1}(0)u + \varphi''_{X_1}(0)u^2/2 + R(u),$$

where $|R(u)|/u^2 \to 0$ as $|u| \to 0$. Hence using (21.22)

$$\varphi_{X_1}(u) = 1 - u^2/2 + R(u).$$

Then

$$\varphi_{S_n/\sqrt{n}}(u) = \mathbb{E}\, e^{iuS_n/\sqrt{n}} = \varphi_{S_n}(u/\sqrt{n}) = (\varphi_{X_1}(u/\sqrt{n}))^n$$

$$= \left[1 - \frac{u^2}{2n} + R(u/\sqrt{n})\right]^n,$$

where we used (21.17). Since $u/\sqrt{n}$ converges to zero as $n \to \infty$, we have

$$\varphi_{S_n/\sqrt{n}}(u) \to e^{-u^2/2}.$$

Now apply Theorem 21.44. $\square$

If the X_i are i.i.d., but don't necessarily have mean 0 with variance 1, we have

$$\frac{S_n - \mathbb{E}\, S_n}{\sqrt{\mathrm{Var}\, X_1}\sqrt{n}}$$

converges weakly to a standard normal random variable. This follows from Theorem 21.46 by looking at the random variables $(X_i - \mathbb{E}\, X_1)/\sqrt{\mathrm{Var}\, X_1}$.

We give another example of the use of characteristic functions to obtain a limit theorem.

Proposition 21.47 *Suppose that $\{X_n\}$ is a sequence of independent random variables such that X_n is a binomial random variable with parameters n and p_n. If $np_n \to \lambda$, then X_n converges weakly to a Poisson random variable with parameter λ.*

Proof. We write

$$\varphi_{X_n}(u) = \left(1 + p_n(e^{iu} - 1)\right)^n$$

$$= \left(1 + \frac{np_n}{n}(e^{iu} - 1)\right)^n \to e^{\lambda(e^{iu} - 1)}.$$

Now apply Theorem 21.44. □

21.10 Kolmogorov extension theorem

The goal of this section is to show how to construct probability measures on $\mathbb{R}^{\mathbb{N}} = \mathbb{R} \times \mathbb{R} \times \cdots$. We may view $\mathbb{R}^{\mathbb{N}}$ as the set of sequences $(x_1, x_2, \ldots)$ of elements of $\mathbb{R}$. Given an element $x = (x_1, x_2, \ldots)$ of $\mathbb{R}^{\mathbb{N}}$, we define $\tau_n(x) = (x_1, \ldots, x_n) \in \mathbb{R}^n$. A *cylindrical set* in $\mathbb{R}^{\mathbb{N}}$ is a set of the form $A \times \mathbb{R}^{\mathbb{N}}$, where A is a Borel subset of $\mathbb{R}^n$ for some $n \geq 1$. Another way of phrasing this is to say a cylindrical set is one of the form $\tau_n^{-1}(A)$, where $n \geq 1$ and A is a Borel subset of $\mathbb{R}^n$. We furnish $\mathbb{R}^n$ with the product topology; see Section 20.1. Recall that this means we take the smallest topology that contains all cylindrical sets. We use the σ-field on $\mathbb{R}^{\mathbb{N}}$ generated by the cylindrical sets. Thus the σ-field we use is the same as the Borel σ-field on $\mathbb{R}^{\mathbb{N}}$. We use $\mathcal{B}_n$ to denote the Borel σ-field on $\mathbb{R}^n$.

We suppose that for each n we have a probability measure μ_n defined on $(\mathbb{R}^n, \mathcal{B}_n)$. The μ_n are *consistent* if $\mu_{n+1}(A \times \mathbb{R}) = \mu_n(A)$ whenever $A \in \mathcal{B}_n$. The *Kolmogorov extension theorem* is the following.

Theorem 21.48 *Suppose for each n we have a probability measure μ_n on $(\mathbb{R}^n, \mathcal{B}_n)$. Suppose the μ_n are consistent. Then there exists a probability measure μ on $\mathbb{R}^{\mathbb{N}}$ such that $\mu(A \times \mathbb{R}^{\mathbb{N}}) = \mu_n(A)$ for all $A \in \mathcal{B}_n$.*

Proof. Define μ on cylindrical sets by $\mu(A \times \mathbb{R}^{\mathbb{N}}) = \mu_m(A)$ if $A \in \mathcal{B}_m$. By the consistency assumption, μ is well defined. If

$\mathcal{A}_0$ is the collection of cylindrical sets, it is easy to see that $\mathcal{A}_0$ is an algebra of sets and that μ is finitely additive on $\mathcal{A}_0$. If we can show that μ is countably additive on $\mathcal{A}_0$, then by the Carathéodory extension theorem, Theorem 4.16, we can extend μ to the σ-field generated by the cylindrical sets. By Exercise 3.2 it suffices to show that whenever $A_n \downarrow \emptyset$ with $A_n \in \mathcal{A}_0$, then $\mu(A_n) \to 0$.

Suppose that A_n are cylindrical sets decreasing to $\emptyset$ but $\mu(A_n)$ does not tend to 0; by taking a subsequence we may assume without loss of generality that there exists $\varepsilon > 0$ such that $\mu(A_n) \geq \varepsilon$ for all n. We will obtain a contradiction.

It is possible that A_n might depend on fewer or more than n coordinates. It will be more convenient if we arrange things so that A_n depends on exactly n coordinates. We want $A_n = \tau_n^{-1}(\widetilde{A}_n)$ for some $\widetilde{A}_n$ a Borel subset of $\mathbb{R}^n$. Suppose A_n is of the form

$$A_n = \tau_{j_n}^{-1}(D_n)$$

for some $D_n \subset \mathbb{R}^{j_n}$; in other words, A_n depends on j_n coordinates. By letting $A_0 = \mathbb{R}^{\mathbb{N}}$ and replacing our original sequence by $A_0, \ldots, A_0, A_1, \ldots, A_1, A_2, \ldots, A_2, \ldots$, where we repeat each A_i sufficiently many times, we may without loss of generality suppose that $j_n \leq n$. On the other hand, if $j_n < n$ and $A_n = \tau_{j_n}^{-1}(D_n)$, we may write $A_n = \tau_n^{-1}(\widehat{D}_n)$ with $\widehat{D}_n = D_n \times \mathbb{R}^{n-j_n}$. Thus we may without loss of generality suppose that A_n depends on exactly n coordinates.

We set $\widetilde{A}_n = \tau_n(A_n)$. For each n, choose $\widetilde{B}_n \subset \widetilde{A}_n$ so that $\widetilde{B}_n$ is compact and $\mu(\widetilde{A}_n - \widetilde{B}_n) \leq \varepsilon/2^{n+1}$. To do this, first we choose M such that $\mu_n(([-M, M]^n)^c) < \varepsilon/2^{n+2}$, and then we use Proposition 17.6 to find a compact subset $\widetilde{B}_n$ of $\widetilde{A}_n \cap [-M, M]^n$ such that $\mu(\widetilde{A}_n \cap [-M, M]^n - \widetilde{B}_n) \leq \varepsilon/2^{n+2}$. Let $B_n = \tau_n^{-1}(\widetilde{B}_n)$ and let $C_n = B_1 \cap \ldots \cap B_n$. Hence $C_n \subset B_n \subset A_n$, and $C_n \downarrow \emptyset$, but

$$\mu(C_n) \geq \mu(A_n) - \sum_{i=1}^{n} \mu(A_i - B_i) \geq \varepsilon/2,$$

and $\widetilde{C}_n = \tau_n(C_n)$, the projection of C_n onto $\mathbb{R}^n$, is compact.

We will find $x = (x_1, \ldots, x_n, \ldots) \in \cap_n C_n$ and obtain our contradiction. For each n choose a point $y(n) \in C_n$. The first coordinates of $\{y(n)\}$, namely, $\{y_1(n)\}$, form a sequence contained in $\widetilde{C}_1$, which is compact, hence there is a convergent subsequence

$\{y_1(n_k)\}$. Let x_1 be the limit point. The first and second coordinates of $\{y(n_k)\}$ form a sequence contained in the compact set $\widetilde{C}_2$, so a further subsequence $\{(y_1(n_{k_j}), y_2(n_{k_j}))\}$ converges to a point in $\widetilde{C}_2$. Since $\{n_{k_j}\}$ is is a subsequence of $\{n_k\}$, the first coordinate of the limit is x_1. Therefore the limit point of $\{(y_1(n_{k_j}), y_2(n_{k_j}))\}$ is of the form (x_1, x_2), and this point is in $\widetilde{C}_2$. We continue this procedure to obtain $x = (x_1, x_2, \ldots, x_n, \ldots)$. By our construction, $(x_1, \ldots, x_n) \in \widetilde{C}_n$ for each n, hence $x \in C_n$ for each n, or $x \in \cap_n C_n$, a contradiction. $\qquad\square$

A typical application of this theorem is to construct a countable sequence of independent random variables. We construct $X_1, \ldots, X_n$ to be an independent collection of n independent random variables using Exercise 21.39. Let μ_n be the joint law of $(X_1, \ldots, X_n)$; it is easy to check that the μ_n form a consistent family. We use Theorem 21.48 to obtain a probability measure μ on $\mathbb{R}^{\mathbb{N}}$. To get random variables out of this, we let $X_i(\omega) = \omega_i$ if $\omega = (\omega_1, \omega_2, \ldots)$.

21.11 Brownian motion

In this section we construct Brownian motion and define Wiener measure.

Let $(\Omega, \mathcal{F}, \mathbb{P})$ be a probability space and let $\mathcal{B}$ be the Borel σ-field on $[0, \infty)$. A *stochastic process*, denoted $X(t, \omega)$ or $X_t(\omega)$ or just X_t, is a map from $[0, \infty) \times \Omega$ to $\mathbb{R}$ that is measurable with respect to the product σ-field of $\mathcal{B}$ and $\mathcal{F}$.

Definition 21.49 A stochastic process X_t is a one-dimensional *Brownian motion* started at 0 if
(1) $X_0 = 0$ a.s.;
(2) for all $s \leq t$, $X_t - X_s$ is a mean zero normal random variable with variance $t - s$;
(3) the random variables $X_{r_i} - X_{r_{i-1}}$, $i = 1, \ldots, n$, are independent whenever $0 \leq r_0 \leq r_1 \leq \cdots \leq r_n$;
(4) there exists a null set N such that if $\omega \notin N$, then the map $t \to X_t(\omega)$ is continuous.

Let us show that there exists a Brownian motion. We give the

Haar function construction, which is one of the quickest ways to the construction of Brownian motion.

For $i = 1, 2, \ldots$, $j = 1, 2, \ldots, 2^{i-1}$, let φ_{ij} be the function on $[0, 1]$ defined by

$$
\varphi_{ij} = \begin{cases}
2^{(i-1)/2}, & x \in \left[\frac{2j-2}{2^i}, \frac{2j-1}{2^i}\right); \\
-2^{(i-1)/2}, & x \in \left[\frac{2j-1}{2^i}, \frac{2j}{2^i}\right); \\
0, & \text{otherwise.}
\end{cases}
$$

Let φ_{00} be the function that is identically 1. The φ_{ij} are called the Haar functions. If $\langle \cdot, \cdot \rangle$ denotes the inner product in $L^2([0,1])$, that is, $\langle f, g \rangle = \int_0^1 f(x)\overline{g}(x)dx$, note the φ_{ij} are orthogonal and have norm 1. It is also easy to see that they form a complete orthonormal system for L^2: $\varphi_{00} \equiv 1$; $1_{[0,1/2)}$ and $1_{[1/2,1)}$ are both linear combinations of φ_{00} and φ_{11}; $1_{[0,1/4)}$ and $1_{[1/4,1/2)}$ are both linear combinations of $1_{[0,1/2)}$, φ_{21}, and φ_{22}. Continuing in this way, we see that $1_{[k/2^n,(k+1)/2^n)}$ is a linear combination of the φ_{ij} for each n and each $k \leq 2^n$. Since any continuous function can be uniformly approximated by step functions whose jumps are at the dyadic rationals, linear combinations of the Haar functions are dense in the set of continuous functions, which in turn is dense in $L^2([0,1])$.

Let $\psi_{ij}(t) = \int_0^t \varphi_{ij}(r)\, dr$. Let Y_{ij} be a sequence of independent identically distributed standard normal random variables. Set

$$
V_0(t) = Y_{00}\psi_{00}(t), \qquad V_i(t) = \sum_{j=1}^{2^{i-1}} Y_{ij}\psi_{ij}(t), \qquad i \geq 1.
$$

We need one more preliminary. If Z is a standard normal random variable, then Z has density $(2\pi)^{-1/2}e^{-x^2/2}$. Since

$$
\int x^4 e^{-x^2/2}\, dx < \infty,
$$

then $\mathbb{E}\, Z^4 < \infty$. We then have

$$
\mathbb{P}(|Z| > \lambda) = \mathbb{P}(Z^4 > \lambda^4) \leq \frac{\mathbb{E}\, Z^4}{\lambda^4}. \tag{21.23}
$$

Theorem 21.50 $\sum_{i=0}^{\infty} V_i(t)$ *converges uniformly in* t *a.s. If we call the sum* X_t, *then* X_t *is a Brownian motion started at 0.*

Proof. *Step 1.* We first prove convergence of the series. Let

$$A_i = (|V_i(t)| > i^{-2} \text{ for some } t \in [0,1]).$$

We will show $\sum_{i=1}^{\infty} \mathbb{P}(A_i) < \infty$. Then by the Borel–Cantelli lemma, except for ω in a null set, there exists $i_0(\omega)$ such that if $i \geq i_0(\omega)$, we have $\sup_t |V_i(t)(\omega)| \leq i^{-2}$. This will show $\sum_{i=0}^{I} V_i(t)(\omega)$ converges as $I \to \infty$, uniformly over $t \in [0,1]$. Moreover, since each $\psi_{ij}(t)$ is continuous in t, then so is each $V_i(t)(\omega)$, and we thus deduce that $X_t(\omega)$ is continuous in t.

Now for $i \geq 1$ and $j_1 \neq j_2$, for each t at least one of $\psi_{ij_1}(t)$ and $\psi_{ij_2}(t)$ is zero. Also, the maximum value of ψ_{ij} is $2^{-(i+1)/2}$. Hence

$$\mathbb{P}(|V_i(t)| > i^{-2} \text{ for some } t \in [0,1])$$
$$\leq \mathbb{P}(|Y_{ij}|\psi_{ij}(t) > i^{-2} \text{ for some } t \in [0,1], \text{ some } 0 \leq j \leq 2^{i-1})$$
$$\leq \mathbb{P}(|Y_{ij}|2^{-(i+1)/2} > i^{-2} \text{ for some } 0 \leq j \leq 2^{i-1})$$
$$\leq \sum_{j=0}^{2^{i-1}} \mathbb{P}(|Y_{ij}|2^{-(i+1)/2} > i^{-2})$$
$$= (2^{i-1} + 1)\mathbb{P}(|Z| > 2^{(i+1)/2}i^{-2})$$

where Z is a standard normal random variable. Using (21.23), we conclude $\mathbb{P}(A_i)$ is summable in i.

Step 2. Next we show that the limit, X_t, satisfies the definition of Brownian motion. It is obvious that each X_t has mean zero and that $X_0 = 0$. In this step we show that $X_t - X_s$ is a mean zero normal random variable with variance $t - s$.

If $f \in L^2([0,1])$, Parseval's identity says that

$$\langle f, f \rangle = \sum_{i,j} \langle \varphi_{ij}, f \rangle^2.$$

Let

$$W_t^k = \sum_{i=0}^{k} V_i(t).$$

Fix $s < t$ and set $d_k^2 = \text{Var}\,(W_t^k - W_s^k)$. We will use the notation $\sum_{i=0}^{k} \sum_j$ to mean that the sum over j is from 1 to 2^{i-1} when $i > 0$ and the sum over j is from 0 to 0, i.e., a single summand with $j = 0$, when $i = 0$. Since

$$\psi_{ij}(t) - \psi_{ij}(s) = \langle \varphi_{ij}, 1_{[s,t]} \rangle,$$

then

$$W_t^k - W_s^k = \sum_{i=0}^{k} \sum_{j} Y_{ij} \langle \varphi_{ij}, 1_{[s,t]} \rangle.$$

Since the Y_{ij} are independent with mean zero and variance one, $\mathbb{E}[Y_{ij}Y_{\ell m}] = 1$ if $i = \ell$ and $j = m$ and otherwise is equal to 0. Thus

$$d_k^2 = \mathbb{E}(W_t^k - W_s^k)^2 \tag{21.24}$$

$$= \mathbb{E}\left[\sum_{i=0}^{k} \sum_{j} Y_{ij} \langle \varphi_{ij}, 1_{[s,t]} \rangle \sum_{\ell=0}^{k} \sum_{m} Y_{\ell m} \langle \varphi_{\ell m}, 1_{[s,t]} \rangle \right]$$

$$= \sum_{i=0}^{k} \sum_{j} \langle \varphi_{ij}, 1_{[s,t]} \rangle^2 \to \sum_{i=0}^{\infty} \sum_{j} \langle \varphi_{ij}, 1_{[s,t]} \rangle^2$$

$$= \langle 1_{[s,t]}, 1_{[s,t]} \rangle = t - s.$$

Since $W_t^k - W_s^k$ is a finite linear combination of standard normal random variables, it is normal random variable with mean zero and variance d_k^2, and therefore its characteristic function is $e^{-d_k^2 u^2/2}$. Since $W_t^k - W_s^k \to X_t - X_s$ a.s. and $d_k^2 \to t - s$, then by (21.18), the characteristic function of $X_t - X_s$ is $e^{-(t-s)u^2/2}$. This proves that $X_t - X_s$ is a normal random variable with mean zero and variance $t - s$.

Step 3. We prove that if $0 \le r_0 < r_1 < \cdots < r_n$, then the random variables $X_{r_1} - X_{r_0}, \ldots, X_{r_n} - X_{r_{n-1}}$ are independent.

For $f, g \in L^2([0,1])$ we have $f = \sum_{i,j} \langle \varphi_{ij}, f \rangle \varphi_{ij}$ and $g = \sum_{i,j} \langle \varphi_{ij}, g \rangle \varphi_{ij}$, hence

$$\langle f, g \rangle = \sum_{i,j} \langle \varphi_{ij}, f \rangle \langle \varphi_{ij}, g \rangle.$$

Therefore for $1 \le I, J \le n$,

$$\mathbb{E}[X_{r_I} - X_{r_{I-1}})(X_{r_J} - X_{r_{J-1}})]$$

$$= \mathbb{E}\left[\left(\sum_{i,j} Y_{ij} \langle \varphi_{ij}, 1_{[r_{I-1}, r_I]} \rangle \right) \left(\sum_{k,\ell} Y_{k\ell} \langle \varphi_{k\ell}, 1_{[r_{J-1}, r_J]} \rangle \right) \right]$$

$$= \sum_{i,j} \langle \varphi_{ij}, 1_{[r_{I-1}, r_I]} \rangle \langle \varphi_{ij}, 1_{[r_{J-1}, r_J]} \rangle$$

$$= \langle 1_{[r_{I-1}, r_I]}, 1_{[r_{J-1}, r_J]} \rangle.$$

This last inner product is 0, and the covariance of $X_{r_I} - X_{r_{I-1}}$ and $X_{r_J} - X_{r_{J-1}}$ is zero, unless $I = J$, in which case the covariance is the same as the variance and is equal to $r_I - r_{I-1}$.

Let $U^k = (W_{r_1}^k - W_{r_0}^k, \ldots, W_{r_n}^k - W_{r_{n-1}}^k)$. This is a collection of jointly normal random variables, each with mean zero. Its joint characteristic function will be

$$\varphi_{U^k}(u) = e^{-u^T \Sigma_k u / 2},$$

where Σ_k is the covariance matrix for U^k. As in Step 2, we see that $\Sigma_k \to \Sigma$, where Σ is a diagonal matrix whose (j,j) entry is $r_j - r_{j-1}$. Since U^k converges almost surely to

$$U = (X_{r_1} - X_{r_0}, \ldots, X_{r_n} - X_{r_{n-1}}),$$

then the joint characteristic function of U is

$$\varphi_U(u) = e^{-u^T \Sigma u / 2}.$$

Since $\varphi_U(u)$ factors as $\prod_{j=1}^n \varphi_{X_{r_j} - X_{r_{j-1}}}(u_j)$, then the components of U are independent by Proposition 21.45. □

The stochastic process X_t induces a measure on $C([0,1])$. We say $A \subset C([0,1])$ is a *cylindrical set* if

$$A = \{f \in C([0,1]) : (f(r_1), \ldots, f(r_n)) \in B\}$$

for some $n \geq 1$, $r_1 \leq \cdots \leq r_n$, and B a Borel subset of $\mathbb{R}^n$. For A a cylindrical set, define $\mu(A) = \mathbb{P}(\{X.(\omega) \in A\})$, where X is a Brownian motion and $X.(\omega)$ is the function $t \to X_t(\omega)$. We extend μ to the σ-field generated by the cylindrical sets. If B is in this σ-field, then $\mu(B) = \mathbb{P}(X. \in B)$. The probability measure μ is called *Wiener measure*.

We defined Brownian motion for $t \in [0,1]$. To define Brownian motion for $t \in [0,\infty)$, take a sequence $\{X^n\}$ of independent Brownian motions on $[0,1]$ and piece them together as follows. Define $X_t = X_t^1$ for $0 \leq t \leq 1$. For $1 < t \leq 2$, define $X_t = X_1 + X_{t-1}^2$. For $2 < t \leq 3$, let $X_t = X_2 + X_{t-2}^3$, and so on.

21.12 Exercises

Exercise 21.1 Show that if X has a continuous distribution function F_X and $Y = F_X(X)$, then Y has a density $f_Y(x) = 1_{[0,1]}(x)$.

Exercise 21.2 Find an example of a probability space and three events A, B, and C such that $\mathbb{P}(A \cap B \cap C) = \mathbb{P}(A)\mathbb{P}(B)\mathbb{P}(C)$, but A, B, and C are not independent events.

Exercise 21.3 Suppose that

$$\mathbb{P}(X \leq x, Y \leq y) = \mathbb{P}(X \leq x)\mathbb{P}(Y \leq y)$$

for all $x, y \in \mathbb{R}$. Prove that X and Y are independent random variables.

Exercise 21.4 Find a sequence of events $\{A_n\}$ such that

$$\sum_{n=1}^{\infty} \mathbb{P}(A_n) = \infty$$

but $\mathbb{P}(A_n \text{ i.o.}) = 0$.

Exercise 21.5 A random vector $X = (X_1, \ldots, X_n)$ has a *joint density* f_X if $\mathbb{P}(X \in A) = \int_A f_X(x)\,dx$ for all Borel subsets A of $\mathbb{R}^n$. Here the integral is with respect to n dimensional Lebesgue measure.
(1) Prove that if the joint density of X factors into the product of densities of the X_j, i.e., $f_X(x) = \prod_{j=1}^n f_{X_j}(x_j)$, for almost every $x = (x_1, \ldots, x_n)$, then the X_j are independent.
(2) Prove that if X has a joint density and the X_j are independent, then each X_j has a density and the joint density of X factors into the product of the densities of the X_j.

Exercise 21.6 Suppose $\{A_n\}$ is a sequence of events, not necessarily independent, such that $\sum_{n=1}^{\infty} \mathbb{P}(A_n) = \infty$. Suppose in addition that there exists a constant c such that for each $N \geq 1$,

$$\sum_{i,j=1}^{N} \mathbb{P}(A_i \cap A_j) \leq c \left(\sum_{i=1}^{N} \mathbb{P}(A_i) \right)^2.$$

Prove that $\mathbb{P}(A_n \text{ i.o.}) > 0$.

Exercise 21.7 Suppose X and Y are independent, $\mathbb{E}\,|X|^p < \infty$ for some $p \in [1, \infty)$, $\mathbb{E}\,|Y| < \infty$, and $\mathbb{E}\,Y = 0$. Prove that

$$\mathbb{E}\left(|X + Y|^p\right) \geq \mathbb{E}\,|X|^p.$$

Exercise 21.8 Suppose that X_i are independent random variables such that $\operatorname{Var} X_i / i \to 0$ as $i \to \infty$. Suppose also that $\mathbb{E} X_i \to a$. Prove that S_n / n converges in probability to a, where $S_n = \sum_{i=1}^n X_i$. We do not assume that the X_i are identically distributed.

Exercise 21.9 Suppose $\{X_i\}$ is a sequence of independent mean zero random variables, not necessarily identically distributed. Suppose that $\sup_i \mathbb{E} X_i^4 < \infty$.
(1) If $S_n = \sum_{i=1}^n X_i$, prove there is a constant c such that $\mathbb{E} S_n^4 \le cn^2$.
(2) Prove that $S_n / n \to 0$ a.s.

Exercise 21.10 Suppose $\{X_i\}$ is an i.i.d. sequence of random variables such that S_n / n converges a.s., where $S_n = \sum_{i=1}^n X_i$.
(1) Prove that $X_n / n \to 0$ a.s.
(2) Prove that $\sum_n \mathbb{P}(|X_n| > n) < \infty$.
(3) Prove that $\mathbb{E} |X_1| < \infty$.

Exercise 21.11 Suppose $\{X_i\}$ is an i.i.d. sequence of random variables with $\mathbb{E} |X_1| < \infty$. Prove that the sequence $\{S_n / n\}$ is uniformly integrable; see Exercise 7.16 for the definition of uniformly integrable. Conclude by Theorem 7.17 that $\mathbb{E} S_n / n$ converges to $\mathbb{E} X_1$.

Exercise 21.12 Suppose $\{X_i\}$ is an i.i.d. sequence of random variables with $\mathbb{E} |X_1| < \infty$ and $\mathbb{E} X_1 = 0$. Prove that

$$\frac{\max_{1 \le k \le n} |S_k|}{n} \to 0, \qquad \text{a.s.}$$

Exercise 21.13 Suppose that $\{X_i\}$ is a sequence of independent random variables with mean zero such that $\sum_i \operatorname{Var} X_i < \infty$. Prove that S_n converges a.s. as $n \to \infty$, where $S_n = \sum_{i=1}^n X_i$.

Exercise 21.14 Let $\{X_i\}$ be a sequence of random variables. The *tail σ-field* is defined to be $\mathcal{T} = \cap_{n \ge 1} \sigma(X_n, X_{n+1}, \dots)$. Let $S_n = \sum_{i=1}^n X_i$.
(1) Prove that the event $(S_n \text{ converges})$ is in $\mathcal{T}$.
(2) Prove that the event $(S_n / n > a)$ is in $\mathcal{T}$ for each real number a.

Exercise 21.15 Let $\{X_i\}$ be a sequence of independent random variables and let $\mathcal{T}$ be the tail σ-field.
(1) Prove that if $A \in \mathcal{T}$, then A is independent of $\sigma(X_1, \ldots, X_n)$ for each n.
(2) Prove that if $A \in \mathcal{T}$, then A is independent of itself, and hence $\mathbb{P}(A)$ is either 0 or 1. This is known as the *Kolmogorov 0-1 law*.

Exercise 21.16 Let $\{X_i\}$ be an i.i.d. sequence. Prove that if $\mathbb{E}\, X_1^+ = \infty$ and $\mathbb{E}\, X_1^- < \infty$, then $S_n/n \to +\infty$ a.s., where $S_n = \sum_{i=1}^{n} X_i$.

Exercise 21.17 Let $\mathcal{F} \subset \mathcal{G}$ be two σ-fields. Let H be the Hilbert space of $\mathcal{G}$ measurable random variables Y such that $\mathbb{E}\, Y^2 < \infty$ and let M be the subspace of H consisting of the $\mathcal{F}$ measurable random variables. Prove that if $Y \in H$, then $\mathbb{E}\,[Y \mid \mathcal{F}]$ is equal to the projection of Y onto the subspace M.

Exercise 21.18 Suppose $\mathcal{F} \subset \mathcal{G}$ are two σ-fields and X and Y are bounded $\mathcal{G}$ measurable random variables. Prove that

$$\mathbb{E}\,[X\mathbb{E}\,[Y \mid \mathcal{F}]] = \mathbb{E}\,[Y\mathbb{E}\,[X \mid \mathcal{F}]].$$

Exercise 21.19 Let $\mathcal{F} \subset \mathcal{G}$ be two σ-fields and let X be a bounded $\mathcal{G}$ measurable random variable. Prove that if

$$\mathbb{E}\,[XY] = \mathbb{E}\,[X\mathbb{E}\,[Y \mid \mathcal{F}]]$$

for all bounded $\mathcal{G}$ measurable random variables Y, then X is $\mathcal{F}$ measurable.

Exercise 21.20 Suppose $\mathcal{F} \subset \mathcal{G}$ are two σ-fields and that X is $\mathcal{G}$ measurable with $\mathbb{E}\, X^2 < \infty$. Set $Y = \mathbb{E}\,[X \mid \mathcal{F}]$. Prove that if $\mathbb{E}\, X^2 = \mathbb{E}\, Y^2$, then $X = Y$ a.s.

Exercise 21.21 Suppose $\mathcal{F}_1 \subset \mathcal{F}_2 \subset \cdots \subset \mathcal{F}_N$ are σ-fields. Suppose A_i is a sequence of random variables adapted to $\{\mathcal{F}_i\}$ such that $A_1 \le A_2 \le \cdots$ and $A_{i+1} - A_i \le 1$ a.s. for each i. Prove that if $\mathbb{E}\,[A_N - A_i \mid \mathcal{F}_i] \le 1$ a.s. for each i, then $\mathbb{E}\, A_N^2 < \infty$.

Exercise 21.22 Let $\{X_i\}$ be an i.i.d. sequence of random variables with $\mathbb{P}(X_1 = 1) = \mathbb{P}(X_1 = -1) = \frac{1}{2}$. Let $S_n = \sum_{i=1}^{n} X_i$. The sequence $\{S_n\}$ is called a *simple random walk*. Let

$$L = \max\{k \le 9 : S_k = 1\} \wedge 9.$$

Prove that L is *not* a stopping time with respect to the family of σ-fields $\mathcal{F}_n = \sigma(S_1, \ldots, S_n)$.

Exercise 21.23 Let $\mathcal{F}_1 \subset \mathcal{F}_2 \subset \cdots$ be an increasing family of σ-fields and let $\mathcal{F}_\infty = \sigma(\cup_{n=1}^\infty \mathcal{F}_n)$. If N is a stopping time, define

$$\mathcal{F}_N = \{A \in \mathcal{F}_\infty : A \cap (N \le n) \in \mathcal{F}_n \text{ for all } n\}.$$

(1) Prove that $\mathcal{F}_N$ is a σ-field.
(2) If M is another stopping time with $M \le N$ a.s., and we define $\mathcal{F}_M$ analogously, prove that $\mathcal{F}_M \subset \mathcal{F}_N$.
(3) If X_n is a martingale with respect to $\{\mathcal{F}_n\}$ and N is a stopping time bounded by the real number K, prove that $\mathbb{E}[X_n \mid \mathcal{F}_N] = X_N$.

Exercise 21.24 Let $\{X_i\}$ be a sequence of bounded i.i.d. random variables with mean 0. Let $S_n = \sum_{i=1}^n X_i$.
(1) Prove that there exists a constant c_1 such that $M_n = e^{S_n - c_1 n}$ is a martingale.
(2) Show there exists a constant c_2 such that

$$\mathbb{P}(\max_{1 \le k \le n} S_n > \lambda) \le 2e^{-c_2 \lambda^2 / n}$$

for all $\lambda > 0$.

Exercise 21.25 Let $\{X_i\}$ be a sequence of i.i.d. standard normal random variables. Let $S_n = \sum_{i=1}^n X_i$.
(1) Prove that for each $a > 0$ $M_n = e^{aS_n - a^2 n/2}$ is a martingale.
(2) Show

$$\mathbb{P}(\max_{1 \le k \le n} S_n > \lambda) \le e^{-\lambda^2 / 2n}$$

for all $\lambda > 0$.

Exercise 21.26 Let $\{X_n\}$ be a submartingale. Let

$$A_n = \sum_{i=2}^n (X_i - \mathbb{E}[X_i \mid \mathcal{F}_{i-1}]).$$

Prove that $M_n = X_n - A_n$ is a martingale. This is known as the *Doob decomposition* of a submartingale.

Exercise 21.27 Suppose M_n is a martingale. Prove that if

$$\sup_n \mathbb{E}\, M_n^2 < \infty,$$

then M_n converges a.s. and also in L^2.

Exercise 21.28 Set $(\Omega, \mathcal{F}, \mathbb{P})$ equal to $([0, 1], \mathcal{B}, m)$, where $\mathcal{B}$ is the Borel σ-field on $[0, 1]$ and m is Lebesgue measure. Define

$$X_n(\omega) = \begin{cases} 1, & \omega \in \left[\frac{2k}{2^n}, \frac{2k+1}{2n}\right) \text{ for some } k \leq 2^{n-1}; \\[2mm] -1, & \omega \in \left[\frac{2k+1}{2^n}, \frac{2k+2}{2n}\right) \text{ for some } k \leq 2^{n-1}. \end{cases}$$

(1) Prove that X_n converges weakly (in the probabilistic sense) to a non-zero random variable.
(2) Prove that X_n converges to 0 with respect to weak convergence in $L^2(m)$, that is, $\mathbb{E}\,[X_n Y] \to 0$ for all $Y \in L^2$.

Exercise 21.29 Suppose X_n is a sequence of random variables that converges weakly to a random variable X. Prove that the sequence $\{X_n\}$ is tight.

Exercise 21.30 Suppose $X_n \to X$ weakly and $Y_n \to 0$ in probability. Prove that $X_n Y_n \to 0$ in probability.

Exercise 21.31 Given two probability measures $\mathbb{P}$ and $\mathbb{Q}$ on $[0, 1]$ with the Borel σ-field, define

$$d(\mathbb{P}, \mathbb{Q}) = \inf\{\left| \int f\, d\mathbb{P} - \int f\, d\mathbb{Q} \right| : f \in C^1, \|f\|_\infty \leq 1, \|f'\|_\infty \leq 1\}.$$

Here C^1 is the collection of continuously differentiable functions and f' is the derivative of f.
(1) Prove that d is a metric.
(2) Prove that $\mathbb{P}_n \to \mathbb{P}$ weakly if and only if $d(\mathbb{P}_n, \mathbb{P}) \to 0$.
This metric makes sense only for probabilities defined on $[0, 1]$. There are other metrics for weak convergence that work in more general situations.

Exercise 21.32 Suppose $F_n \to F$ weakly and every point of F is a continuity point. Prove that F_n converges to F uniformly over $x \in \mathbb{R}$:

$$\sup_{x \in \mathbb{R}} |F_n(x) - F(x)| \to 0.$$

Exercise 21.33 Suppose $X_n \to X$ weakly. Prove that φ_{X_n} converges uniformly to φ_X on each bounded interval.

Exercise 21.34 Suppose $\{X_n\}$ is a collection of random variables that is tight. Prove that $\{\varphi_{X_n}\}$ is equicontinuous on $\mathbb{R}$.

Exercise 21.35 Suppose $X_n \to X$ weakly, $Y_n \to Y$ weakly, and X_n and Y_n are independent for each n. Prove that $X_n + Y_n \to X + Y$ weakly.

Exercise 21.36 X is said to be a *gamma* random variable with parameters λ and t if X has density

$$\frac{1}{\Gamma(t)} \lambda^t x^{t-1} e^{-\lambda x} 1_{(0,\infty)}(x),$$

where $\Gamma(t) = \int_0^\infty y^{t-1} e^{-y} \, dy$ is the Gamma function.
(1) Prove that an exponential random variable with parameter λ is also a gamma random variable with parameters 1 and λ.
(2) Prove that if X is a standard normal random variable, then X^2 is a gamma random variable with parameters $1/2$ and $1/2$.
(3) Find the characteristic function of a gamma random variable.
(4) Prove that if X is a gamma random variable with parameters s and λ, Y is a gamma random variable with parameters t and λ, and X and Y are independent, then $X + Y$ is also a gamma random variable; determine the parameters of $X + Y$.

Exercise 21.37 Suppose X_n is a sequence of independent random variables, not necessarily independent, with $\sup_n \mathbb{E} |X_n|^3 < \infty$ and $\mathbb{E} X_n = 0$ and $\text{Var}\, X_n = 1$ for each n. Prove that $S_n / \sqrt{n}$ converges weakly to a standard normal random variable, where $S_n = \sum_{i=1}^n X_i$.

Exercise 21.38 Suppose that X_n a Poisson random variable with parameter n for each n. Prove that $(X_n - n)/\sqrt{n}$ converges weakly to a standard normal random variable as $n \to \infty$.

Exercise 21.39 In this exercise we show how to construct a random vector whose law is a given probability measure on $\mathbb{R}^n$.
(1) Let $\mathbb{P}$ be a probability measure on the Borel subsets of $\mathbb{R}^n$. If $\omega = (\omega_1, \ldots, \omega_n) \in \mathbb{R}^n$, define $X_n(\omega) = \omega_n$. Let $X = (X_1, \ldots, X_n)$.

Prove that the law $\mathbb{P}_X$ of X is equal to $\mathbb{P}$.
(2) If $\mathbb{P}$ is a product measure, prove that the components of X are independent.

Exercise 21.40 Prove that if X_t is a Brownian motion and a is a non-zero real number, then $Y_t = aX_{a^2t}$ is also a Brownian motion.

Exercise 21.41 Let X_t be a Brownian motion. Let $n \geq 1$ and let $M_k = X_{k/2^n}$.
(1) Prove that M_k is a martingale.
(2) Prove that $e^{aM_k - a^2(k/2^n)/2}$ is a martingale.
(3) Prove that

$$\mathbb{P}(\sup_{t \leq r} X_t \geq \lambda) \leq e^{-\lambda^2/2r}.$$

Exercise 21.42 Let X_t be a Brownian motion. Let

$$A_n = (\sup_{t \leq 2^{n+1}} X_t > \sqrt{4 \cdot 2^n \log \log 2^n}).$$

(1) Prove that $\sum_{n=1}^{\infty} \mathbb{P}(A_n) < \infty$.
(2) Prove that

$$\limsup_{t \to \infty} \frac{X_t}{\sqrt{t \log \log t}} < \infty, \qquad \text{a.s.}$$

This is part of what is known as the *law of the iterated logarithm* or *LIL* for Brownian motion.

Exercise 21.43 Let X_t be a Brownian motion. Let $M > 0$, $t_0 > 0$, and

$$B_n = (X_{t_0+2^{-n}} - X_{t_0+2^{-n-1}} > M2^{-n-1}).$$

(1) Prove that $\sum_{n=1}^{\infty} \mathbb{P}(B_n) = \infty$.
(2) Prove that the function $t \to X_t(\omega)$ is not differentiable at $t = t_0$.
(3) Prove that except for ω in a null set, the function $t \to X_t(\omega)$ is not differentiable at almost every t (with respect to Lebesgue measure on $[0, \infty)$.)
This can actually be strengthened, via a different proof, to the fact that except for a set of ω in a null set, the function $t \to X_t(\omega)$ is nowhere differentiable.

Exercise 21.44 Let X_t be a Brownian motion and let $h > 0$. Prove that except for ω in a null set, there are times $t \in (0, h)$ for which $X_t(\omega) > 0$. (The times will depend on ω.) This says that Brownian motion started at 0 cannot stay negative for a time. Similarly it cannot stay positive for a time; the path of X_t must oscillate quite a bit near 0.

Exercise 21.45 Let X_t be a Brownian motion on $[0, 1]$. Prove that $Y_t = X_1 - X_{1-t}$ is also a Brownian motion.

Chapter 22

Harmonic functions

Harmonic functions are important in complex analysis, partial differential equations, mathematical physics, and probability theory, as well as in real analysis. In this chapter we present some of their basic properties.

22.1 Definitions

Recall that a C^2 function is one whose second partial derivatives are continuous and a domain is an open set in $\mathbb{R}^n$. If f is a C^2 function in a domain D, the Laplacian of f, written Δf, is the function

$$\Delta f(x) = \sum_{i=1}^{n} \frac{\partial^2 f}{\partial x_i^2}(x), \qquad x \in D.$$

A real-valued function h is *harmonic* on a domain D if h is C^2 in D and $\Delta h(x) = 0$ for all $x \in D$.

In one dimension, the linear functions $h(x) = ax + b$ are harmonic in any interval, and any harmonic function is linear in each open interval on which it is defined, since $h''(x) = 0$.

When we turn to two dimensions, we can identity $\mathbb{R}^2$ with the complex plane $\mathbb{C}$. If D is a domain in $\mathbb{C}$ and f is analytic in D, that is, f is differentiable at each point of D, then the real and imaginary parts of f are harmonic in D. This is a consequence of

303

the Cauchy-Riemann equations. If $f = u + iv$, then the Cauchy-Riemann equations from complex analysis say that $\partial u/\partial x = \partial v/\partial y$ and $\partial u/\partial y = -\partial v/\partial x$. Since an analytic function is infinitely differentiable, it follows that

$$\Delta u = \frac{\partial^2 u}{\partial x^2} + \frac{\partial^2 u}{\partial y^2} = \frac{\partial^2 v}{\partial x \partial y} - \frac{\partial^2 v}{\partial y \partial x} = 0,$$

and similarly $\Delta v(x) = 0$.

In particular, the real part of the logarithm function is harmonic as long as we are not at 0. If $z = re^{i\theta} \in \mathbb{C} - \{0\}$, then $\log z = \log r + i\theta$, so the real part of $f(z) = \log z$ is $u(z) = \log r = \log |z|$. We conclude that the function $f(x) = \log |x|$ is harmonic for $x \in \mathbb{R}^2 - \{0\}$, where $|x| = (x_1^2 + x_2^2)^{1/2}$. Alternatively, this can be verified by computing the Laplacian of $\log |x|$.

When the dimension n is greater than or equal to 3, the function $h(x) = |x|^{2-n}$ is seen by a direct calculation of the partial derivatives to be harmonic in $\mathbb{R}^n - \{0\}$, where $|x| = (x_1^2 + \cdots + x_n^2)^{1/2}$. When doing the calculation, it is helpful to write

$$\frac{\partial}{\partial x_i} |x| = \frac{\partial}{\partial x_i} (x_1^2 + \cdots + x_n^2)^{1/2} = \frac{2x_i}{2(x_1^2 + \cdots + x_n^2)^{1/2}} = \frac{x_i}{|x|}.$$

22.2 The averaging property

Recall the *divergence theorem* from undergraduate analysis: if D is a nice domain such as a ball, then

$$\int_D \operatorname{div} F(x) \, dx = \int_{\partial D} F \cdot n(y) \, \sigma(dy), \tag{22.1}$$

where $F = (F_1, \ldots, F_n) : \mathbb{R}^n \to \mathbb{R}^n$ is a vector field, $\operatorname{div} F = \sum_{i=1}^n \partial F_i/\partial x_i$ is the divergence of F, ∂D is the boundary of D, $n(y)$ is the outward pointing unit normal vector at y, and $\sigma(dy)$ is surface measure on ∂D.

If u and v are two real-valued functions on $\mathbb{R}^n$ and we let $F = u\nabla v$, then

$$\operatorname{div} F = \sum_{i=1}^n \frac{\partial}{\partial x_i} \left(u \frac{\partial v}{\partial x_i} \right) = u \sum_{i=1}^n \frac{\partial^2 v}{\partial x_i^2} + \sum_{i=1}^n \frac{\partial u}{\partial x_i} \frac{\partial v}{\partial x_i}$$

$$= u\Delta v + \nabla u \cdot \nabla v$$

and

$$F \cdot n = u \frac{\partial v}{\partial n},$$

where $\partial v / \partial n$ is the normal derivative. Substituting into the divergence theorem we get *Green's first identity*:

$$\int_D u \Delta v \, dx + \int_D \nabla u \cdot \nabla v \, dx = \int_{\partial D} u \frac{\partial v}{\partial n} \, d\sigma. \qquad (22.2)$$

If we reverse the roles of u and v in Green's first identity and take the difference, we get *Green's second identity*:

$$\int_D (u \Delta v - v \Delta u) \, dx = \int_{\partial D} \left(u \frac{\partial v}{\partial n} - v \frac{\partial u}{\partial n} \right) d\sigma. \qquad (22.3)$$

Each of the two following theorems are known as the *mean value property* or the *averaging property* of harmonic functions.

Theorem 22.1 *Suppose h is harmonic in a domain D, $x_0 \in D$, and $r < \text{dist}\,(x_0, D^c)$. Then*

$$h(x_0) = \int_{\partial B(x_0, r)} h(y) \, \sigma(dy), \qquad (22.4)$$

where σ is surface measure on $\partial B(x_0, r)$.

Proof. By looking instead at $h(x - x_0) - h(x_0)$, we may suppose without loss of generality that $x_0 = 0$ and $h(x_0) = 0$.

If $s \leq r$ and we apply Green's first identity with $v = h$ and u identically equal to one, we see that

$$\int_{\partial B(0, s)} \frac{\partial h}{\partial n}(y) \, \sigma(dy) = 0, \qquad (22.5)$$

since $\nabla u = 0$ and $\Delta v = 0$ in $B(0, s)$.

Now let $\varepsilon > 0$ and choose δ such that $|h(x)| < \varepsilon$ if $|x| \leq \delta$. This can be done because h is continuous at 0 and $h(0) = 0$. If $n \geq 3$, let v be a C^2 function on $\mathbb{R}^n$ such that $v(x) = |x|^{2-n}$ if $|x| \geq \delta/2$. If $n = 2$, let v be a C^2 function such that $v(x) = \log|x|$ if $|x| \geq \delta/2$. We now apply Green's second identity with $u = h$

and v as just described in each of the balls $B(0, r)$ and $B(0, \delta)$ and take the difference:

$$
0 = \int_{B(0,r)-B(0,\delta)} (u\Delta v - v\Delta u)
$$

$$
= \int_{\partial B(0,r)} u(y)\frac{\partial v}{\partial n}(y)\,\sigma(dy) - \int_{\partial B(0,r)} v(y)\frac{\partial u}{\partial n}(y)\,dy
$$

$$
- \int_{\partial B(0,\delta)} u(y)\frac{\partial v}{\partial n}(y)\,\sigma(dy) + \int_{\partial B(0,\delta)} v(y)\frac{\partial u}{\partial n}(y)\,\sigma(dy)
$$

$$
= I_1 - I_2 - I_3 + I_4.
$$

We used that $\Delta u = 0$ and $\Delta v = 0$ in $B(0, r) - B(0, \delta)$. We then have

$$
I_1 = I_2 + I_3 - I_4.
$$

For $y \in \partial B(0, r)$ and for $y \in \partial B(0, \delta)$, we see that $n(y) = y/|y|$. A calculation shows that $\nabla v(y) = c_1 y/|y|^n$ for y on the boundary of either of those sets, where c_1 is a constant depending only on the dimension n, and we conclude

$$
\frac{\partial v}{\partial n}(y) = \frac{c_1}{|y|^{n-1}}
$$

on the boundary of either of those sets. Therefore

$$
I_1 = \frac{c_2}{r^{n-1}} \int_{\partial B(0,r)} u(y)\,\sigma(dy)
$$

is equal to a constant times the right hand side of (22.4). We also have

$$
|I_3| \leq \sup_{y \in \partial B(0,\delta)} |u(y)|\, \frac{c_2}{\delta^{n-1}}\sigma(\partial B(0,\delta)) \leq c_3\varepsilon.
$$

I_2 and I_4 are both zero by (22.5) and the fact that v is constant on $\partial B(0, r)$ and is constant on $\partial B(0, \delta)$. We conclude that the right hand side of (22.4) is bounded in absolute value by a constant times ε. Since ε is arbitrary, this proves that $\int_{\partial B(0,r)} h(y)\,\sigma(dy) = 0$, which yields the theorem.   □

The previous theorem says that the value of a harmonic function at the center of a ball contained in the domain is equal to the average of the values of the harmonic function on the boundary of the ball. The next theorem says that the value at the center is equal to the average of the values inside the ball.

Theorem 22.2 *Suppose h is harmonic in D, $x_0 \in D$, and $r <$ dist (x_0, D^c). Then*

$$h(x_0) = \frac{1}{m(B(0,r))} \int_{B(x_0,r)} h(y)\, dy, \qquad (22.6)$$

where m is Lebesgue measure.

Proof. This result follows easily from Theorem 22.1 by changing to polar coordinates. Again we may suppose $x_0 = 0$. If $y \in B(0, r)$, we may write $y = sv$, where $s = |y| \in (0, r)$ and $v = y/|y| \in \partial B(0, 1)$. If $\sigma_s(dy)$ is surface measure on $\partial B(0, s)$, then

$$\int_{B(x_0,r)} h(y)\, dy = \int_0^r \int_{\partial B(0,s)} h(vs)\, \sigma_s(dv)\, ds$$
$$= \int_0^r h(0)\, \sigma_s(\partial B(0, s))\, ds$$
$$= h(0) m(B(0, r)),$$

where we used Theorem 22.1 for the second equality. $\qquad \square$

22.3 Maximum principle

The following theorem is known as the *maximum principle* for harmonic functions.

Theorem 22.3 *Suppose D is a connected domain and h is harmonic in D. If h takes its maximum inside D, then h is constant in D.*

Proof. Let $M = \sup_{x \in D} h(x)$. Suppose $h(x_0) = M$ for some $x_0 \in D$ and let $r < $ dist (x, D^c). If $h(x) < M$ for some $x \in B(x_0, r)$, then by the continuity of h, we see that $h < M$ for a ball contained in $B(x_0, r)$. Then

$$h(x_0) = M > \frac{1}{m(B(x_0, r))} \int_{B(x_0,r)} h(y)\, dy = h(x_0),$$

a contradiction. Therefore h is identically equal to M on $B(x_0, r)$ if $h(x_0) = M$ and $\overline{B(x_0, r)} \subset D$. Thus $\{y \in D : h(y) = M\}$ is open. Since h is continuous, then $\{y \in D : h(y) < M\}$ is also open. Since D is connected, either $\{y \in D : h(y) = M\}$ must be empty or must be equal to D.   □

If D is a bounded connected domain, h is harmonic in D, and h is continuous on $\overline{D}$, then Theorem 22.3 says that h takes its maximum on ∂D. The hypothesis that D be bounded is essential. If we consider

$$D = \{(x, y) \in \mathbb{R}^2 : y > 0\}$$

and let $h(x, y) = y$, then h is harmonic, but does not takes its maximum on ∂D.

22.4   Smoothness of harmonic functions

In this section we prove that harmonic functions are C^∞ in the domain in which they are defined, and then show that functions satisfying the averaging property are harmonic.

Theorem 22.4 *Suppose D is a domain and h is bounded on D and satisfies the averaging property (22.6) for each $x_0 \in D$ and each $r < \text{dist}\,(x_0, D^c)$. Then h is C^∞ in D.*

Remark 22.5 Suppose for each $x \in D$ there is an open subset N_x of D containing x on which h is bounded. We can apply the above theorem to N_x and conclude that h is C^∞ on each set N_x, and hence is C^∞ on D.

Since harmonic functions are C^2 functions and satisfy the averaging property, they are C^∞ in their domain.

Proof. Suppose $z_0 \in D$, $8r < \text{dist}\,(z_0, D^c)$, $x \in B(z_0, 2r)$, $x' \in B(z_0, 3r)$, and $r > |\varepsilon| > |x - x'|$. Suppose $|h|$ is bounded by M. Applying the averaging property,

$$h(x) = \frac{1}{m(B(0, r))} \int_{B(x, r)} h(y) \, dy$$

and similarly with x replaced by x'. Taking the difference,

$$|h(x) - h(x')| \leq \frac{1}{m(B(0,r))} \int_{B(x,r) \triangle B(x',r)} |h(y)| \, dy$$
$$\leq \frac{c_1 M}{r^n} m(B(x,r) \triangle B(x',r)),$$

where $A \triangle B = (A - B) \cup (B - A)$. Some easy geometry shows that

$$B(x,r) \triangle B(x',r) \subset B(x, r + \varepsilon) - B(x, r - \varepsilon),$$

so

$$|h(x) - h(x')| \leq c_1 M r^{-n} m(B(x, r + \varepsilon) - B(x, r - \varepsilon))$$
$$= c_2 M r^{-n} [(r + \varepsilon)^n - (r - \varepsilon)^n],$$

which in turn is bounded by

$$c_3 M r^{-n} (r^{n-1} \varepsilon).$$

We used here the inequality

$$(a - b)^n = (a - b)(a^{n-1} + a^{n-2}b + \cdots + ab^{n-2} + b^{n-1})$$
$$\leq n(a - b)(a \vee b)^{n-1}.$$

This is true for each $|\varepsilon| > |x - x'|$, and therefore

$$\frac{|h(x) - h(x')|}{|x - x'|} \leq c_3 \frac{M}{r}. \tag{22.7}$$

One conclusion we draw from this is that h is continuous.

Now let $e_1 = (1, 0, \ldots, 0)$ be the unit vector in the x_1 direction. Let

$$F_\varepsilon(x) = \frac{h(x + \varepsilon e_1) - h(x)}{\varepsilon}.$$

We have seen that $|F_\varepsilon(x)|$ is bounded by $c_3 M/r$ if $x \in B(z_0, 2r)$ and $|\varepsilon| < r$. Applying the averaging property and doing some algebra,

$$F_\varepsilon(x) = \frac{1}{m(B(0,r))} \int_{B(x,r)} F_\varepsilon(y) \, dy. \tag{22.8}$$

Just as in the derivation of (22.7),

$$|F_\varepsilon(x) - F_\varepsilon(x')| \leq c_4 \frac{M}{r} |x - x'|.$$

This implies that $\{F_\varepsilon(x)\}$ is an equicontinuous family of functions of x on $\overline{B(z_0, r)}$.

Fix $x_2, \ldots, x_n$. In view of (22.7),

$$H(x_1) = h(x_1, \ldots, x_n)$$

is of bounded variation in the x_1 variable, and hence is differentiable for almost every x_1. Therefore $G_\varepsilon(x_1) = F_\varepsilon(x_1, \ldots, x_n)$ has a limit as $\varepsilon \to 0$ for almost every x_1 such that $(x_1, \ldots, x_n) \in B(z_0, r)$. This and the equicontinuity of the family $\{F_\varepsilon\}$ imply that $G_\varepsilon(x_1)$ has a limit for every such x_1. Thus, for each $(x_1, \ldots, x_n) \in B(z_0, r)$, the partial derivative of h with respect to x_1 exists. Moreover, $\partial h / \partial x_1$ is bounded in $B(z_0, r)$. Since $z_0 \in D$ is arbitrary, we see that $\partial h / \partial x_1$ exists at each point of D and a compactness argument shows that it is bounded on each bounded subdomain D' of D such that $\overline{D'} \subset D$.

Passing to the limit in (22.8), we obtain

$$\frac{\partial h}{\partial x_1}(x) = \frac{1}{m(B(0, r))} \int_{B(x, r)} \frac{\partial f}{\partial x_1}(y)\, dy.$$

Thus $\partial h / \partial x_1$ also satisfies the averaging property and is bounded in each bounded subdomain D' of D such that $\overline{D'} \subset D$. Hence it is continuous. These facts also apply to each of the first partial derivatives of h.

Repeating the argument and using Remark 22.5, we see each second partial derivative $\partial^2 h / \partial x_i\, \partial x_j$ satisfies the averaging property, hence is continuous, and so on. Therefore h is a C^∞ function in D. $\qquad\square$

We now have the following converse of the averaging property.

Theorem 22.6 *If D is a domain and h is bounded on D and satisfies (22.6), then h is harmonic in D.*

Proof. Let $x_0 \in D$. We may take $x_0 = 0$ and $h(0) = 0$ without loss of generality. By Taylor's theorem,

$$h(y) = h(0) + \sum_{i=1}^n \frac{\partial h}{\partial x_i}(0)y_i + \frac{1}{2} \sum_{i,j=1}^n \frac{\partial^2 h}{\partial x_i \partial x_j}(0)y_i y_j + R(y),$$

where the remainder R satisfies $|R(y)|/|y|^2 \to 0$ as $|y| \to 0$. Integrating over $B(0, r)$ and using that the integrals of y_i and of $y_i y_j$ over $B(0, r)$ are zero unless $i = j$, we obtain

$$0 = h(0) = \sum_{i=1}^{n} \int_{B(0,r)} \frac{\partial^2 h}{\partial x_i^2}(0) y_i^2 \, dy + \int_{B(0,r)} R(y) \, dy.$$

Therefore given ε,

$$c_1 r^3 |\Delta h(0)| \leq \varepsilon \int_{B(0,r)} |y|^2 \, dy$$

if r is small enough. Dividing both sides by r^3, we have

$$|\Delta h(0)| \leq c_2 \varepsilon,$$

and since ε is arbitrary, then $\Delta h(0) = 0$. □

Now that we know that harmonic functions are C^∞ in their domain, then $\partial h/\partial x_i \in C^2$ and

$$\Delta \left(\frac{\partial h}{\partial x_i} \right)(x) = \frac{\partial(\Delta h)}{\partial x_i}(x) = 0,$$

so $\partial h/\partial x_i$ is also harmonic. This could also be deduced from the fact that $\partial h/\partial x_i$ satisfies the averaging property by the proof of Theorem 22.4.

22.5 Poisson kernels

Let $H \subset \mathbb{R}^{n+1}$ be defined by $H = \mathbb{R}^n \times (0, \infty)$ and denote points of H by (x, y). Define

$$P(x, y) = \frac{cy}{(|x|^2 + y^2)^{(n+1)/2}},$$

where

$$c_n = \frac{\Gamma((n+1)/2)}{\pi^{(n+1)/2}}$$

and Γ is the Gamma function:

$$\Gamma(x) = \int_0^\infty t^{x-1} e^{-t} \, dt.$$

$P(x, y)$ is called the *Poisson kernel* for the half space H.

Proposition 22.7 *The function* $(x, y) \to P(x, y)$ *is harmonic in* H. *Moreover, for each* y,

$$\int_{\mathbb{R}^n} P(x, y) \, dx = 1.$$

Proof. This is just calculus. Calculating derivatives shows that the Laplacian is zero. A trigonometric substitution shows that the integral is equal to one. □

If $1 \leq p \leq \infty$ and $f \in L^p(\mathbb{R}^n)$, define

$$u(x, y) = \int P(x - t, y) f(t) \, dt.$$

u is called the *Poisson integral* of f and also the *harmonic extension* of f.

Proposition 22.8 *(1) If* $1 \leq p \leq \infty$ *and* $f \in L^p(\mathbb{R}^n)$, *then the harmonic extension of* f *in* H *is harmonic.*
(2) If f *is bounded and continuous, then*

$$\lim_{y \to 0} u(x_0, y) = f(x_0)$$

for each $x_0 \in \mathbb{R}^n$.

Proof. The first follows from Proposition 22.7 and the dominated convergence theorem. To prove (2), by looking at $f(x - x_0) - f(x_0)$ and using the fact that $\int P(x, y) \, dx = 1$, it suffices to prove this when $x_0 = 0$ and $f(0) = 0$.

Given ε, choose δ such that $|f(x)| \leq \varepsilon$ if $|x| \leq \delta$. We have

$$u(0, y) = \int_{|t| \leq \delta} P(t, y) f(t) \, dt + \int_{|t| > \delta} P(t, y) f(t) \, dt.$$

The first integral on the right is bounded in absolute value by

$$\varepsilon \int P(t, y) \, dt = \varepsilon.$$

The second integral on the right is bounded in absolute value by

$$\sup_{t \in \mathbb{R}^n} |f(t)| \int_{|t| > \delta} P(t, y) \, dt.$$

By a change of variables,

$$\int_{|t|>\delta} P(t,y)\, dt = \int_{|t|>\delta/y} \frac{c}{(1+t^2)^{(n+1)/2}}\, dt,$$

which tends to 0 as $y \to 0$ by the dominated convergence theorem. Since ε is arbitrary,

$$\limsup_{y\to 0} |u(0,y)| = 0,$$

which proves (2). □

If $D = B(0,r)$, the *Poisson kernel* for the ball D is given by

$$P_r(x,y) = c\frac{r^2 - |x|^2}{r|x-y|^n}, \qquad x \in D, \quad y \in \partial D,$$

where $c = 1/\sigma(\partial B(0,1))$ and $\sigma(dy)$ is surface area on ∂D.

If f is a continuous function on $\partial B(0,r)$, then

$$u(x) = \int P_r(x,y)f(y)\, \sigma(dy)$$

is harmonic in D, u has a continuous extension to $\overline{B(0,r)}$ and

$$\lim_{x\to y, u\in D} u(x) = f(y), \qquad y \in \partial B(0,r).$$

These facts can be shown by some not-so-easy calculus, and an argument similar to the proof of (2) of Proposition 22.8.

How does one arrive at the formula for the Poisson kernel for the ball? If you are good at calculations, you can show by tedious calculations that if h is harmonic in a domain E not containing zero, then $|x|^{2-n}h(x/|x|^2)$ is harmonic in the domain $\{y \in \mathbb{R}^n : y/|y|^2 \in E\}$. The Poisson kernel formula is obvious when $x = 0$ and $r = 1$. By a simple change of variables, one can get the Poisson kernel for $E = B(e_1, 1)$, where e_1 is the unit vector in the x_1 direction. We then apply the transformation $y \to y/|y|^2$ to get the Poisson kernel for the half space $H' = \{y : y_1 > 1/2\}$ with $x = e_1$. By another simple change of variables we get the Poisson kernel for H' with x any point of H'. Finally we do another inversion $x \to x/|x|^2$ to get the Poisson kernel for the unit ball, and do yet another change of variables to get the Poisson kernel for the ball of radius r. See [2] or [3] for details.

Remark 22.9 The Dirichlet problem for a ball $D = B(0,r)$ from partial differential equations is the following: given a continuous function f on the boundary of the ball, find a function u that is C^2 in D and continuous on $\overline{D}$ such that $\Delta u(x) = 0$ for $x \in D$ and $u(x) = f(x)$ for $x \in \partial D$. By the above,

$$u(x) = \int_{\partial B(0,r)} P_r(x,y) f(y) \, \sigma(dy)$$

provides the solution.

22.6   Harnack inequality

The following theorem is known as the *Harnack inequality*.

Theorem 22.10 *Suppose $h \geq 0$ is harmonic in $B(0,R)$ and $r < R$. There exists a constant c_1 depending only on r and R such that*

$$h(x) \leq c_1 h(x'), \qquad x, x' \in B(0,r).$$

Proof. If $\varepsilon < R - r$, then

$$g(x) = \int_{\partial B(0,R-\varepsilon)} P_{R-\varepsilon}(x,y) h(y) \, \sigma(dy)$$

is harmonic in $B(0, R - \varepsilon)$, agrees with h on $\partial B(0, R - \varepsilon)$, and so $g - h$ is harmonic in $B(0, R - \varepsilon)$ and equal to 0 on the boundary of $B(0, R - \varepsilon)$. By the maximum principle, g is identically equal to h.

If $x, x' \in B(0,r)$ and $y \in \partial B(0,s)$ with $s > r$, then

$$c_2(r,s) P_s(x,y) \leq P_s(x',y) \leq c_3(r,s) P_s(x,y)$$

with

$$c_2(r,s) = \frac{s^2 - r^2}{s(s+r)^n}, \qquad c_3(r,s) = \frac{s^2}{s(s-r)^n}.$$

Setting $s = R - \varepsilon$, multiplying the Poisson kernel by $h(y)$, and integrating over $y \in \partial B(0, R - \varepsilon)$ proves our result for the balls $B(0,r)$ and $B(0, R - \varepsilon)$. Letting $\varepsilon \to 0$ yields our inequality for the balls $B(0,r)$ and $B(0,R)$.   □

22.7 Exercises

Exercise 22.1 Suppose h is harmonic in a domain and $g(x) = x \cdot \nabla h(x)$. Prove that g is harmonic in the domain.

Exercise 22.2 Prove that if u, v are harmonic in a domain D, then uv harmonic in harmonic in D if and only if $\nabla u(x) \cdot \nabla v(x) = 0$ in D.

Exercise 22.3 Suppose D connected and h and h^2 are harmonic in D. Prove that h is constant in D.

Exercise 22.4 Let D be a bounded connected domain. Suppose that h is harmonic in D and C^1 in $\overline{D}$. Prove that if $\partial h / \partial n = 0$ everywhere on the boundary of D, then h is constant.

Exercise 22.5 Suppose that h is bounded and harmonic in a domain D, $x_0 \in D$, and $r > \text{dist}(x_0, D^c)$. Prove there exists a constant c_k depending only on k such that if g is any of the kth partial derivatives of h, then

$$|g(x_0)| \leq \frac{c_k}{r^k} \sup_{x \in D} |h(x)|.$$

Exercise 22.6 Prove that if h is harmonic in a domain D not containing 0 and

$$g(x) = |x|^{2-n} h(x/|x|^2),$$

then g is harmonic in $\{y : y/|y|^2 \in D\}$.

Exercise 22.7 Prove that if f is continuous on $\partial B(0, r)$ and

$$h(x) = \int_{\partial B(0,r)} P_r(x, y) f(y) \, \sigma(dy),$$

where $P_r(x, y)$ is the Poisson kernel for the ball, then

$$\lim_{x \to z, x \in B(0,r)} h(x) = f(z), \qquad z \in \partial B(0, r).$$

Exercise 22.8 Prove that if h is harmonic in $\mathbb{R}^n$, then h has a Taylor series that converges everywhere.

Exercise 22.9 Suppose that D is a bounded connected domain, $x_0 \in D$, and that h and all of its partial derivatives are equal to 0 at x_0. Prove that h is identically zero.

Exercise 22.10 (1) Show that the constant c_1 in Theorem 22.10 can be taken to be equal to

$$c_1 = \frac{R^2}{R^2 - r^2} \left(\frac{R + r}{R - r} \right)^n.$$

(2) Prove *Liouville's theorem*: if h is harmonic and non-negative in $\mathbb{R}^n$, then h is constant.

Chapter 23

Sobolev spaces

For some purposes, particularly when studying partial differential equations, one wants to study functions which only have a derivative in the weak sense. We look at spaces of such functions in this chapter, and prove the important Sobolev inequalities.

23.1 Weak derivatives

Let C_K^∞ be the set of functions on $\mathbb{R}^n$ that have compact support and have partial derivatives of all orders. For $j = (j_1, \ldots, j_n)$, write

$$D^j f = \frac{\partial^{j_1 + \cdots + j_n} f}{\partial_{x_1}^{j_1} \cdots \partial_{x_n}^{j_n}},$$

and set $|j| = j_1 + \cdots + j_n$. We use the convention that $\partial^0 f / \partial x_i^0$ is the same as f.

Let f, g be locally integrable. We say that $D^j f = g$ in the *weak sense* or g is the *weak j^{th} order partial derivative* of f if

$$\int f(x) \, D^j \varphi(x) \, dx = (-1)^{|j|} \int g(x) \varphi(x) \, dx$$

for all $\varphi \in C_K^\infty$. Note that if $g = D^j f$ in the usual sense, then integration by parts shows that g is also the weak derivative of f.

Let

$$W^{k,p}(\mathbb{R}^n) = \{f : f \in L^p, D^j f \in L^p \text{ for each } j \text{ such that } |j| \leq k\}.$$

Set

$$\|f\|_{W^{k,p}} = \sum_{\{j:0\le|j|\le k\}} \|D^j f\|_p,$$

where we set $D^0 f = f$.

Theorem 23.1 *The space $W^{k,p}$ is complete.*

Proof. Let f_m be a Cauchy sequence in $W^{k,p}$. For each j such that $|j| \le k$, we see that $D^j f_m$ is a Cauchy sequence in L^p. Let g_j be the L^p limit of $D^j f_m$. Let f be the L^p limit of f_m. Then

$$\int f_m D^j \varphi = (-1)^{|j|} \int (D^j f_m)\varphi \to (-1)^{|j|} \int g_j \varphi$$

for all $\varphi \in C_K^\infty$. On the other hand, $\int f_m D^j \varphi \to \int f D^j \varphi$. Therefore

$$(-1)^{|j|} \int g_j \varphi = \int f D^j \varphi$$

for all $\varphi \in C_K^\infty$. We conclude that $g_j = D^j f$ a.e. for each j such that $|j| \le k$. We have thus proved that $D^j f_m$ converges to $D^j f$ in L^p for each j such that $|j| \le k$, and that suffices to prove the theorem. $\qquad\square$

23.2 Sobolev inequalities

Lemma 23.2 *If $k \ge 1$ and $f_1, \dots, f_k \ge 0$, then*

$$\int f_1^{1/k} \cdots f_k^{1/k} \le \left(\int f_1\right)^{1/k} \cdots \left(\int f_k\right)^{1/k}.$$

Proof. We will prove

$$\left(\int f_1^{1/k} \cdots f_k^{1/k}\right)^k \le \left(\int f_1\right) \cdots \left(\int f_k\right). \qquad (23.1)$$

We will use induction. The case $k = 1$ is obvious. Suppose (23.1) holds when k is replaced by $k - 1$ so that

$$\left(\int f_1^{1/(k-1)} \cdots f_{k-1}^{1/(k-1)}\right)^{k-1} \le \left(\int f_1\right) \cdots \left(\int f_{k-1}\right). \qquad (23.2)$$

Let $p = k/(k-1)$ and $q = k$ so that $p^{-1} + q^{-1} = 1$. Using Hölder's inequality,

$$\int (f_1^{1/k} \cdots f_{k-1}^{1/k}) f_k^{1/k}$$

$$\leq \left(\int f_1^{1/(k-1)} \cdots f_{k-1}^{1/(k-1)} \right)^{(k-1)/k} \left(\int f_k \right)^{1/k}.$$

Taking both sides to the kth power, we obtain

$$\left(\int (f_1^{1/k} \cdots f_{k-1}^{1/k}) f_k^{1/k} \right)^k$$

$$\leq \left(\int f_1^{1/(k-1)} \cdots f_{k-1}^{1/(k-1)} \right)^{(k-1)} \left(\int f_k \right).$$

Using (23.2), we obtain (23.1). Therefore our result follows by induction. $\qquad \square$

Let C_K^1 be the continuously differentiable functions with compact support. The following theorem is sometimes known as the *Gagliardo-Nirenberg inequality.*

Theorem 23.3 *There exists a constant c_1 depending only on n such that if $u \in C_K^1$, then*

$$\|u\|_{n/(n-1)} \leq c_1 \| |\nabla u| \|_1.$$

We observe that u having compact support is essential; otherwise we could just let u be identically equal to one and get a contradiction. On the other hand, the constant c_1 does not depend on the support of u.

Proof. For simplicity of notation, set $s = 1/(n-1)$. Let $K_{j_1 \cdots j_m}$ be the integral of $|\nabla u(x_1, \ldots, x_n)|$ with respect to the variables $x_{j_1}, \ldots, x_{j_m}$. Thus

$$K_1 = \int |\nabla u(x_1, \ldots, x_n)| \, dx_1$$

and

$$K_{23} = \int \int |\nabla u(x_1, \ldots, x_n)| \, dx_2 \, dx_3.$$

Note K_1 is a function of $(x_2, \ldots, x_n)$ and K_{23} is a function of $(x_1, x_4, \ldots, x_n)$.

If $x = (x_1, \ldots, x_n) \in \mathbb{R}^n$, then since u has compact support,

$$|u(x)| = \left| \int_{-\infty}^{x_1} \frac{\partial u}{\partial x_1}(y_1, x_2, \ldots, x_n)\, dy_1 \right|$$
$$\leq \int_{\mathbb{R}} |\nabla u(y_1, x_2, \ldots, x_n)|\, dy_1$$
$$= K_1.$$

The same argument shows that $|u(x)| \leq K_i$ for each i, so that

$$|u(x)|^{n/(n-1)} = |u(x)|^{ns} \leq K_1^s K_2^s \cdots K_n^s.$$

Since K_1 does not depend on x_1, Lemma 23.2 shows that

$$\int |u(x)|^{ns}\, dx_1 \leq K_1^s \int K_2^s \cdots K_n^s\, dx_1$$
$$\leq K_1^s \left(\int K_2\, dx_1 \right)^s \cdots \left(\int K_n\, dx_1 \right)^s.$$

Note that

$$\int K_2\, dx_1 = \int \left(\int |\nabla u(x_1, \ldots, x_n)|\, dx_2 \right) dx_1 = K_{12},$$

and similarly for the other integrals. Hence

$$\int |u|^{ns}\, dx_1 \leq K_1^s K_{12}^s \cdots K_{1n}^s.$$

Next, since K_{12} does not depend on x_2,

$$\int |u(x)|^{ns}\, dx_1\, dx_2 \leq K_{12}^s \int K_1^s K_{13}^s \cdots K_{1n}^s\, dx_2$$
$$\leq K_{12}^s \left(\int K_1\, dx_2 \right)^s \left(\int K_{13}\, dx_2 \right)^s \cdots \left(\int K_{1n}\, dx_2 \right)^s$$
$$= K_{12}^s K_{12}^s K_{123}^s \cdots K_{12n}^s.$$

We continue, and get

$$\int |u(x)|^{ns}\, dx_1\, dx_2\, dx_3 \leq K_{123}^s K_{123}^s K_{123}^s K_{1234}^s \cdots K_{123n}^s$$

and so on, until finally we arrive at

$$\int |u(x)|^{ns} \, dx_1 \cdots dx_n \leq \left(K^s_{12\cdots n}\right)^n = K^{ns}_{12\cdots n}.$$

If we then take the $ns = n/(n-1)$ roots of both sides, we get the inequality we wanted. $\qquad\square$

From this we can get the *Sobolev inequalities*.

Theorem 23.4 *Suppose $1 \leq p < n$ and $u \in C^1_K$. Then there exists a constant c_1 depending only on n such that*

$$\|u\|_{np/(n-p)} \leq c_1 \| \, |\nabla u| \, \|_p.$$

Proof. The case $p = 1$ is the case above, so we assume $p > 1$. The case when u is identically equal to 0 is obvious, so we rule that case out. Let

$$r = \frac{p(n-1)}{n-p},$$

and note that $r > 1$ and

$$r - 1 = \frac{np - n}{n - p}.$$

Let $w = |u|^r$. We observe that

$$|\nabla w| \leq c_2 |u|^{r-1} |\nabla u|.$$

Applying Theorem 23.3 to w and using Hölder's inequality with $q = \frac{p}{p-1}$, we obtain

$$\left(\int |w|^{n/(n-1)}\right)^{\frac{n-1}{n}} \leq c_3 \int |\nabla w|$$

$$\leq c_4 \int |u|^{(np-n)/(n-p)} |\nabla u|$$

$$\leq c_5 \left(\int |u|^{np/(n-p)}\right)^{\frac{p-1}{p}} \left(\int |\nabla u|^p\right)^{1/p}.$$

The left hand side is equal to

$$\left(\int |u|^{np/(n-p)}\right)^{\frac{n-1}{n}}.$$

Divide both sides by

$$\left(\int |u|^{np/(n-p)} \right)^{\frac{p-1}{p}}.$$

Since

$$\frac{n-1}{n} - \frac{p-1}{p} = \frac{1}{p} - \frac{1}{n} = \frac{n-p}{pn},$$

we get our result. $\square$

We can iterate to get results on the L^p norm of f in terms of the L^q norm of $D^k f$ when $k > 1$. The proof of the following theorem is left as Exercise 23.8.

Theorem 23.5 *Suppose $k \geq 1$. Suppose $p < n/k$ and we define q by $\frac{1}{q} = \frac{1}{p} - \frac{k}{n}$. Then there exists c_1 such that*

$$\|f\|_q \leq c \left\| \sum_{\{j:|j|=k\}} |D^k f| \right\|_p.$$

Remark 23.6 It is possible to show that if $p > n/k$, then f is Hölder continuous.

23.3   Exercises

Exercise 23.1 Prove that if $p_1, \ldots, p_n > 1$,

$$\sum_{i=1}^{n} \frac{1}{p_i} = 1,$$

and μ is a σ-finite measure, then

$$\int |f_1 \ldots f_n| \, d\mu \leq \|f_1\|_{p_1} \cdots \|f_n\|_{p_n}.$$

This is known as the *generalized Hölder inequality*.

Exercise 23.2 Suppose $1 \leq p < \infty$. Prove that if there exist f_m such that
(1) $f_m \in C_K^\infty$;
(2) $\|f - f_m\|_p \to 0$;
(3) for all $|j| \leq k$, $D^j f_m$ converges in L^p,
then $f \in W^{k,p}$.

Exercise 23.3 Suppose $1 \leq p < \infty$. Prove that if $f \in W^{k,p}$, then there exist f_m such that

(1) $f_m \in C_K^\infty$;

(2) $\|f - f_m\|_p \to 0$;

(3) for all $|j| \leq k$, $D^j f_m$ converges in L^p.

Exercise 23.4 Suppose $\frac{1}{r} = \frac{1}{p} + \frac{1}{q} - 1$. Prove that

$$\|f * g\|_r \leq \|f\|_p \|g\|_q.$$

This is known as *Young's inequality.*

Exercise 23.5 (1) Prove that $W^{k,2}$ can be regarded as a Hilbert space. (It is common to write H^k for $W^{k,2}$.)

(2) Suppose $k \geq 1$. Prove that $f \in W^{k,2}$ if and only if

$$\int (1 + |u|^2)^k |\widehat{f}(u)|^2 \, du < \infty.$$

Exercise 23.6 If s is a real number, define

$$H^s = \left\{ f : \int (1 + |u|^2)^s |\widehat{f}(u)|^2 \, du < \infty \right\}.$$

Prove that if $s > n/2$, then $\widehat{f}$ is in L^1. Conclude that f is continuous.

Exercise 23.7 Does the product formula hold for weak derivatives? That is, if $p \geq 2$ and $f, g \in W^{1,p}$, is $fg \in W^{1,p/2}$ with $D(fg) = f(Dg) + (Df)g$? Prove or give a counterexample.

Exercise 23.8 Prove Theorem 23.5.

Exercise 23.9 Let ψ be a C_K^1 function on $\mathbb{R}^2$ that is equal to one on $B(0,1)$ and let

$$f(x_1, x_2) = \psi(x_1, x_2) \frac{x_1^2}{x_1^2 + x_2^2}.$$

Prove that $f \in W^{1,p}(\mathbb{R}^2)$ for $1 \leq p < 2$, but that f is not continuous. (The function ψ is introduced only to make sure f has compact support.)

Exercise 23.10 Suppose the dimension $n = 1$.
(1) Prove that if $f \in W^{1,1}(\mathbb{R})$, then f is continuous.
(2) Prove that if $f \in W^{1,p}(\mathbb{R})$ for some $p > 1$, then f is Hölder continuous, that is, there exist $c_1 > 0$ and $\alpha \in (0,1)$ such that $|f(x) - f(y)| \le c_1 |x - y|^\alpha$ for all x and y.

Exercise 23.11 Prove that if $f \in C_K^1$, then

$$f(x) = c_1 \sum_{j=1}^{n} \int \frac{\partial f}{\partial x_j}(x - y) \frac{y_j}{|y|^n} \, dy,$$

where c_1^{-1} is equal to the surface measure of $\partial B(0,1)$.

Exercise 23.12 Suppose $n \ge 3$. Prove the *Nash inequality*:

$$\left(\int |f|^2 \right)^{1 + 2/n} \le c_1 \left(\int |\nabla f|^2 \right) \left(\int |f| \right)^{4/n}$$

if $f \in C_K^1(\mathbb{R}^n)$, where the constant c_1 depends only on n. (The Nash inequality is also true when $n = 2$.)

Chapter 24

Singular integrals

This chapter is concerned with the Hilbert transform, which is the prototype for more general singular integrals. The Hilbert transform of a function f is defined by

$$Hf(x) = \lim_{\varepsilon \to 0, N \to \infty} \frac{1}{\pi} \int_{\varepsilon < |y| < N} \frac{f(x-y)}{y} \, dy,$$

and is thus a principal value integral. Remarkably, H is a bounded operator on $L^p(\mathbb{R})$ for $1 < p < \infty$.

In preparation for the study of the Hilbert transform, we also investigate the Marcinkiewicz interpolation theorem and delve more deeply into properties of the maximal function, which was defined in Chapter 14.

24.1 Marcinkiewicz interpolation theorem

Let $(X, \mathcal{A}, \mu)$ be a measure space. An operator T mapping a collection of real-valued measurable functions on X to real-valued measurable functions on X is *sublinear* if

$$|T(f+g)(x)| \leq |Tf(x)| + |Tg(x)|$$

for all $x \in X$ and for all measurable functions f and g in the collection. Recall that the L^p norm of a function is given by

$$\|f\|_p = \left(\int_X |f(x)|^p \, \mu(dx) \right)^{1/p}$$

if $1 \leq p < \infty$. We say that an operator T is bounded on L^p or is of *strong-type p-p* if there exists a constant c_1 such that

$$\|Tf\|_p \leq c_1 \|f\|_p$$

for every $f \in L^p$. We say that an operator T is of *weak-type p-p* if there exists a constant c_2 such that

$$\mu(\{x : |Tf(x)| \geq \lambda\}) \leq c_2 \frac{\|f\|_p^p}{\lambda^p}$$

for all $\lambda > 0$. An operator that is bounded on L^p is automatically of weak-type p-p. This follows by Chebyshev's inequality (Lemma 10.4):

$$\mu(\{x : |Tf(x)| > \lambda\}) = \mu(\{x : |Tf(x)|^p > \lambda^p\})$$

$$\leq \frac{1}{\lambda^p} \int |Tf(x)|^p \, \mu(dx)$$

$$= \frac{1}{\lambda^p} \|Tf\|_p^p \leq \frac{c_1^p}{\lambda^p} \|f\|_p^p.$$

The *Marcinkiewicz interpolation theorem* says that if $1 \leq p < r < q \leq \infty$ and a sublinear operator T is of weak-type p-p and of weak-type q-q, then T is a bounded operator on L^r. A more general version considers operators that are what are known as weak-type p-q, but we do not need this much generality.

Theorem 24.1 *Suppose $1 \leq p < r < q \leq \infty$. Let T be a sublinear operator defined on $\{f : f = f_1 + f_2, f_1 \in L^p, f_2 \in L^q\}$.*
(1) If T is of weak-type p-p and T is a bounded operator on L^∞, then T is a bounded operator on L^r.
(2) If $q < \infty$, T is of weak-type p-p, and T is of weak-type q-q, then T is a bounded operator on L^r.

Proof. (1) Suppose $\|Tg\|_\infty \leq c_1 \|g\|_\infty$ if $g \in L^\infty$. Let $f \in L^r$ and define $f_1(x) = f(x)$ if $|f(x)| > \lambda/(2c_1)$ and 0 otherwise. Let

$f_2 = f - f_1$. This implies that $|f_2(x)|$ is bounded by $\lambda/2c_1$ and

$$\int |f_1(x)|^p \, dx = \int_{|f(x)|>\lambda/2c_1} |f(x)|^p \, dx$$
$$\leq \left(\frac{\lambda}{2c_1}\right)^{r-p} \int |f(x)|^r \, dx < \infty.$$

Because T is a bounded operator on L^∞, then $|Tf_2(x)|$ is bounded by $\lambda/2$. By the sublinearity of T,

$$|Tf(x)| \leq |Tf_1(x)| + |Tf_2(x)| \leq |Tf_1(x)| + \lambda/2,$$

and hence

$$\{x : |Tf(x)| > \lambda\} \subset \{x : |Tf_1(x)| > \lambda/2\}.$$

We therefore have

$$\mu(\{x : |Tf(x)| > \lambda\}) \leq \mu(\{x : |Tf_1(x)| > \lambda/2\}).$$

Since T is of weak-type p-p, there exists a constant c_2 not depending on f such that the right hand side is bounded by

$$c_2 \frac{\|f_1\|_p^p}{(\lambda/2)^p} = \frac{c_2 2^p}{\lambda^p} \int |f_1(x)|^p \, \mu(dx) = \frac{c_2 2^p}{\lambda^p} \int_{|f|>\lambda/2c_1} |f(x)|^p \, \mu(dx).$$

We then write, using Exercise 15.3 and the Fubini theorem,

$$\int |Tf(x)|^r \, \mu(dx) = \int_0^\infty r\lambda^{r-1} \mu(\{x : |Tf(x)| > \lambda\}) \, d\lambda$$
$$\leq \int_0^\infty r\lambda^{r-1} \frac{c_2 2^p}{\lambda^p} \int_{|f|>\lambda/(2c_1)} |f(x)|^p \, \mu(dx) \, d\lambda$$
$$= c_2 2^p r \int_0^\infty \int \lambda^{r-p-1} \chi_{(|f|>\lambda/2c_1)}(x) |f(x)|^p \, \mu(dx) \, d\lambda$$
$$= c_2 2^p r \int \int_0^{2c_1|f(x)|} \lambda^{r-p-1} \, d\lambda \, |f(x)|^p \, \mu(dx)$$
$$= c_2 \frac{2^p r}{r-p} \int |f(x)|^{r-p} |f(x)|^p \, \mu(dx)$$
$$= c_2 \frac{2^p r}{r-p} \int |f(x)|^r \, \mu(dx).$$

This is exactly what we want.

(2) Let $\lambda > 0$, let $f \in L^r$, and let $f_1 = f(x)$ if $|f(x)| > \lambda$ and 0 otherwise, and let $f_2(x) = f(x) - f_1(x)$. Since $|Tf(x)| \leq |Tf_1(x)| + |Tf_2(x)|$, we have

$$\{|Tf(x)| > \lambda\} \subset \{|Tf_1(x)| > \lambda/2\} \cup \{|Tf_2(x)| > \lambda/2\}.$$

Since T is of weak-type p-p and weak-type q-q, there exist constants c_3 and c_4 so that

$$\mu(\{|Tf(x)| > \lambda\}) \leq \mu(\{|Tf_1(x)| > \lambda/2\}) + \mu(\{|Tf_2(x)| > \lambda/2\})$$

$$\leq \frac{c_3}{(\lambda/2)^p} \int |f_1|^p + \frac{c_4}{(\lambda/2)^q} \int |f_2|^q$$

$$= c_3 2^p \lambda^{-p} \int_{|f|>\lambda} |f|^p + c_4 2^q \lambda^{-q} \int_{|f|\leq\lambda} |f|^q.$$

Therefore

$$\int |Tf(x)|^r \, \mu(dx) = \int_0^\infty r\lambda^{r-1} \mu(\{|Tf(x)| > \lambda\}) \, d\lambda$$

$$\leq c_3 2^p r \int_0^\infty \lambda^{r-p-1} \int_{|f|>\lambda} |f|^p \, \mu(dx) \, d\lambda$$

$$+ c_4 2^q r \int_0^\infty \lambda^{r-q-1} \int_{|f|\leq\lambda} |f|^q \, \mu(dx) \, d\lambda$$

$$= c_3 2^p r \int |f|^p \int_0^{|f(x)|} \lambda^{r-p-1} \, d\lambda \, \mu(dx)$$

$$+ c_4 2^q r \int |f|^q \int_{|f(x)|}^\infty \lambda^{r-q-1} \, d\lambda \, \mu(dx)$$

$$\leq c_3 \frac{2^p r}{r-p} \int |f|^p |f|^{r-p} \, \mu(dx) + c_4 \frac{2^q r}{q-r} \int |f|^q |f|^{r-q} \, \mu(dx)$$

$$= c_5 \int |f|^r \, \mu(dx),$$

where $c_5 = c_3 2^p r/(r-p) + c_4 2^q r/(q-r)$.   □

An application of the Marcinkiewicz interpolation theorem is the following, although a proof using Hölder's inequality is also possible; cf. Exercise 15.11.

Theorem 24.2 *Suppose $1 \leq p \leq \infty$. There exists a constant c such that if $g \in L^1$ and $f \in L^p$, then $f * g \in L^p$ and $\|f * g\|_p \leq c\|f\|_p\|g\|_1$.*

Proof. By linearity we may suppose $\|g\|_1 = 1$. The case $p = 1$ is Proposition 15.7. If $f \in L^\infty$, then

$$|f * g(x)| \le \int |f(x - y)| \, |g(y)| \, dy \le \|f\|_\infty \int |g(y)| \, dy = \|f\|_\infty \|g\|_1,$$

which takes care of the case $p = \infty$. If we define the operator $Tf = f * g$, then we have shown T is a bounded operator on L^1 and on L^∞. Therefore it is a bounded operator on L^p for all $1 < p < \infty$ by the Marcinkiewicz interpolation theorem. $\square$

24.2 Maximal functions

In Chapter 14 we defined

$$Mf(x) = \sup_{r > 0} \frac{1}{m(B(x,r))} \int_{B(x,r)} |f(y)| \, dy \qquad (24.1)$$

for locally integrable functions on $\mathbb{R}^n$ and called Mf the *maximal function* of f. Here m is n-dimensional Lebesgue measure. Note $Mf \ge 0$ and $Mf(x)$ might be infinite.

The main goal in this section is to relate the size of Mf as measured by the L^p norm to the size of f.

Theorem 24.3 *The operator M is of weak-type 1-1 and bounded on L^p for $1 < p \le \infty$. More precisely,*
(1) $m(\{x : Mf(x) > \lambda\}) \le c_1 \|f\|_1 / \lambda$ for $\lambda > 0$ and $f \in L^1$. The constant c_1 depends only on the dimension n.
(2) If $1 < p \le \infty$, then $\|Mf\|_p \le c_2 \|f\|_p$. The constant c_2 depends only on p and the dimension n.

In this theorem and the others that we will consider, it is important to pay attention to the range of p for which it holds. Frequently theorems hold only for $1 < p < \infty$. In this theorem, we have boundedness on L^p for $p > 1$ and including $p = \infty$. For the $p = 1$ case we only have a weak-type 1-1 estimate.

In the course of the proof of Theorem 24.3 we will show that $Mf(x)$ exists for almost every x if $f \in L^p$ for some $1 \le p \le \infty$.

Recall from Section 14.1 that M is not a bounded operator on L^1.

Proof. (1) This is just Theorem 14.2.

(2) It is obvious from the definition that $Mf(x) \leq \|f\|_\infty$, and so M is a bounded operator on L^∞. It is clear that M is sublinear. If we write $f \in L^p$ as $f\chi_{(|f|>1)} + f\chi_{(|f|\leq 1)}$, then the first summand is in L^1 and the second is in L^∞; the sublinearity then shows that Mf is finite almost everywhere.

By Theorem 24.1(1), assertion (2) follows.   □

24.3   Approximations to the identity

Let φ be integrable on $\mathbb{R}^n$ and let $\varphi_r(x) = r^{-d}\varphi(x/r)$. Let $\psi : [0, \infty) \to [0, \infty)$ be a decreasing function and suppose

$$c_1 = \int_{\mathbb{R}^n} \psi(|x|)\, dx < \infty.$$

Suppose also that $|\varphi(x)| \leq \psi(|x|)$ for all x. Recall that the convolution of f and g is defined by $f * g(x) = \int f(x - y)g(y)\, dy$. We continue to let m be n-dimensional Lebesgue measure.

Theorem 24.4 (Approximation to the identity) *(1) If $f \in L^p$, $1 \leq p \leq \infty$, then*

$$\sup_{r>0} |f * \varphi_r(x)| \leq c_1 Mf(x).$$

(2) If $f \in L^p$, $1 \leq p < \infty$, and $\int \varphi(x)\, dx = 1$, then $\|f*\varphi_r - f\|_p \to 0$ as $r \to 0$.
(3) If $f \in L^p$, $1 \leq p \leq \infty$ and $\int \varphi(x)\, dx = 1$, then

$$\lim_{r \to 0}(f * \varphi_r)(x) = f(x), \qquad \text{a.e.}$$

Proof. In proving (1), by a change of variables, we need only show that $|f * \varphi(0)| \leq c_1 Mf(0)$. First suppose ψ is piecewise constant: there exist $a_1 \leq a_2 \leq \ldots \leq a_k$ and $A_1 \geq A_2 \geq \ldots \geq A_k$ such that $\psi(y) = A_1$ for $y \in [0, a_1]$, $\psi(y) = A_i$ for $y \in (a_{i-1}, a_i]$, and $\psi(y) = 0$ for $|x| > a_k$. Then

$$|f*\varphi(0)| = \left| \int f(x)\varphi(-x)\, dx \right|$$

$$\leq \int |f(x)| \psi(|x|)\, dx$$

$$= A_1 \int_{B(0,a_1)} |f| + A_2 \int_{B(0,a_2)-B(0,a_1)} |f| + \cdots$$

$$+ A_k \int_{B(0,a_k)-B(0,a_{k-1})} |f|$$

$$= A_1 \int_{B(0,a_1)} |f| + \left[A_2 \int_{B(0,a_2)} |f| - A_2 \int_{B(0,a_1)} |f| \right]$$

$$+ \cdots + \left[A_k \int_{B(0,a_k)} |f| - A_k \int_{B(0,a_{k-1})} |f| \right]$$

$$= (A_1 - A_2) \int_{B(0,a_1)} |f| + (A_2 - A_3) \int_{B(0,a_2)} |f|$$

$$+ \cdots + (A_{k-1} - A_k) \int_{B(0,a_{k-1})} |f| + A_k \int_{B(0,a_k)} |f|$$

$$\leq \Big[(A_1 - A_2) m(B(0, a_1)) + \cdots + (A_{k-1} - A_k) m(B(0, a_{k-1}))$$

$$+ A_k m(B(0, a_k)) \Big] M f(0)$$

$$= \Big[A_1 m(B(0, a_1)) + A_2 m(B(0, a_2) - B(0, a_1)) + \cdots$$

$$+ A_k m(B(0, a_k) - B(0, a_{k-1})) \Big] M f(0).$$

Observe that the coefficient of $Mf(0)$ in the last expression is just $\int \psi(|x|)\, dx$. To handle the general case where ψ is not piecewise constant, we approximate ψ by piecewise constant ψ_j of the above form and take a limit.

Turning to (2), by a change of variables

$$f * \varphi_r(x) - f(x) = \int [f(x - y) - f(x)] \varphi_r(y)\, dy, \qquad (24.2)$$

$$= \int [f(x - ry) - f(x)] \varphi(y)\, dy.$$

Let $\varepsilon > 0$ and write $f = g + h$ where g is continuous with compact support and $\|h\|_p < \varepsilon$. By (24.2) with f replaced by g, $g * \varphi_r(x) - g(x) \to 0$ as $r \to 0$ by the dominated convergence theorem. Using that g is bounded with compact support and the dominated convergence theorem again, $\|g * \varphi_r - g\|_p \to 0$. By

Theorem 24.2

$$\|h * \varphi_r - h\|_p \le \|h\|_p \|\varphi_r\|_1 + \|h\|_p$$
$$= \|h\|_p \|\varphi\|_1 + \|h\|_p$$
$$\le \varepsilon(1 + c_1).$$

Therefore

$$\limsup_{r \to 0} \|f * \varphi_r - f\|_p \le \varepsilon(1 + c_1).$$

Since ε is arbitrary, (2) follows.

Finally we prove (3). If $p < \infty$, we proceed exactly as in the proof of Theorem 14.3, using part (1). We let $\beta > 0$ and let $\varepsilon > 0$. We write $f = g + h$, where g is continuous with compact support and $\|h\|_p < \varepsilon$. As in the proof of (2), $g * \varphi_r(x) - g(x) \to 0$. For each r we have

$$\sup_r |h * \varphi_r(x) - h(x)| \le \sup_r |h * \varphi_r(x)| + |h(x)| \le c_1 M h(x) + |h(x)|$$

by (1). Therefore by Theorem 14.2 and Chebyshev's inequality (Lemma 10.4),

$$m(\{x : \limsup_{r \to 0} |h * \varphi_r(x) - h(x)| > \beta\})$$
$$\le m(\{x : c_1 M h(x) > \beta/2\}) + m(\{x : |h(x)| > \beta/2\})$$
$$\le c_1 c_2 \frac{\|h\|_1}{\beta/2} + \frac{\|h\|_1}{\beta/2}$$
$$\le (2c_1 c_2 + 2)\varepsilon/\beta,$$

where c_2 is a constant depending only on the dimension n. Since

$$\limsup_{r \to 0} |f * \varphi_r(x) - f(x)| \le \limsup_{r \to 0} |h * \varphi_r(x) - h(x)|$$

and ε is arbitrary, then $\limsup_{r \to 0} |f * \varphi_r(x) - f(x)| \le \beta$ for almost every x. Since β is arbitrary, we conclude $f * \varphi_r(x) \to f(x)$ a.e.

There remains the case $p = \infty$. It suffices to let R be arbitrary and to show that $f * \varphi_r(x) \to f(x)$ a.e. for $x \in B(0, R)$. Write $f = f\chi_{B(0,2R)} + f\chi_{B(0,2R)^c}$. Since f is bounded, $f\chi_{B(0,2R)}$ is in L^1 and we obtain our result for this function by the $p = 1$ result. Set $h = f\chi_{B(0,2R)^c}$. If $x \in B(0, R)$, then $h(x) = 0$, and

$$|h * \varphi_r(x)| = \left| \int h(x - y)\varphi_r(y) \, dy \right| \le \int_{|y| \ge R} \varphi_r(y) \, dy \, \|h\|_\infty,$$

since $h(x - y) = 0$ if $x \in B(0, R)$ and $|y| < R$. Note now that $\|h\|_\infty \leq \|f\|_\infty$ and

$$\int_{|y| \geq R} \varphi_r(y) \, dy = \int_{|y| \geq R/r} \varphi(y) \, dy \to 0$$

as $r \to 0$ by the dominated convergence theorem. □

We will need the following in Section 24.4.

For each integer k, let $\mathcal{R}_k$ be the collection of closed cubes with side length 2^{-k} such that each coordinate of each vertex is an integer multiple of 2^{-k}. If $x \in \mathbb{R}^n$, $x = (x_1, \ldots, x_n)$, and $j_i/2^k \leq x_i < (j_i + 1)/2^k$ for each i, let

$$S_k(x) = [j_1/2^k, (j + 1)/2^k] \times \cdots [j_n/2^k, (j_n + 1)/2^k].$$

Thus $S_k(x)$ is an element of $\mathcal{R}_k$ containing x.

Theorem 24.5 *If $f \in L^1$, then*

$$\frac{1}{m(S_k(x))} \int_{S_k(x)} f(y) \, dy \to f(x)$$

as $k \to \infty$ for a.e. x.

Proof. The proof is similar to that of Theorems 14.3 and 24.4. First we show that there exists c_1 not depending on f such that

$$\frac{1}{m(S_k(x))} \int_{S_k(x)} |f(y)| \, dy \leq c_1 M f(x). \tag{24.3}$$

Note $S_k(x) \subset B(x, 2^{-k}\sqrt{n})$. Hence the left hand side of (24.3) is bounded by

$$\frac{m(B(x, 2^{-k}\sqrt{n}))}{m(S_k(x))} \cdot \frac{1}{m(B(x, 2^{-k}\sqrt{n}))} \int_{B(x, 2^{-k}\sqrt{n})} |f(y)| \, dy$$
$$\leq c_1 M f(x),$$

where $c_1 = m(B(x, 2^{-k}\sqrt{n}))/m(S_k(x))$ does not depend on x or k.

Once we have (24.3), we proceed as in Theorems 14.3 and 24.4. We let $\beta > 0$ and $\varepsilon > 0$ and we write $f = g + h$, where g is

continuous with compact support and $\|h\|_1 < \varepsilon$. The average of
g over $S_k(x)$ converges to $g(x)$ since g is continuous, while (24.3)
guarantees that the averages of h are small. Since the proof is so
similar to that of Theorems 14.3 and 24.4 we leave the details to
the reader. $\square$

Here is an important application of Theorem 24.4. Recall from
Chapter 22 that for $x \in \mathbb{R}^n$ and $y > 0$, we defined

$$P(x,y) = c_n \frac{y}{(|x|^2 + y^2)^{(n+1)/2}}, \qquad c_n = \frac{\Gamma((n+1)/2)}{\pi^{(n+1)/2}}. \qquad (24.4)$$

We will also write $P_y(x)$ for $P(x,y)$. We called P_y the *Poisson
kernel*. If $f \in L^p$ for some $1 \le p \le \infty$, define

$$u(x,y) = \int_{\mathbb{R}^n} P_y(t) f(x-t)\, dt.$$

u is called the *Poisson integral* of f and is sometime denoted by

$$P_y f(x).$$

u is also sometimes called the *harmonic extension* of f.

Obviously $P_1(x)$ as a function of x is radially symmetric, de-
creasing as a function of $|x|$, and $P_y(x) = y^{-d} P_1(x/y)$. Therefore
by Theorem 24.4 we see that $P_y f(x) \to f(x)$ a.e. if $x \in L^p$ for some
$1 \le p \le \infty$.

24.4   The Calderon-Zygmund lemma

The following theorem, known as the *Calderon-Zygmund lemma*, is
very important to the theory of singular integrals.

Theorem 24.6 *Suppose $f \ge 0$, f is integrable, and $\lambda > 0$. There
exists a closed set F such that*
(1) $f(x) \le \lambda$ almost everywhere on F.
*(2) F^c is the union of open disjoint cubes $\{Q_j\}$ such that for each
j,*

$$\lambda < \frac{1}{m(Q_j)} \int_{Q_j} f(x)\, dx \le 2^n \lambda. \qquad (24.5)$$

(3) $m(F^c) \le \|f\|_1/\lambda$.

Proof. Let $\mathcal{R}_k$ be defined as in Section 24.3. Choose k_0 to be a negative integer such that $\|f\|_1 2^{nk_0} \leq \lambda$. Then if $Q \in \mathcal{R}_{k_0}$,

$$\frac{1}{m(Q)} \int_Q f \leq \frac{\|f\|_1}{m(Q)} \leq \lambda.$$

The idea is that we look at each cube in $\mathcal{R}_{k_0}$ and divide it into 2^n equal subcubes. If the average of f over a subcube is greater than λ, then we include that subcube in our collection $\{Q_j\}$. If the average of a subcube is less than or equal to λ, we divide that subcube into 2^n further subcubes and look at the averages over these smaller subcubes.

To be precise, we proceed as follows. Let $\mathcal{Q}_{k_0} = \emptyset$. For $k > k_0$ we define $\mathcal{Q}_k$ inductively. Suppose we have defined $\mathcal{Q}_{k_0}, \ldots, \mathcal{Q}_{k-1}$. We let $\mathcal{Q}_k$ consist of those cubes R in $\mathcal{R}_k$ such that
(1) the average of f over R is greater than λ:

$$\frac{1}{m(R)} \int_R f > \lambda;$$

(2) R is not contained in any element of $\mathcal{Q}_{k_0}, \mathcal{Q}_{k_0+1}, \ldots, \mathcal{Q}_{k-1}$.

We then let $\{Q_j\}$ consist of the interiors of the cubes that are in $\cup_{k \geq k_0} \mathcal{Q}_k$. The Q_j are open cubes. For each k the interiors of the cubes in $\mathcal{R}_k$ are disjoint, while two cubes, one in $\mathcal{R}_k$ and the other in $\mathcal{R}_{k'}$ with $k \neq k'$, either have disjoint interiors or else one is contained in the other. The fact that we never chose a cube R in $\mathcal{R}_k$ that was contained in any of the cubes in $\cup_{i=k_0}^{k-1} \mathcal{Q}_i$ implies that the Q_j are disjoint.

Suppose R is one of the Q_j and its closure is in $\mathcal{R}_k$. Let S be the cube in $\mathcal{R}_{k-1}$ that contains R. Since R is one of the Q_j, then

$$\frac{1}{m(R)} \int_R f > \lambda.$$

Since S is not one of the Q_j (otherwise we would not have chosen R), then

$$\frac{1}{m(S)} \int_S f \leq \lambda.$$

From this we deduce

$$\frac{1}{m(R)} \int_R f = \frac{m(S)}{m(R)} \cdot \frac{1}{m(S)} \int_S f \leq 2^n \lambda.$$

Consequently (24.5) holds.

Let G be the union of the Q_j's and let $F = G^c$. G is open, so F is closed. If $x \in F$ and x is not on the boundary of any of the cubes in any of the $\mathcal{R}_k$, then there exists a sequence of cubes $S_i(x)$ decreasing to x with $S_i(x) \in \mathcal{R}_i$ for which $\frac{1}{m(S_i(x))} \int_{S_i(x)} f \leq \lambda$. By Theorem 24.5, for almost every such x we have

$$f(x) = \lim_{i \to \infty} \frac{1}{S_i(x)} \int_{S_i(x)} f(y)\, dy \leq \lambda.$$

Since the union of the boundaries of all the cubes has Lebesgue measure 0, (2) is proved.

We have $\int_{Q_j} f / m(Q_j) > \lambda$ for each j, hence

$$m(Q_j) < \frac{1}{\lambda} \int_{Q_j} f.$$

Since the Q_j are disjoint, then

$$m(F^c) = \sum_j m(Q_j) \leq \frac{1}{\lambda} \int_{\cup_j Q_j} f \leq \frac{\|f\|_1}{\lambda}.$$

This proves (3). $\square$

24.5   Hilbert transform

A function is in C^1 if it has a continuous derivative. If in addition f has compact support, we say that $f \in C_K^1$.

The *Hilbert transform* is an operator on functions defined by

$$Hf(x) = \lim_{\varepsilon \to 0, N \to \infty} \frac{1}{\pi} \int_{N > |y| > \varepsilon} \frac{f(x - y)}{y}\, dy. \qquad (24.6)$$

Of course, $1/y$ is not absolutely integrable, so even when f is identically constant, $\int_{N_1 > y > \varepsilon_1} dy/y$ will not have a limit as $\varepsilon_1 \to 0$ or $N_1 \to \infty$. It is important, therefore, to take integrals over symmetric intervals. Let us show the limit exists for each x if $f \in C_K^1$.

Proposition 24.7 *If $f \in C_K^1$,*

$$\lim_{\varepsilon \to 0, N \to \infty} \int_{N > |y| > \varepsilon} \frac{f(x-y)}{y} \, dy$$

exists for every x.

Proof. Fix x. Since f has compact support, $f(x-y)$ will be 0 if $|y|$ is large enough. Hence for each fixed ε we see that

$$\lim_{N \to \infty} \int_{N > |y| > \varepsilon} f(x-y)/y \, dy$$

exists. We now consider

$$\lim_{\varepsilon \to 0} \int_{|y| > \varepsilon} \frac{f(x-y)}{y} \, dy. \tag{24.7}$$

Observe that if $\varepsilon_1 < \varepsilon_2$, then

$$\int_{\varepsilon_2 \geq |y| > \varepsilon_1} \frac{f(x-y)}{y} \, dy = \int_{\varepsilon_2 \geq |y| > \varepsilon_1} \frac{f(x-y) - f(x)}{y} \, dy,$$

using the fact that $\int_{\varepsilon_2 \geq |y| > \varepsilon_1} dy/y = 0$ because $1/y$ is an odd function. By the mean value theorem, $|f(x-y) - f(x)| \leq \|f'\|_\infty |y|$, and so

$$\left| \int_{|y| > \varepsilon_1} \frac{f(x-y)}{y} \, dy - \int_{|y| > \varepsilon_2} \frac{f(x-y)}{y} \, dy \right|$$

$$\leq \int_{\varepsilon_2 \geq |y| > \varepsilon_1} \frac{|f(x-y) - f(x)|}{|y|} \, dy$$

$$\leq \|f'\|_\infty \int_{\varepsilon_2 \geq |y| > \varepsilon_1} dy$$

$$\leq 2|\varepsilon_2 - \varepsilon_1| \, \|f'\|_\infty.$$

Hence $\int_{|y| > \varepsilon} f(x-y)/y \, dy$ is a Cauchy sequence in ε. This implies that the limit in (24.7) exists. $\square$

The Hilbert transform turns out to be related to conjugate harmonic functions. To see the connection, let us first calculate the Fourier transform of Hf.

Remark 24.8 We will need the fact that the absolute value of $\int_a^b \frac{\sin x}{x}\,dx$ is bounded uniformly over $0 \le a < b < \infty$. To see this, since $\sin x / x \to 1$ as $x \to 0$, there is no difficulty as $a \to 0$ and it is enough to bound

$$\left| \int_0^b \frac{\sin x}{x}\,dx \right|.$$

If $A_k = \int_{k\pi}^{(k+1)\pi} \frac{\sin x}{x}\,dx$, then $\sum_k A_k$ is an alternating series with $A_k \to 0$ as $k \to \infty$, hence the series converges. This implies $\sum_{k=1}^N A_k$ is bounded in absolute value independently of N. Now if N is the largest integer less than or equal to b/π, write

$$\int_0^b \frac{\sin x}{x}\,dx = \sum_{k=1}^{N-1} A_k + \int_{(N-1)\pi}^b \frac{\sin x}{x}\,dx.$$

The last term is bounded in absolute value by A_N, and this proves the assertion.

Proposition 24.9 *If $f \in C_K^1$, then*

$$\widehat{Hf}(u) = i\,\mathrm{sgn}\,(u)\widehat{f}(u).$$

Proof. Let

$$H_{\varepsilon N}(x) = \frac{1}{\pi x}\chi_{(N>|x|>\varepsilon)}. \tag{24.8}$$

and let us look at $\widehat{H}_{\varepsilon N}$. Since $1/x$ is an odd function,

$$\int_{N>|x|>\varepsilon} \frac{e^{iux}}{x}\,dx = 2i \int_{N>x>\varepsilon} \frac{\sin(ux)}{x}\,dx.$$

This is 0 if u is 0, and is equal to $-2i \int_{N>x>\varepsilon} \sin(|u|x)/x\,dx$ if $u < 0$. Also

$$\int_{N>x>\varepsilon} \frac{\sin(|u|x)}{x}\,dx = \int_{|u|N>x>|u|\varepsilon} \frac{\sin x}{x}\,dx.$$

This converges to the value $\pi/2$ as $N \to \infty$ and $\varepsilon \to 0$; see Exercise 11.12. Moreover,

$$\sup_{\varepsilon,N} \left| \int_{N>|x|>\varepsilon} \frac{\sin(|u|x)}{x}\,dx \right| < \infty$$

by Remark 24.8. Therefore $\widehat{H}_{\varepsilon N}(u) \to i\,\mathrm{sgn}\,(u)$ pointwise and boundedly.

By the Plancherel theorem, Theorem 16.8,

$$\|H_{\varepsilon_1 N_1} f - H_{\varepsilon_2 N_2} f\|_2^2 = (2\pi)^{-1} \int |\widehat{H}_{\varepsilon_1 N_1}(u) - \widehat{H}_{\varepsilon_2 N_2}(u)|^2 |\widehat{f}(u)|^2 du.$$

This tends to 0 as $\varepsilon_1, \varepsilon_2 \to 0$ and $N_1, N_2 \to \infty$ by the dominated convergence theorem and the fact that $\|\widehat{f}\|_2 = (2\pi)^{1/2} \|f\|_2 < \infty$. Therefore $H_{\varepsilon N} f$ converges in L^2 as $\varepsilon \to 0$ and $N \to \infty$. Since $H_{\varepsilon N} f$ converges pointwise to Hf by Proposition 24.7, it converges to Hf in L^2. By the Plancherel theorem again, the Fourier transform of $H_{\varepsilon N} f$ converges in L^2 to the Fourier transform of Hf. The Fourier transform of $H_{\varepsilon N} f$ is $\widehat{H}_{\varepsilon N}(u) \widehat{f}(u)$, which converges pointwise to $i \operatorname{sgn}(u) \widehat{f}(u)$. □

Proposition 24.10 *Suppose $f \in C_K^1$. Let U be the harmonic extension of f and let V be the harmonic extension of Hf. Then U and V are conjugate harmonic functions.*

Proof. We will show that U and V satisfy the Cauchy-Riemann conditions by looking at their Fourier transforms. By Exercise 24.4, $\widehat{P}_y(u)$, the Fourier transform of the Poisson kernel in x with y held fixed, is $e^{-y|u|}$.

In each of the formulas below the Fourier transform is in the x variable only, with y considered to be fixed. We have $U(x, y) = (P_y f)(x)$. Then the Fourier transform of U is

$$\widehat{U}(u, y) = \widehat{P}_y(u) \widehat{f}(u) = e^{-y|u|} \widehat{f}(u). \tag{24.9}$$

Also, by Exercise 16.4,

$$\widehat{\frac{\partial U}{\partial x}}(u, y) = iu\widehat{U}(u, y) = iue^{-y|u|} \widehat{f}(u) \tag{24.10}$$

and

$$\widehat{\frac{\partial U}{\partial y}}(u, y) = -|u|e^{-y|u|} \widehat{f}(u). \tag{24.11}$$

We obtain (24.11) by differentiating (24.9). Similarly,

$$\widehat{V}(u, y) = e^{-y|u|} \widehat{Hf}(u) = i \operatorname{sgn}(u) e^{-y|u|} \widehat{f}(u),$$

hence

$$\widehat{\frac{\partial V}{\partial x}}(u, y) = iue^{-y|u|} i \operatorname{sgn}(u) \widehat{f}(u) \tag{24.12}$$

and

$$\widehat{\frac{\partial V}{\partial y}}(u, y) = -|u|e^{-y|u|}i\operatorname{sgn}(u)\widehat{f}(u) \qquad (24.13)$$

Comparing (24.10) with (24.13) and (24.10) with (24.12) and using the inversion theorem for Fourier transforms (Theorem 16.7), we see that the Cauchy-Riemann equations hold for almost all pairs (x, y). Since $P_y(x)$ is continuous in x and y for $y > 0$, then U and V are both continuous, and hence the Cauchy-Riemann equations hold everywhere.   □

24.6 L^p boundedness

Throughout this section we let m be one-dimensional Lebesgue measure. We say a function K satisfies the *Hörmander condition* if

$$\int_{|x|\geq 2|y|} |K(x - y) - K(x)|\,dx \leq c_1, \qquad |y| > 0,$$

where c_1 does not depend on y.

As an example, consider the Hilbert transform. Here $K(x) = 1/\pi x$, and

$$|K(x - y) - K(x)| = \frac{1}{\pi}\left|\frac{1}{x - y} - \frac{1}{x}\right| = \frac{1}{\pi}\left|\frac{y}{x(x - y)}\right|.$$

If $|x| > 2|y|$, then $|x - y| \geq |x|/2$, so the right hand side is bounded by $2|y|/\pi|x|^2$. Then

$$\int_{|x|\geq 2|y|} |K(x - y) - K(x)|\,dx \leq \frac{2}{\pi}|y| \int_{|x|>2|y|} \frac{1}{|x|^2}\,dx \leq \frac{2}{\pi}.$$

Theorem 24.11 *Define $Hf(x)$ by (24.6) when $f \in C_K^1$. If $1 < p < \infty$, then there exists a constant c_p such that*

$$\|Hf\|_p \leq c_p\|f\|_p$$

for $f \in C_K^1$.

Since the C_K^1 functions are dense in L^p, we can use this theorem to extend the definition of H to all of L^p as follows. If $f \in L^p$, choose $f_m \in C_K^1$ such that $\|f - f_m\|_p \to 0$. Then

$$\|Hf_m - Hf_k\|_p \le c_p\|f_m - f_k\|_p \to 0,$$

and we let Hf be the limit of the Cauchy sequence Hf_m. If $\{f_m\}$ and $\{g_m\}$ are two sequences of C_K^1 functions converging to f in L^p, then

$$\|Hf_m - Hg_m\|_p \le c_p\|f_m - g_m\|_p \to 0.$$

Thus the definition of Hf is independent of which sequence we choose to approximate f by.

This theorem is not true for $p = \infty$. Let $f = \chi_{[0,1]}$. If $x < 0$,

$$\int \frac{f(x-y)}{y}\, dy = \int_{x-1}^{x} \frac{dy}{y} = \log\left|\frac{x}{x-1}\right|,$$

which is not bounded on $[-1, 0)$. Exercise 24.6 shows that the theorem is not true for $p = 1$ either.

Proof. *Step 1: $p = 2$.* Let

$$K(x) = \frac{1}{\pi x}\chi_{(N > |x| > \varepsilon)}$$

and define $Tf(x) = \int K(x-y)f(y)\, dy$. By the proof of Proposition 24.9 we saw that $\widehat{K}$ is bounded in absolute value by a constant c_1 not depending on N or ε. Moreover, Exercise 24.8 asks you to show that K satisfies Hörmander's condition with a constant c_2 not depending on ε or N. By the Plancherel theorem, Theorem 16.8,

$$\|Tf\|_2 = (2\pi)^{-1/2}\|\widehat{Tf}\|_2 = (2\pi)^{-1/2}\|\widehat{K}\widehat{f}\|_2$$
$$\le c_1(2\pi)^{-1/2}\|\widehat{f}\|_2 = c_1\|f\|_2.$$

Step 2: $p = 1$. We want to show that T is of weak-type 1-1. Fix λ and use the Calderon-Zygmund lemma for $|f|$. We thus have disjoint open intervals Q_j with $G = \cup_j Q_j$ and a closed set $F = G^c$ on which $|f| \le \lambda$ a.e. Also, $m(G) \le \|f\|_1/\lambda$.

Define

$$g(x) = \begin{cases} f(x), & x \in F; \\ \frac{1}{m(Q_j)}\int_{Q_j} f(x)\, dx, & x \in Q_j. \end{cases}$$

Note $|g(x)| \leq \lambda$ a.e. on F and

$$|g(x)| \leq \frac{1}{m(Q_j)} \int_{Q_j} |f(y)|\,dy \leq 2\lambda$$

on Q_j. Let $h = f - g$. Then $h(x) = 0$ on F and

$$\int_{Q_j} h(x)\,dx = \int_{Q_j} f(x)\,dx - \int_{Q_j} \left[\frac{1}{m(Q_j)} \int_{Q_j} f(y)\,dy \right] dx = 0$$

for each Q_j.

We have $Tf = Tg + Th$, so

$$m(\{x : |Tf(x)| > \lambda\}) \leq m(\{x : |Tg(x)| > \lambda/2\})$$
$$+ m(\{x : |Th(x)| > \lambda/2\}).$$

If we show

$$m(\{x : |Tg(x)| > \lambda/2\}) \leq \frac{c_3}{\lambda} \|f\|_1$$

for some constant c_3 with a similar estimate for Th, then we will have that T is of weak-type 1-1.

Step 3. We look at Tg. Since $|g| = |f| \leq \lambda$ on F and

$$\frac{1}{m(Q_j)} \int_{Q_j} |f(x)|\,dx \leq 2\lambda$$

by Theorem 24.6, we have

$$\begin{aligned}
\|g\|_2^2 &= \int_F g^2 + \int_G g^2 \\
&= \int_F f^2 + \sum_j \int_{Q_j} \left[\frac{1}{m(Q_j)} \int_{Q_j} f(y)\,dy \right]^2 dx \\
&\leq \lambda \int_F |f(x)|\,dx + 4 \sum_j m(Q_j)\lambda^2 \\
&\leq \lambda\|f\|_1 + 4\lambda^2 m(G) \\
&\leq \left(\lambda + 4\lambda^2 \frac{1}{\lambda}\right) \|f\|_1 \\
&= 5\lambda\|f\|_1.
\end{aligned}$$

Therefore

$$m(\{x : |Tg(x)| > \lambda/2\}) \leq m(\{x : |Tg(x)|^2 > \lambda^2/4\})$$
$$\leq \frac{4}{\lambda^2}\|Tg\|_2^2 \leq \frac{c_1}{\lambda_2}\|g\|_2^2$$
$$\leq \frac{c_2\lambda}{\lambda^2}\|f\|_1 = \frac{c_2\|f\|_1}{\lambda}$$

for some constants c_1, c_2. This provides the required inequality for Tg.

Step 4. We now turn to Th. Define $h_j(x)$ to be equal to $h(x)$ if $x \in Q_j$ and to be zero otherwise. Let Q_j^* be the interval with the same center as Q_j but with length twice as long. Let $G^* = \cup_j Q_j^*$ and $F^* = (G^*)^c$. Note that

$$m(G^*) \leq \sum_j m(Q_j^*) = 2\sum_j m(Q_j) = 2m(G).$$

If y_j is the center of Q_j, r_j is the length of Q_j, $x \notin Q_j^*$, and $y \in Q_j$, then

$$|x - y_j| \geq r_j \geq 2|y - y_j|.$$

Since $\int h_j(y)\,dy = \int_{Q_j} h(y)\,dy = 0$,

$$|Th_j(x)| = \left|\int K(x-y)h_j(y)\,dy\right|$$
$$= \left|\int [K(x-y) - K(x-y_j)]h_j(y)\,dy\right|$$
$$\leq \int_{Q_j} |K(x-y) - K(x-y_j)|\,|h_j(y)|\,dy.$$

Therefore, since $F^* = (G^*)^c = \cap_j(Q_j^*)^c \subset (Q_j^*)^c$ for each j,

$$\int_{F^*} |Th(x)|\,dx \leq \sum_j \int_{F^*} |Th_j(x)|\,dx$$
$$\leq \sum_j \int_{(Q_j^*)^c} |Th_j(x)|\,dx$$
$$\leq \sum_j \int_{(Q_j^*)^c} \int_{Q_j} |K(x-y) - K(x-y_j)|\,|h_j(y)|\,dy\,dx$$
$$= \sum_j \int_{Q_j} \int_{(Q_j^*)^c} |K(x-y) - K(x-y_j)|\,dx\,|h_j(y)|\,dy$$

$$\leq \sum_j \int_{Q_j} \int_{|x'| \geq 2|y - y_j|} |K(x' - (y - y_j)) - K(x')| \, dx' |h_j(y)| \, dy$$

$$\leq c_2 \sum_j \int_{Q_j} |h_j(y)| \, dy.$$

In the next to the last inequality we made the substitution $x' = x - y_j$ and in the last inequality we used the fact that K satisfies Hörmander's condition.

Now $h = h_j$ on Q_j and $h = f - g$, so

$$\int_{Q_j} |h(y)| \, dy \leq \int_{Q_j} |f(y)| \, dy + \int_{Q_j} \left[\frac{1}{m(Q_j)} \int_{Q_j} |f(x)| \, dx \right] dy$$

$$= 2 \int_{Q_j} |f(y)| \, dy.$$

We therefore conclude

$$\int_{F^*} |Th(x)| \leq 2c_2 \sum_j \int_{Q_j} |f(y)| dy \leq 2c_2 \|f\|_1.$$

By the Chebyshev inequality,

$$m(\{x \in F^* : |Th(x)| > \lambda/2\}) = m(\{x : |Th(x)|\chi_{F^*}(x) > \lambda/2\})$$

$$\leq \frac{\int |Th(x)|\chi_{F^*}(x) \, dx}{\lambda/2}$$

$$= \frac{2}{\lambda} \int_{F^*} |Th(x)| \, dx$$

$$\leq \frac{2c_2}{\lambda} \|f\|_1.$$

We also have

$$m(\{x \in G^* : |Th(x)| > \lambda/2\}) \leq m(G^*) \leq 2m(G) \leq \frac{2}{\lambda} \|f\|_1.$$

Combining gives the required inequality for Th.

Step 5: $1 < p < 2$. The operator T is linear, is of weak-type 1-1 and is bounded on L^2. By the Marcinkiewicz interpolation theorem, T is bounded on $L^1 \cap L^p$ for $1 < p < 2$.

Step 6: $2 < p < \infty$. We obtain the boundedness of T on L^p for $p > 2$ by a duality argument. Suppose $p > 2$ and choose q so that

$p^{-1} + q^{-1} = 1$. If $f \in L^1 \cap L^p$,

$$\|Tf\|_p = \sup \left\{ \int g(y)(Tf)(y)\, dy : \|g\|_q, g \in C_K^1 \right\}$$

since the C_K^1 functions are dense in L^q. But

$$\int g(Tf)dy = \int g(y)K(y-x)f(x)\, dx\, dy = -\int f(x)(Tg)(x)\, dx.$$

By Hölder's inequality, the absolute value of the last integral is less than or equal to

$$\|f\|_p \|Tg\|_q \le c_4 \|f\|_p \|g\|_q$$

for a constant c_4. The inequality here follows from Step 5 since $1 < q < 2$. Taking the supremum over $g \in L^q \cap C_K^1$ with $\|g\|_q \le 1$, we obtain by the duality of L^p and L^q that

$$\|Tf\|_p \le c_4 \|f\|_p.$$

Step 7. We have proved the boundedness of Tf where we chose ε and N and then left them fixed, and obtained

$$\|Tf\|_p \le c_p \|f\|_p, \tag{24.14}$$

where c_p does not depend on ε or N. If we apply (24.14) for $f \in C_K^1$ and let $\varepsilon \to 0$ and $N \to \infty$, we obtain by Fatou's lemma that

$$\|Hf\|_p \le c_p \|f\|_p.$$

This is what we were required to prove. □

Remark 24.12 An examination of the proof shows that the essential properties of the Hilbert transform that were used are that its Fourier transform is bounded, that it satisfies Hörmander's condition, and that Hf exists for a sufficiently large class of functions. Almost the same proof shows L^p boundedness for a much larger class of operators. See [9] for an exposition of further results in this theory.

24.7   Exercises

Exercise 24.1 Let $f \in L^p(\mathbb{R}^n)$ for some $1 \leq p \leq \infty$. Define $\Gamma(x)$ to be the cone in $\mathbb{R}^n \times [0, \infty)$ defined by

$$\Gamma(x) = \{(z, y) : z \in \mathbb{R}^n, y \in (0, \infty), |z - x| < y\}.$$

(1) Prove that there exists a constant c_1 such that

$$\sup_{(z,y)\in\Gamma(x)} |P_y f(x)| \leq c_1 M f(x).$$

(2) Prove that

$$\lim_{(z,y)\in\Gamma(x),(z,y)\to x} P_y f(x) = f(x), \qquad \text{a.e.}$$

The assertion in (2) is known as *Fatou's theorem* and the convergence is called *nontangential convergence*.

Exercise 24.2 Let A be a bounded open set in $\mathbb{R}^n$ and let $rA+x = \{ry + x : y \in A\}$ for $r > 0$ and $x \in \mathbb{R}^n$. Suppose f is an integrable function on $\mathbb{R}^n$ and m is Lebesgue measure. Prove that

$$\lim_{r\to 0} \frac{1}{m(rA)} \int_{rA+x} f(y)\,dy = f(x), \qquad \text{a.e.}$$

Exercise 24.3 Suppose f is in $L^p(\mathbb{R}^2)$ for $1 < p < \infty$. Let R_{hk} be a rectangle whose sides are parallel to the x and y axes, whose base is h, and whose height is h. Prove that

$$\lim_{h,k\to 0} \frac{1}{hk} \int_{R_{hk}+x} f(y)\,dy = f(x), \qquad \text{a.e.,}$$

where $R_{hk} + x = \{y + x : y \in R_{hk}\}$.

Exercise 24.4 Let P_y be defined by (24.4) where we take the dimension n to be 1.
(1) Prove that the Fourier transform of $e^{-|x|}$ is a constant times $1/(1 + u^2)$.
(2) Fix $y > 0$. Prove that the Fourier transform of $h(x) = P(x, y)$ is $e^{-|u|y}$.

Exercise 24.5 Prove that if $y_1, y_2 > 0$ and $f \in L^p$ for some $1 \le p \le \infty$, then $P_{y_1}(P_{y_2} f) = P_{y_1+y_2} f$. This shows that the family $\{P_y\}$ is an example of what is known as a *semigroup* of operators.

Exercise 24.6 Let P_y be defined by (24.4) where we take the dimension n to be 1.
(1) Prove that the Hilbert transform of P_y is equal to the function $x/(x^2 + y^2)$.
(2) Conclude that the Hilbert transform is not a bounded operator on $L^1(\mathbb{R})$.

Exercise 24.7 Suppose that $f \in C_K^1$ and $f \ge 0$. Prove that the Hilbert transform of f is not in $L^1(\mathbb{R})$.

Exercise 24.8 Let $H_{\varepsilon N}$ be defined by (24.8). Prove that $H_{\varepsilon N}$ satisfies Hörmander's condition with a constant not depending on ε or N.

Exercise 24.9 Let f be a C_K^1 function on $\mathbb{R}^n$ and for $1 \le j \le n$ prove that the limit

$$R_j f = \lim_{\varepsilon \to 0, N \to \infty} \int_{\varepsilon < |x| < N} \frac{y_j}{|y|} f(x - y) \, dy$$

exists. R_j is a constant multiple of the jth *Riesz transform*.

Exercise 24.10 Show that the Fourier transform of R_j defined in Exercise 24.9 is a constant multiple of $u_j/|u|$.

Exercise 24.11 Prove that R_j defined in Exercise 24.9 satisfies

$$\|R_f\|_p \le c_p \|f\|_p$$

for all C_K^1 functions f, where $1 \le j \le n$, $1 < p < \infty$, and c_p depends only on n and p.

Exercise 24.12 Suppose $1 < p < \infty$ and $1 \le i, j \le n$.
(1) Prove there exists a constant c_0 such that if f is C^∞ with compact support, then

$$\frac{\partial^2 f}{\partial x_i \partial x_j} = c_0 R_i R_j \Delta f,$$

where R_i and R_j are Riesz transforms, and $\Delta f = \sum_{k=1}^{d} \frac{\partial^2 f}{\partial x_k^2}$ is the Laplacian.

(2) Prove there exists a constant c_p such that

$$\left\| \frac{\partial^2 f}{\partial x_i \partial x_j} \right\|_p \leq c_p \|\Delta f\|_p$$

for $f \in C^2$ with compact support.

Exercise 24.13 Suppose K is an odd function, $K \in L^2(\mathbb{R})$, we have $|K(x)| \leq 1/|x|$ for each $x \neq 0$, and K satisfies Hörmander's condition. Prove that $\widehat{K}$ is bounded.

Exercise 24.14 Let $f \in L^2(\mathbb{R})$ and set $u(x, y) = P_y f(x)$, where $P_y f$ is the Poisson integral of f. Define

$$g(f)(x) = \left(\int_0^\infty y |\nabla u(x, y)|^2 \, dy \right)^{1/2}.$$

Here

$$|\nabla u|^2 = \left| \frac{\partial u}{\partial y} \right|^2 + \sum_{k=1}^{n} \left| \frac{\partial u}{\partial x_k} \right|^2.$$

$g(f)$ is one of a class of functions known as *Littlewood-Paley functions*. Prove there exists a constant c_1 such that

$$\|f\|_2 = c_1 \|g(f)\|_2.$$

Chapter 25

Spectral theory

An important theorem from undergraduate linear algebra says that a symmetric matrix is diagonalizable. Our goal in this chapter is to give the infinite dimensional analog of this theorem.

We first consider compact symmetric operators on a Hilbert space. We prove in this case that there is an orthonormal basis of eigenvectors. We apply this to an example that arises from partial and ordinary differential equations.

We then turn to general bounded symmetric operators on a Hilbert space. We derive some properties of the spectrum of such operators, give the spectral resolution, and prove the spectral theorem.

Throughout this chapter we assume our Hilbert space is separable, that is, there exists a countable dense subset. We also only consider Hilbert spaces whose scalar field is the complex numbers.

25.1   Bounded linear operators

Let H be a Hilbert space over the complex numbers. Recall that a linear operator $A : H \to H$ is bounded if

$$\|A\| = \sup\{\|Ax\| : \|x\| \leq 1\}$$

is finite. Given two linear operators A and B and a complex number c, we define $(A + B)(x) = Ax + Bx$ and $(cA)(x) = cA(x)$. The set

$\mathcal{L}$ of all bounded linear operators from H into H is a linear space, and by Exercise 18.7, this linear space is a Banach space. We can define the composition of two operators A and B by

$$(AB)(x) = A(Bx).$$

Note that

$$\|(AB)(x)\| = \|A(Bx)\| \le \|A\| \, \|Bx\| \le \|A\| \, \|B\| \, \|x\|,$$

and we conclude that

$$\|AB\| \le \|A\| \, \|B\|. \tag{25.1}$$

In particular, $\|A^2\| \le (\|A\|)^2$ and $\|A^i\| \le (\|A\|)^i$.

The operator of composition is not necessarily commutative (think of the case where H is $\mathbb{C}^n$ and A acts by matrix multiplication), but it is associative and the distributive laws hold; see Exercise 25.1. The space of bounded linear operators from a Banach space to itself is an example of what is known as a *Banach algebra*, but that is not important for what follows.

Let A be a bounded linear operator. If z is a complex number and I is the identity operator, then $zI - A$ is a bounded linear operator on H which might or might not be invertible. We define the *spectrum* of A by

$$\sigma(A) = \{z \in \mathbb{C} : zI - A \text{ is not invertible}\}.$$

We sometimes write $z - A$ for $zI - A$. The *resolvent set* for A is the set of complex numbers z such that $z - A$ is invertible. We define the *spectral radius* of A by

$$r(A) = \sup\{|z| : z \in \sigma(A)\}.$$

We say $x \in H$ with $x \ne 0$ is an *eigenvector* corresponding to an *eigenvalue* $\lambda \in \mathbb{C}$ if $Ax = \lambda x$. If λ is an eigenvalue, then $\lambda \in \sigma(A)$. This follows since $(\lambda - A)x = 0$ while $x \ne 0$, so $\lambda - A$ is not one-to-one.

The converse is not true, that is, not every element of the spectrum is necessarily an eigenvalue.

Example 25.1 Let $H = \ell^2$ and let e_n be the sequence in ℓ^2 which has all coordinates equal to 0 except for the nth one, where the coordinate is equal to 1. Define

$$A(a_1, a_2, a_3, \ldots) = (a_1, a_2/2, a_3/3, \ldots).$$

Clearly A is a bounded operator. On the one hand, A does not have an inverse, for if it did, then $A^{-1}e_n = ne_n$, and A^{-1} would not be a bounded operator. Therefore $0I - A$ is not invertible, or 0 is in the spectrum. On the other hand, we do not have $Ax = 0x$ for any non-zero x, so 0 is not an eigenvalue.

Proposition 25.2 *If B is a bounded linear operator from H to H with $\|B\| < 1$, then $I - B$ is invertible and*

$$(I - B)^{-1} = \sum_{i=0}^{\infty} B^i, \tag{25.2}$$

with the convention that $B^0 = I$.

Proof. If $\|B\| < 1$, then

$$\left\| \sum_{i=m}^{n} B^i \right\| \leq \sum_{i=m}^{n} \|B^i\| \leq \sum_{i=m}^{n} \|B\|^i,$$

which shows that $S_n = \sum_{i=0}^{n} B^i$ is a Cauchy sequence in the Banach space $\mathcal{L}$ of linear operators from H to H. By the completeness of this space, S_n converges to $S = \sum_{i=0}^{\infty} B^i$. We see that $BS = \sum_{i=1}^{\infty} B^i = S - I$, so $(I - B)S = I$. Similarly we show that $S(I - B) = I$. $\qquad \square$

Proposition 25.3 *If A is an invertible bounded linear operator from H to H and B is a bounded linear operator from H to H with $\|B\| < 1/\|A^{-1}\|$, then $A - B$ is invertible.*

Proof. We have $\|A^{-1}B\| \leq \|A^{-1}\| \|B\| < 1$, so by Proposition 25.2 we know that $I - A^{-1}B$ is invertible. If M and N are invertible, then

$$(N^{-1}M^{-1})(MN) = I = (MN)(N^{-1}M^{-1}),$$

so MN is invertible. We set $M = A$ and $N = I - A^{-1}B$, and conclude that $A - B$ is invertible. $\qquad \square$

Proposition 25.4 *If A is a bounded linear operator from H into H, then $\sigma(A)$ is a closed and bounded subset of $\mathbb{C}$ and $r(A) \leq \|A\|$.*

Proof. If $z \notin \sigma(A)$, then $zI - A$ is invertible. By Proposition 25.3, if $|w - z|$ is small enough, then

$$wI - A = (zI - A) - (z - w)I$$

will be invertible, and hence $w \notin \sigma(A)$. This proves that $\sigma(A)^c$ is open, and we conclude that $\sigma(A)$ is closed.

We know from (25.2) that

$$(zI - A)^{-1} = z^{-1}(I - Az^{-1})^{-1} = \sum_{n=0}^{\infty} A^n z^{-n-1}$$

converges if $\|Az^{-1}\| < 1$, or equivalently, $|z| > \|A\|$. In other words, if $|z| > \|A\|$, then $z \notin \sigma(A)$. Hence the spectrum is contained in the closed ball in H of radius $\|A\|$ centered at 0. $\quad\square$

If A is a bounded operator on H, the *adjoint* of A, denoted A^*, is the operator on H such that $\langle Ax, y \rangle = \langle x, A^*y \rangle$ for all x and y.

It follows from the definition that the adjoint of cA is $\bar{c}A^*$ and the adjoint of A^n is $(A^*)^n$. If $P(x) = \sum_{j=0}^{n} a_j x^j$ is a polynomial, the adjoint of $P(A) = \sum_{j=0}^{n} a_j A^j$ will be

$$\overline{P}(A^*) = \sum_{j=0}^{n} \bar{a}_j P(A^*).$$

The adjoint operator always exists.

Proposition 25.5 *If A is a bounded operator on H, there exists a unique operator A^* such that $\langle Ax, y \rangle = \langle x, A^*y \rangle$ for all x and y.*

Proof. Fix y for the moment. The function $f(x) = \langle Ax, y \rangle$ is a linear functional on H. By the Riesz representation theorem for Hilbert space, Theorem 19.9, there exists z_y such that $\langle Ax, y \rangle = \langle x, z_y \rangle$ for all x. Since

$$\langle x, z_{y_1+y_2} \rangle = \langle Ax, y_1 + y_2 \rangle = \langle Ax, y_1 \rangle + \langle Ax, y_2 \rangle = \langle x, z_{y_1} \rangle + \langle x, z_{y_2} \rangle$$

for all x, then $z_{y_1+y_2} = z_{y_1} + z_{y_2}$ and similarly $z_{cy} = cz_y$. If we define $A^*y = z_y$, this will be the operator we seek.

If A_1 and A_2 are two operators such that $\langle x, A_1y \rangle = \langle Ax, y \rangle = \langle x, A_2y \rangle$ for all x and y, then $A_1y = A_2y$ for all y, so $A_1 = A_2$. Thus the uniqueness assertion is proved. $\qquad \square$

25.2 Symmetric operators

A bounded linear operator A mapping H into H is called *symmetric* if

$$\langle Ax, y \rangle = \langle x, Ay \rangle \tag{25.3}$$

for all x and y in H. Other names for symmetric are *Hermitian* or *self-adjoint*. When A is symmetric, then $A^* = A$, which explains the name "self-adjoint."

Example 25.6 For an example of a symmetric bounded linear operator, let $(X, \mathcal{A}, \mu)$ be a measure space with μ a σ-finite measure, let $H = L^2(X)$, and let $F(x, y)$ be a jointly measurable function from $X \times X$ into $\mathbb{C}$ such that $F(y, x) = \overline{F(x, y)}$ and

$$\int \int F(x, y)^2 \, \mu(dx) \, \mu(dy) < \infty. \tag{25.4}$$

Define $A : H \to H$ by

$$Af(x) = \int F(x, y)f(y) \, \mu(dy). \tag{25.5}$$

Exercise 25.4 asks you to verify that A is a bounded symmetric operator.

We have the following proposition.

Proposition 25.7 *Suppose A is a bounded symmetric operator.*
(1) (Ax, x) is real for all $x \in H$.
(2) The function $x \to (Ax, x)$ is not identically 0 unless $A = 0$.
(3) $\|A\| = \sup_{\|x\|=1} |\langle Ax, x \rangle|$.

Proof. (1) This one is easy since

$$\langle Ax, x \rangle = \langle x, Ax \rangle = \overline{\langle Ax, x \rangle},$$

where we use $\bar{z}$ for the complex conjugate of z.

(2) If $\langle Ax, x \rangle = 0$ for all x, then

$$0 = \langle A(x+y), x+y \rangle = \langle Ax, x \rangle + \langle Ay, y \rangle + \langle Ax, y \rangle + \langle Ay, x \rangle$$
$$= \langle Ax, y \rangle + \langle y, Ax \rangle = \langle Ax, y \rangle + \overline{\langle Ax, y \rangle}.$$

Hence $\mathrm{Re}\,\langle Ax, y \rangle = 0$. Replacing x by ix and using linearity,

$$\mathrm{Im}\,(\langle Ax, y \rangle) = -\mathrm{Re}\,(i\langle Ax, y \rangle) = -\mathrm{Re}\,(\langle A(ix), y \rangle) = 0.$$

Therefore $\langle Ax, y \rangle = 0$ for all x and y. We conclude $Ax = 0$ for all x, and thus $A = 0$.

(3) Let $\beta = \sup_{\|x\|=1} |\langle Ax, x \rangle|$. By the Cauchy-Schwarz inequality,

$$|\langle Ax, x \rangle| \le \|Ax\|\,\|x\| \le \|A\|\,\|x\|^2,$$

so $\beta \le \|A\|$.

To get the other direction, let $\|x\| = 1$ and let $y \in H$ such that $\|y\| = 1$ and $\langle y, Ax \rangle$ is real. Then

$$\langle y, Ax \rangle = \tfrac{1}{4}(\langle x+y, A(x+y) \rangle - \langle x-y, A(x-y) \rangle).$$

We used that $\langle y, Ax \rangle = \langle Ay, x \rangle = \langle Ax, y \rangle = \langle x, Ay \rangle$ since $\langle y, Ax \rangle$ is real and A is symmetric. Then

$$16|\langle y, Ax \rangle|^2 \le \beta^2 (\|x+y\|^2 + \|x-y\|^2)^2$$
$$= 4\beta^2 (\|x\|^2 + \|y\|^2)^2$$
$$= 16\beta^2.$$

We used the parallelogram law (equation (19.1)) in the first equality. We conclude $|\langle y, Ax \rangle| \le \beta$.

If $\|y\| = 1$ but $\langle y, Ax \rangle = re^{i\theta}$ is not real, let $y' = e^{-i\theta}y$ and apply the above with y' instead of y. We then have

$$|\langle y, Ax \rangle| = |\langle y', Ax \rangle| \le \beta.$$

Setting $y = Ax/\|Ax\|$, we have $\|Ax\| \le \beta$. Taking the supremum over x with $\|x\| = 1$, we conclude $\|A\| \le \beta$. □

25.3 Compact symmetric operators

Let H be a separable Hilbert space over the complex numbers and let B_1 be the open unit ball in H. We say that K is a *compact operator* from H to itself if the closure of $K(B_1)$ is compact in H.

Example 25.8 The identity operator on $H = \ell^2$ is not compact; see Exercise 19.6.

Example 25.9 Let $H = \ell^2$, let $n > 0$, let $\lambda_1, \ldots, \lambda_n$ be complex numbers, and define

$$K(a_1, a_2, \ldots,) = (\lambda_1 a_1, \lambda_2 a_2, \ldots, \lambda_n a_n, 0, 0, \ldots).$$

Then $\overline{K(B_1)}$ is contained in $F = E \times \{(0, 0, \ldots)\}$, where $E = \prod_{i=1}^{n} \overline{B(0, \lambda_i)}$. The set F is homeomorphic (in the topological sense) to E, which is a closed and bounded subset of $\mathbb{C}^n$. Since $\mathbb{C}^n$ is topologically the same as $\mathbb{R}^{2n}$, the Heine-Borel theorem says that E is compact, hence F is also. Closed subsets of compact sets are compact, so the closure of $K(B_1)$ is compact.

Before we can give other examples of compact operators, we need a few facts.

Proposition 25.10 *(1) If K_1 and K_2 are compact operators and c is a complex number, then $cK_1 + K_2$ is a compact operator.*
(2) If L is a bounded linear operator from H to H and K is a compact operator, then KL and LK are compact operators.
(3) If K_n are compact operators and $\lim_{n \to \infty} \|K_n - K\| = 0$, then K is a compact operator.

Proof. (1) Since $(K_1 + K_2)(B_1) \subset K_1(B_1) + K_2(B_1)$, (1) follows by Exercise 25.5. The proof for cK_1 is even easier.

(2) The closure of $ML(B_1)$ will be compact because the closure of $L(B_1)$ is compact and M is a continuous function, hence M maps compact sets into compact sets.

$L(B_1)$ will be contained in the ball $B(0, \|L\|)$. Then $ML(B_1)$ will be contained in $\|L\| M(B_1)$, and the closure of this set is compact.

(3) Recall from Theorem 20.23 that a subset A of a metric space is compact if and only if it complete and totally bounded. Saying A is totally bounded means that given $\varepsilon > 0$, A can be covered by finitely many balls of radius ε. Let $\varepsilon > 0$. Choose n such that $\|K_n - K\| < \varepsilon/2$. Since K_n is compact, the closure of $K_n(B_1)$ can be covered by finitely many balls of radius $\varepsilon/2$. Hence the closure of $K(B_1)$ can be covered by the set of balls with the same centers but with radius ε. Therefore the closure of $K(B_1)$ is totally bounded. Since H is a Hilbert space, it is complete. We know that closed subsets of complete metric spaces are complete. Hence the closure of $K(B_1)$ is complete and totally bounded, so is compact.   □

We now give an example of a non-trivial compact operator.

Example 25.11 Let $H = \ell^2$ and let

$$K(a_1, a_2, a_3, \ldots) = (a_1/2, a_2/2^2, a_3/2^3, \ldots).$$

Note K is the limit in norm of K_n, where

$$K_n(a_1, a_2, \ldots) = (a_1/2, a_2/2^2, \ldots, a_n/2^n, 0, 0, \ldots).$$

Each K_n is compact by Example 25.9. By Proposition 25.10, K is a compact operator.

Here is another interesting example of a compact symmetric operator.

Example 25.12 Let $H = L^2([0, 1])$ and let $F : [0, 1]^2 \to \mathbb{R}$ be a continuous function with $F(x, y) = F(y, x)$ for all x and y. Define $K : H \to H$ by

$$Kf(x) = \int_0^1 F(x, y)f(y)\, dy.$$

We discussed in Example 25.6 the fact that K is a bounded symmetric operator. Let us show that it is compact.

If $f \in L^2([0, 1])$ with $\|f\| \le 1$, then

$$|Kf(x) - Kf(x')| = \left| \int_0^1 [F(x, y) - F(x', y)]f(y)\, dy \right|$$

$$\le \left(\int_0^1 |F(x, y) - F(x', y)|^2\, dy \right)^{1/2} \|f\|,$$

using the Cauchy-Schwarz inequality. Since F is continuous on $[0,1]^2$, which is a compact set, then it is uniformly continuous there. Let $\varepsilon > 0$. There exists δ such that

$$\sup_{|x-x'|<\delta} \sup_y |F(x,y) - F(x',y)| < \varepsilon.$$

Hence if $|x - x'| < \delta$, then $|Kf(x) - Kf(x')| < \varepsilon$ for every f with $\|f\| \leq 1$. In other words, $\{Kf : \|f\| \leq 1\}$ is an equicontinuous family.

Since F is continuous, it is bounded, say by N, and therefore

$$|Kf(x)| \leq \int_0^1 N|f(y)|\, dy \leq N\|f\|,$$

again using the Cauchy-Schwarz inequality. If Kf_n is a sequence in $K(B_1)$, then $\{Kf_n\}$ is a bounded equicontinuous family of functions on $[0,1]$, and by the Ascoli-Arzelà theorem, there is a subsequence which converges uniformly on $[0,1]$. It follows that this subsequence also converges with respect to the L^2 norm. Since every sequence in $K(B_1)$ has a subsequence which converges, the closure of $K(B_1)$ is compact. Thus K is a compact operator.

We will use the following easy lemma repeatedly.

Lemma 25.13 *If K is a compact operator and $\{x_n\}$ is a sequence with $\|x_n\| \leq 1$ for each n, then $\{Kx_n\}$ has a convergent subsequence.*

Proof. Since $\|x_n\| \leq 1$, then $\{\frac{1}{2}x_n\} \subset B_1$. Hence $\{\frac{1}{2}Kx_n\} = \{K(\frac{1}{2}x_n)\}$ is a sequence contained in $\overline{K(B_1)}$, a compact set and therefore has a convergent subsequence.   □

We now prove the *spectral theorem* for compact symmetric operators.

Theorem 25.14 *Suppose H is a separable Hilbert space over the complex numbers and K is a compact symmetric linear operator. There exist a sequence $\{z_n\}$ in H and a sequence $\{\lambda_n\}$ in $\mathbb{R}$ such that*

(1) $\{z_n\}$ is an orthonormal basis for H,

(2) each z_n is an eigenvector with eigenvalue λ_n, that is, $Kz_n = \lambda_n z_n$,
(3) for each $\lambda_n \neq 0$, the dimension of the linear space $\{x \in H : Kx = \lambda_n x\}$ is finite,
(4) the only limit point, if any, of $\{\lambda_n\}$ is 0; if there are infinitely many distinct eigenvalues, then 0 is a limit point of $\{\lambda_n\}$.

Note that part of the assertion of the theorem is that the eigenvalues are real. (3) is usually phrased as saying the non-zero eigenvalues have finite *multiplicity*.

Proof. If $K = 0$, any orthonormal basis will do for $\{z_n\}$ and all the λ_n are zero, so we suppose $K \neq 0$. We first show that the eigenvalues are real, that eigenvectors corresponding to distinct eigenvalues are orthogonal, the multiplicity of non-zero eigenvalues is finite, and that 0 is the only limit point of the set of eigenvalues. We then show how to sequentially construct a set of eigenvectors and that this construction yields a basis.

If λ_n is an eigenvalue corresponding to a eigenvector $z_n \neq 0$, we see that

$$\lambda_n \langle z_n, z_n \rangle = \langle \lambda_n z_n, z_n \rangle = \langle K z_n, z_n \rangle = \langle z_n, K z_n \rangle$$
$$= \langle z_n, \lambda_n z_n \rangle = \overline{\lambda}_n \langle z_n, z_n \rangle,$$

which proves that λ_n is real.

If $\lambda_n \neq \lambda_m$ are two distinct eigenvalues corresponding to the eigenvectors z_n and z_m, we observe that

$$\lambda_n \langle z_n, z_m \rangle = \langle \lambda_n z_n, z_m \rangle = \langle K z_n, z_m \rangle = \langle z_n, K z_m \rangle$$
$$= \langle z_n, \lambda_m z_m \rangle = \lambda_m \langle z_n, z_m \rangle,$$

using that λ_m is real. Since $\lambda_n \neq \lambda_m$, we conclude $\langle z_n, z_m \rangle = 0$.

Suppose $\lambda_n \neq 0$ and that there are infinitely many orthonormal vectors x_k such that $Kx_k = \lambda_n x_k$. Then

$$\|x_k - x_j\|^2 = \langle x_k - x_j, x_k - x_j \rangle = \|x_k\|^2 - 2\langle x_k, x_j \rangle + \|x_j\|^2 = 2$$

if $j \neq k$. But then no subsequence of $\lambda_n x_k = K x_k$ can converge, a contradiction to Lemma 25.13. Therefore the multiplicity of λ_n is finite.

Suppose we have a sequence of distinct non-zero eigenvalues converging to a real number $\lambda \neq 0$ and a corresponding sequence of eigenvectors each with norm one. Since K is compact, there is a subsequence $\{n_j\}$ such that Kz_{n_j} converges to a point in H, say w. Then

$$z_{n_j} = \frac{1}{\lambda_{n_j}} K z_{n_j} \to \frac{1}{\lambda} w,$$

or $\{z_{n_j}\}$ is an orthonormal sequence of vectors converging to $\lambda^{-1} w$. But as in the preceding paragraph, we cannot have such a sequence.

Since $\{\lambda_n\} \subset \overline{B(0, r(K))}$, a bounded subset of the complex plane, if the set $\{\lambda_n\}$ is infinite, there will be a subsequence which converges. By the preceding paragraph, 0 must be a limit point of the subsequence.

We now turn to constructing eigenvectors. By Lemma 25.7(3), we have

$$\|K\| = \sup_{\|x\|=1} |\langle Kx, x \rangle|.$$

We claim the maximum is attained. If $\sup_{\|x\|=1} \langle Kx, x \rangle = \|K\|$, let $\lambda = \|K\|$; otherwise let $\lambda = -\|K\|$. Choose x_n with $\|x_n\| = 1$ such that $\langle Kx_n, x_n \rangle$ converges to λ. There exists a subsequence $\{n_j\}$ such that Kx_{n_j} converges, say to z. Since $\lambda \neq 0$, then $z \neq 0$, for otherwise $\lambda = \lim_{j \to \infty} \langle Kx_{n_j}, x_{n_j} \rangle = 0$. Now

$$\|(K - \lambda I)z\|^2 = \lim_{j \to \infty} \|(K - \lambda I)Kx_{n_j}\|^2$$
$$\leq \|K\|^2 \lim_{j \to \infty} \|(K - \lambda I)x_{n_j}\|^2$$

and

$$\|(K - \lambda I)x_{n_j}\|^2 = \|Kx_{n_j}\|^2 + \lambda^2 \|x_{n_j}\|^2 - 2\lambda \langle x_{n_j}, Kx_{n_j} \rangle$$
$$\leq \|K\|^2 + \lambda^2 - 2\lambda \langle x_{n_j}, Kx_{n_j} \rangle$$
$$\to \lambda^2 + \lambda^2 - 2\lambda^2 = 0.$$

Therefore $(K - \lambda I)z = 0$, or z is an eigenvector for K with corresponding eigenvalue λ.

Suppose we have found eigenvalues $z_1, z_2, \ldots, z_n$. Let X_n be the linear subspace spanned by $\{z_1, \ldots, z_n\}$ and let $Y = X^\perp$ be the orthogonal complement of X_n, that is, the set of all vectors orthogonal to every vector in X_n. If $x \in Y$, then

$$\langle Kx, z_k \rangle = \langle x, Kz_k \rangle = \overline{\lambda}_k \langle x, z_k \rangle = 0,$$

or $Kx \in Y$. Hence K maps Y into Y. By Exercise 25.6, $K|_Y$ is a compact symmetric operator. If Y is non-zero, we can then look at $K|_Y$, and find a new eigenvector z_{n+1}.

It remains to prove that the set of eigenvectors forms a basis. Suppose y is orthogonal to every eigenvector. Then

$$\langle Ky, z_k \rangle = \langle y, K z_k \rangle = \langle y, \lambda_k z_k \rangle = 0$$

if z_k is an eigenvector with eigenvalue λ_k, so Ky is also orthogonal to every eigenvector. Suppose X is the closure of the linear subspace spanned by $\{z_k\}$, $Y = X^\perp$, and $Y \neq \{0\}$. If $y \in Y$, then $\langle Ky, z_k \rangle = 0$ for each eigenvector z_k, hence $\langle Ky, z \rangle = 0$ for every $z \in X$, or $K : Y \to Y$. Thus $K|_Y$ is a compact symmetric operator, and by the argument already given, there exists an eigenvector for $K|_Y$. This is a contradiction since Y is orthogonal to every eigenvector.   □

Remark 25.15 If $\{z_n\}$ is an orthonormal basis of eigenvectors for K with corresponding eigenvalues λ_n, let E_n be the projection onto the subspace spanned by z_n, that is, $E_n x = \langle x, z_n \rangle z_n$. A vector x can be written as $\sum_n \langle x, z_n \rangle z_n$, thus $Kx = \sum_n \lambda_n \langle x, z_n \rangle z_n$. We can then write

$$K = \sum_n \lambda_n E_n.$$

For general bounded symmetric operators there is a related expansion where the sum gets replaced by an integral; see Section 25.6.

Remark 25.16 If z_n is an eigenvector for K with corresponding eigenvalue λ_n, then $K z_n = \lambda_n z_n$, so

$$K^2 z_n = K(K z_n) = K(\lambda_n z_n) = \lambda_n K z_n = (\lambda_n)^2 z_n.$$

More generally, $K^j z_n = (\lambda_n)^j z_n$. Using the notation of Remark 25.15, we can write

$$K^j = \sum_n (\lambda_n)^j E_n.$$

If Q is any polynomial, we can then use linearity to write

$$Q(K) = \sum_n Q(\lambda_n) E_n.$$

It is a small step from here to make the definition

$$f(K) = \sum_n f(\lambda_n) E_n$$

for any bounded and Borel measurable function f.

25.4 An application

Consider the ordinary differential equation

$$-f''(x) = g(x) \tag{25.6}$$

with boundary conditions $f(0) = 0$, $f(2\pi) = 0$. We put the minus sign in front of f'' so that later some eigenvalues will be positive.

When g is continuous, we can give an explicit solution as follows. Let

$$G(x, y) = \begin{cases} x(2\pi - y)/2\pi, & 0 \le x \le y \le 2\pi; \\ y(2\pi - x)/2\pi, & 0 \le y \le x \le 2\pi. \end{cases} \tag{25.7}$$

Recall that a function is in C^2 if it has a continuous second derivative. We then have

Proposition 25.17 *If g is continuous on $[0, 2\pi]$, then*

$$f(x) = \int_0^{2\pi} G(x, y) g(y) \, dy$$

is a C^2 function, $f(0) = f(2\pi) = 0$, and $-f''(x) = g(x)$.

Proof. Clearly $f(0) = f(2\pi) = 0$ by the formula for G. We see that

$$\frac{f(x+h) - f(x)}{h} = \int_0^{2\pi} \frac{G(x+h, y) - G(x, y)}{h} g(y) \, dy$$

$$\rightarrow \int_0^{2\pi} H(x, y) \, dy$$

as $h \to 0$ by the dominated convergence theorem, where

$$H(x,y) = \begin{cases} (2\pi - y)/2\pi, & x \le y; \\ -y/2\pi, & x > y. \end{cases}$$

Thus

$$f'(x) = \int_0^x \frac{-y}{2\pi} g(y)\,dy + \int_x^{2\pi} \frac{2\pi - y}{2\pi} g(y)\,dy.$$

By the fundamental theorem of calculus, f' is differentiable and

$$f''(x) = \frac{-x}{2\pi} g(x) - \frac{2\pi - x}{2\pi} g(x) = -g(x)$$

as required. $\square$

The function G is called the *Green's function* for the operator $Lf = f''$. The phrase "the Green's function" is awkward English, but is nevertheless what people use.

By Examples 25.6 and 25.12 the operator

$$Kf(x) = \int G(x,y) f(y)\,dy$$

is a bounded compact symmetric operator on $L^2([0, 2\pi])$, so there exists an orthonormal basis of eigenvectors $\{z_n\}$ with corresponding eigenvalues $\{\lambda_n\}$. If $\lambda_n = 0$, then $z_n = \lambda_n^{-1} K z_n$ is continuous since K maps L^2 functions to continuous ones. By Proposition 25.17 we can say more, namely that $z_n \in C^2$, $z_n(0) = z_n(2\pi) = 0$, and $-z_n'' = \lambda_n^{-1} z_n$. The solutions to this differential equation, or equivalently $z_n'' + \lambda_n^{-1} z_n = 0$, with $z_n(0) = z_n(2\pi)$, are $z_n = \pi^{-1/2} \sin(nx/2)$ with $\lambda_n = 4/n^2$.

We note that 0 is not an eigenvalue for K because if $Kz = 0$, then z is equal to the second derivative of the function that is identically 0, hence z is identically 0. Therefore z is not an eigenvector.

A function $f \in L^2([0, 2\pi])$ can be written as

$$f = \sum_{n=1}^{\infty} \langle f, z_n \rangle z_n,$$

where $z_n(x) = \pi^{-1/2} \sin(nx/2)$ and the sum converges in L^2. This is called a *Fourier sine series*.

We can use the above expansion to solve the partial differential equation

$$\frac{\partial u}{\partial t}(t, x) = \frac{\partial^2 u}{\partial x^2}(t, x)$$

with $u(t, 0) = u(t, 2\pi) = 0$ for all t and $u(0, x) = f(x)$ for all $x \in [0, 2\pi]$, where $f \in L^2([0, 2\pi])$. This partial differential equation is known as the *heat equation*. Write $f = \sum_{n=1}^{\infty} \langle f, z_n \rangle z_n$. It is then a routine matter to show that the solution is

$$u(t, x) = \sum_{n=1}^{\infty} \langle f, z_n \rangle e^{-t/\lambda_n} z_n.$$

This formula may look slightly different from other formulas you may have seen. The reason is that we are using λ_n for the eigenvalues of K; $\{1/\lambda_n\}$ will be the eigenvalues for the operator $Lf = -f''$.

25.5 Spectra of symmetric operators

When we move away from compact operators, the spectrum can become much more complicated. Let us look at an instructive example.

Example 25.18 Let $H = L^2([0, 1])$ and define $A : H \to H$ by $Af(x) = xf(x)$. There is no difficulty seeing that A is bounded and symmetric.

We first show that no point in $[0, 1]^c$ is in the spectrum of A. If z is a fixed complex number and either has a non-zero imaginary part or has a real part that is not in $[0, 1]$, then $z - A$ has the inverse $Bf(x) = \frac{1}{z-x} f(x)$. It is obvious that B is in fact the inverse of $z - A$ and it is a bounded operator because $1/|z - x|$ is bounded on $x \in [0, 1]$.

If $z \in [0, 1]$, we claim $z - A$ does not have a bounded inverse. The function that is identically equal to 1 is in $L^2([0, 1])$. The only function g that satisfies $(z - A)g = 1$ is $g = 1/(z - x)$, but g is not in $L^2([0, 1])$, hence the range of $z - A$ is not all of H.

We conclude that $\sigma(A) = [0, 1]$. We show now, however, that no point in $[0, 1]$ is an eigenvalue for A. If $z \in [0, 1]$ were an eigenvalue, then there would exist a non-zero f such that $(z - A)f = 0$. Since our Hilbert space is L^2, saying f is non-zero means that the set of

x where $f(x) \neq 0$ has positive Lebesgue measure. But $(z - A)f = 0$ implies that $(z - x)f(x) = 0$ a.e., which forces $f = 0$ a.e. Thus A has no eigenvalues.

We have shown that the spectrum of a bounded symmetric operator is closed and bounded. A bit more difficult is the fact that the spectrum is never empty.

Proposition 25.19 *If A is a bounded symmetric operator, then $\sigma(A)$ contains at least one point.*

Proof. Suppose not. Then for each $z \in \mathbb{C}$, the inverse of the operator $z - A$ exists. Let us denote the inverse by R_z. Since $z - A$ has an inverse, then $z - A$ is one-to-one and onto, and by the open mapping theorem, $z - A$ is an open map. This translates to R_z being a continuous operator, hence a bounded operator.

Let $x, y \in H$ and define $f(z) = \langle R_z x, y \rangle$ for $z \in H$. We want to show that f is an analytic function of z. If $w \neq z$,

$$w - A = (z - A) - (z - w)I = (z - A)(I - (z - w)R_z).$$

If $|z - w| < 1/\|R_z\|$, then

$$R_w = (w - A)^{-1} = R_z \left(\sum_{i=0}^{\infty} ((z - w)R_z)^i \right).$$

Therefore

$$\lim_{w \to z} \frac{R_w - R_z}{w - z} = -R_z^2. \tag{25.8}$$

It follows that

$$\lim_{w \to z} \frac{f(w) - f(z)}{w - z} = -\langle R_z^2 x, y \rangle,$$

which proves that f has a derivative at z, and so is analytic.

For $z > \|A\|$ we have

$$R_z = (z - A)^{-1} = z^{-1}(I - z^{-1}A)^{-1},$$

and using Proposition 25.2, we conclude that $f(z) = \langle R_z x, y \rangle \to 0$ as $|z| \to \infty$.

We thus know that f is analytic on $\mathbb{C}$, i.e., it is an entire function, and that $f(z)$ tends to 0 as $|z| \to \infty$. Therefore f is a bounded entire function. By Liouville's theorem from complex analysis (see [1] or [8]), f must be constant. Since f tends to 0 as $|z|$ tends to infinity, that constant must be 0. This holds for all y, so $R_z x$ must be equal to 0 for all x and z. But then for each x we have $x = (z - A)R_z x = 0$, a contradiction. $\qquad\square$

Before proceeding, we need an elementary lemma.

Lemma 25.20 *If M and N are operators on H that commute and MN is invertible, then both M and N are also invertible.*

Proof. If M has a left inverse A and a right inverse B, that is, $AM = I$ and $MB = I$, then $A = A(MB) = (AM)B = B$ and so M has an inverse. It therefore suffices to prove that M has both a left and right inverse, and then to apply the same argument to N.

Let $L = (MN)^{-1}$. Then $M(NL) = (MN)L = I$ and M has a right inverse. Using the commutativity, $(LN)M = L(MN) = I$, and so M has a left inverse. Now use the preceding paragraph. $\square$

Here is the *spectral mapping theorem* for polynomials.

Theorem 25.21 *Suppose A is a bounded linear operator and P is a polynomial. Then $\sigma(P(A)) = P(\sigma(A))$.*

By $P(\sigma(A))$ we mean the set $\{P(\lambda) : \lambda \in \sigma(A)\}$.

Proof. We first suppose $\lambda \in \sigma(P(A))$ and prove that $\lambda \in P(\sigma(A))$. Factor

$$\lambda - P(x) = c(x - a_1) \cdots (x - a_n).$$

Since $\lambda \in \sigma(P(A))$, then $\lambda - P(A)$ is not invertible, and therefore for at least one i we must have that $A - a_i$ is not invertible. That means that $a_i \in \sigma(A)$. Since a_i is a root of the equation $\lambda - P(x) = 0$, then $\lambda = P(a_i)$, which means that $\lambda \in P(\sigma(A))$.

Now suppose $\lambda \in P(\sigma(A))$. Then $\lambda = P(a)$ for some $a \in \sigma(A)$. We can write

$$P(x) = \sum_{i=0}^{n} b_i x^i$$

for some coefficients b_i, and then

$$P(x) - P(a) = \sum_{i=1}^{n} b_i(x^i - a^i) = (x - a)Q(x)$$

for some polynomial Q, since $x - a$ divides $x^i - a^i$ for each $i \geq 1$.
We then have

$$P(A) - \lambda = P(A) - P(a) = (A - a)Q(A).$$

If $P(A) - \lambda$ were invertible, then by Lemma 25.20 we would have
that $A - a$ is invertible, a contradiction. Therefore $P(A) - \lambda$ is not
invertible, i.e., $\lambda \in \sigma(P(A))$. $\square$

A key result is the spectral radius formula. First we need a
consequence of the uniform boundedness principle.

Lemma 25.22 *If $\mathcal{B}$ is a Banach space and $\{x_n\}$ a subset of $\mathcal{B}$
such that $\sup_n |f(x_n)|$ is finite for each bounded linear functional
f, then $\sup_n \|x_n\|$ is finite.*

Proof. For each $x \in \mathcal{B}$, define a linear functional L_x on $\mathcal{B}^*$, the
dual space of $\mathcal{B}$, by

$$L_x(f) = f(x), \qquad f \in \mathcal{B}^*.$$

Note $|L_x(f)| = |f(x)| \leq \|f\| \, \|x\|$, so $\|L_x\| \leq \|x\|$.

To show equality, let $M = \{cx : c \in \mathbb{C}\}$ and define $f(cx) = c\|x\|$. We observe that f is a linear functional on the subspace M,
$|f(x)| = \|x\|$, and $\|f\| = 1$. We use the Hahn-Banach theorem,
Theorem 18.6, to extend f to a bounded linear functional on $\mathcal{B}$,
also denoted by f, with $\|f\| = 1$. Then $|L_x(f)| = |f(x)| = \|x\|$, so
$\|L_x\| \geq \|x\|$. We conclude $\|L_x\| = \|x\|$.

Since $\sup_n |L_{x_n}(f)| = \sup_n |f(x_n)|$ is finite for each $f \in \mathcal{B}^*$, by
the uniform boundedness principle (Theorem 18.8),

$$\sup_n \|L_{x_n}\| < \infty.$$

Since $\|L_{x_n}\| = \|x_n\|$, we obtain our result. $\square$

Here is the *spectral radius formula*.

Theorem 25.23 *If A is a bounded linear operator on H, then*

$$r(A) = \lim_{n \to \infty} \|A^n\|^{1/n}.$$

Proof. By Theorem 25.21 with the polynomial $P(z) = z^n$, we have $(\sigma(A))^n = \sigma(A^n)$. Then

$$
\begin{aligned}
(r(A))^n &= (\sup_{z \in \sigma(A)} |z|)^n = \sup_{z \in \sigma(A)} |z^n| \\
&= \sup_{w \in \sigma(A^n)} |w| = r(A^n) \le \|A^n\| \le \|A\|^n.
\end{aligned}
$$

We conclude

$$r(A) \le \liminf_{n \to \infty} \|A^n\|^{1/n}.$$

For the other direction, if $z \in \mathbb{C}$ with $|z| < 1/r(A)$, then $|1/z| > r(A)$, and thus $1/z \notin \sigma(A)$ by the definition of $r(A)$. Hence $I - zA = z(z^{-1}I - A)$ if invertible if $z \ne 0$. Clearly $I - zA$ is invertible when $z = 0$ as well.

Suppose $\mathcal{B}$ is the set of bounded linear operators on H and f a linear functional on $\mathcal{B}$. The function $F(z) = f((I - zA)^{-1})$ is analytic in $B(0, 1/r(A)) \subset \mathbb{C}$ by an argument similar to that in deriving (25.8). We know from complex analysis that a function has a Taylor series that converges absolutely in any disk on which the function is analytic. Therefore F has a Taylor series which converges absolutely at each point of $B(0, 1/r(A))$.

Let us identify the coefficients of the Taylor series. If $|z| < 1/\|A\|$, then using (25.2) we see that

$$F(z) = f\left(\sum_{n=0}^{\infty} z^n A^n\right) = \sum_{n=0}^{\infty} f(A^n) z^n. \tag{25.9}$$

Therefore $F^{(n)}(0) = n! f(A^n)$, where $F^{(n)}$ is the nth derivative of F. We conclude that the Taylor series for F in $B(0, 1/r(A))$ is

$$F(z) = \sum_{n=0}^{\infty} f(A^n) z^n. \tag{25.10}$$

The difference between (25.9) and (25.10) is that the former is valid in the ball $B(0, 1/\|A\|)$ while the latter is valid in $B(0, 1/r(A))$.

It follows that $\sum_{n=0}^{\infty} f(z^n A^n)$ converges absolutely for z in the ball $B(0, 1/r(A))$, and consequently

$$\lim_{n \to \infty} |f(z^n A^n)| = 0$$

if $|z| < 1/r(A)$. By Lemma 25.22 there exists a real number K such that

$$\sup_n \|z^n A^n\| \leq K$$

for all $n \geq 1$ and all $z \in B(0, 1/r(A))$. This implies that

$$|z| \, \|A^n\|^{1/n} \leq K^{1/n},$$

and hence

$$|z| \limsup_{n \to \infty} \|A^n\|^{1/n} \leq 1$$

if $|z| < 1/r(A)$. Thus

$$\limsup_{n \to \infty} \|A^n\|^{1/n} \leq r(A),$$

which completes the proof. $\square$

We have the following important corollary.

Corollary 25.24 *If A is a symmetric operator, then*

$$\|A\| = r(A).$$

Proof. In view of Theorem 25.23, it suffices to show that $\|A^n\| = \|A\|^n$ when n is a power of 2. We show this for $n = 2$ and the general case follows by induction.

On the one hand, $\|A^2\| \leq \|A\|^2$. On the other hand,

$$\|A\|^2 = (\sup_{\|x\|=1} \|Ax\|)^2 = \sup_{\|x\|=1} \|Ax\|^2$$

$$= \sup_{\|x\|=1} \langle Ax, Ax \rangle = \sup_{\|x\|=1} \langle A^2 x, x \rangle$$

$$\leq \|A^2\|$$

by the Cauchy-Schwarz inequality. $\square$

The following corollary will be important in the proof of the spectral theorem.

Corollary 25.25 *Let A be a symmetric bounded linear operator.*
(1) If P is a polynomial with real coefficients, then

$$\|P(A)\| = \sup_{z \in \sigma(A)} |P(z)|.$$

(2) If P is a polynomial with complex coefficients, then

$$\|P(A)\| \le 2 \sup_{z \in \sigma(A)} |P(z)|.$$

Proposition 25.27 will provide an improvement of assertion (2).

Proof. (1) Since P has real coefficients, then $P(A)$ is symmetric and

$$\|P(A)\| = r(P(A)) = \sup_{z \in \sigma(P(A))} |z|$$
$$= \sup_{z \in P(\sigma(A))} |z| = \sup_{w \in \sigma(A)} |P(w)|,$$

where we used Corollary 25.24 for the first equality and the spectral mapping theorem for the third.

(2) If $P(z) = \sum_{j=0}^{n}(a_j + ib_j)z^j$, let $Q(z) = \sum_{j=0}^{n} a_j z^n$ and $R(z) = \sum_{j=0}^{m} b_j z^n$. By (1),

$$\|P(A)\| \le \|Q(A)\| + \|R(A)\| \le \sup_{z \in \sigma(A)} |Q(z)| + \sup_{z \in \sigma(A)} |R(z)|,$$

and (2) follows. $\qquad\qquad\qquad\qquad\qquad\qquad\qquad\qquad$ □

The last fact we need is that the spectrum of a bounded symmetric operator is real. We know that each eigenvalue of a bounded symmetric operator is real, but as we have seen, not every element of the spectrum is an eigenvalue.

Proposition 25.26 *If A is bounded and symmetric, then $\sigma(A) \subset \mathbb{R}$.*

Proof. Suppose $\lambda = a + ib$, $b \neq 0$. We want to show that λ is not in the spectrum.

If r and s are real numbers, rewriting the inequality $(r - s)^2 \geq 0$ yields the inequality $2rs \leq r^2 + s^2$. By the Cauchy-Schwarz inequality

$$2a\langle x, Ax \rangle \leq 2a\|x\|\,\|Ax\| \leq a^2\|x\|^2 + \|Ax\|^2.$$

We then obtain the inequality

$$
\begin{aligned}
\|(\lambda - A)x\|^2 &= \langle (a + bi - A)x, (a + bi - A)x \rangle \\
&= (a^2 + b^2)\|x\|^2 + \|Ax\|^2 - (a + bi)\langle Ax, x \rangle \\
&\quad - (a - bi)\langle x, Ax \rangle \\
&= (a^2 + b^2)\|x\|^2 + \|Ax\|^2 - 2a\langle Ax, x \rangle \\
&\geq b^2\|x\|^2.
\end{aligned}
\tag{25.11}
$$

This inequality shows that $\lambda - A$ is one-to-one, for if $(\lambda - A)x_1 = (\lambda - A)x_2$, then

$$0 = \|(\lambda - A)(x_1 - x_2)\| \geq b^2\|x_1 - x_2\|^2.$$

Suppose λ is in the spectrum of A. Since $\lambda - A$ is one-to-one but not invertible, it cannot be onto. Let R be the range of $\lambda - A$. We next argue that R is closed.

If $y_k = (\lambda - A)x_k$ and $y_k \to y$, then (25.11) shows that

$$b^2\|x_k - x_m\|^2 \leq \|y_k - y_m\|^2,$$

or x_k is a Cauchy sequence. If x is the limit of this sequence, then

$$(\lambda - A)x = \lim_{n \to \infty} (\lambda - A)x_k = \lim_{n \to \infty} y_k = y.$$

Therefore R is a closed subspace of H but is not equal to H. Choose $z \in R^{\perp}$. For all $x \in H$,

$$0 = \langle (\lambda - A)x, z \rangle = \langle x, (\overline{\lambda} - A)z \rangle.$$

This implies that $(\overline{\lambda} - A)z = 0$, or $\overline{\lambda}$ is an eigenvalue for A with corresponding eigenvector z. However we know that all the eigenvalues of a bounded symmetric operator are real, hence $\overline{\lambda}$ is real. This shows λ is real, a contradiction. $\qquad\square$

25.6 Spectral resolution

Let f be a continuous function on $\mathbb{C}$ and let A be a bounded symmetric operator on a separable Hilbert space over the complex numbers. We describe how to define $f(A)$.

We have shown in Proposition 25.4 that the spectrum of A is a closed and bounded subset of $\mathbb{C}$, hence a compact set. By Exercise 20.46 we can find polynomials P_n (with complex coefficients) such that P_n converges to f uniformly on $\sigma(A)$. Then

$$\sup_{z \in \sigma(A)} |(P_n - P_m)(z)| \to 0$$

as $n, m \to \infty$. By Corollary 25.25

$$\|(P_n - P_m)(A)\| \to 0$$

as $n, m \to \infty$, or in other words, $P_n(A)$ is a Cauchy sequence in the space $\mathcal{L}$ of bounded symmetric linear operators on H. We call the limit $f(A)$.

The limit is independent of the sequence of polynomials we choose. If Q_n is another sequence of polynomials converging to f uniformly on $\sigma(A)$, then

$$\lim_{n \to \infty} \|P_n(A) - Q_n(A)\| \leq 2 \sup_{z \in \sigma(A)} |(P_n - Q_n)(z)| \to 0,$$

so $Q_n(A)$ has the same limit $P_n(A)$ does.

We record the following facts about the operators $f(A)$ when f is continuous.

Proposition 25.27 *Let f be continuous on $\sigma(A)$.*
(1) $\langle f(A)x, y \rangle = \langle x, \overline{f}(A)y \rangle$ for all $x, y \in H$.
(2) If f is equal to 1 on $\sigma(A)$, then $f(A) = I$, the identity.
(3) If $f(z) = z$ on $\sigma(A)$, then $f(A) = A$.
(4) $f(A)$ and A commute.
(5) If f and g are two continuous functions, then $f(A)g(A) = (fg)(A)$.
(6) $\|f(A)\| \leq \sup_{z \in \sigma(A)} |f(z)|$.

Proof. The proofs of (1)-(4) are routine and follow from the corresponding properties of $P_n(A)$ when P_n is a polynomial. Let us prove (5) and (6) and leave the proofs of the others to the reader.

(5) Let P_n and Q_n be polynomials converging uniformly on $\sigma(A)$ to f and g, respectively. Then P_nQ_n will be polynomials converging uniformly to fg. The assertion (6) now follows from

$$(fg)(A) = \lim_{n\to\infty}(P_nQ_n)(A) = \lim_{n\to\infty}P_n(A)Q_n(A) = f(A)g(A).$$

The limits are with respect to the norm on bounded operators on H.

(6) Since f is continuous on $\sigma(A)$, so is $g = |f|^2$. Let P_n be polynomials with real coefficients converging to g uniformly on $\sigma(A)$. By Corollary 25.25(1),

$$\|g(A)\| = \lim_{n\to\infty}\|P_n(A)\| \le \lim_{n\to\infty}\sup_{z\in\sigma(A)}|P_n(z)| = \sup_{z\in\sigma(A)}|g(z)|.$$

If $\|x\| = 1$, using (1) and (5),

$$\|f(A)x\|^2 = \langle f(A)x, f(A)x\rangle = \langle x, \overline{f}(A)f(A)x\rangle = \langle x, g(A)x\rangle$$
$$\le \|x\|\,\|g(A)x\| \le \|g(A)\| \le \sup_{z\in\sigma(A)}|g(z)|$$
$$= \sup_{z\in\sigma(A)}|f(z)|^2.$$

Taking the supremum over the set of x with $\|x\| = 1$ yields

$$\|f(A)\|^2 \le \sup_{z\in\sigma(A)}|f(z)|^2,$$

and (6) follows.   □

We now want to define $f(A)$ when f is a bounded Borel measurable function on $\mathbb{C}$. Fix $x, y \in H$. If f is a continuous function on $\mathbb{C}$, let

$$L_{x,y}f = \langle f(A)x, y\rangle. \tag{25.12}$$

It is easy to check that $L_{x,y}$ is a bounded linear functional on $\mathcal{C}(\sigma(A))$, the set of continuous functions on $\sigma(A)$. By the Riesz representation theorem for complex-valued linear functionals, Exercise 17.10, there exists a complex measure $\mu_{x,y}$ such that

$$\langle f(A)x, y\rangle = L_{x,y}f = \int_{\sigma(A)} f(z)\,\mu_{x,y}(dz)$$

for all continuous functions f on $\sigma(A)$.

We have the following properties of $\mu_{x,y}$.

Proposition 25.28 *(1) $\mu_{x,y}$ is linear in x.*
(2) $\mu_{y,x} = \overline{\mu_{x,y}}$.
(3) The total variation of $\mu_{x,y}$ is less than or equal to $\|x\| \, \|y\|$.

Proof. (1) The linear functional $L_{x,y}$ defined in (25.12) is linear in x and

$$\int f \, d(\mu_{x,y} + \mu_{x',y}) = L_{x,y}f + L_{x',y}f = L_{x+x',y}f = \int f \, d\mu_{x+x',y}.$$

By the uniqueness of the Riesz representation (see Exercise 17.9), $\mu_{x+x',y} = \mu_{x,y} + \mu_{x'+y}$. The proof that $\mu_{cx,y} = c\mu_{x,y}$ is similar.

(2) follows from the fact that if f is continuous on $\sigma(A)$, then

$$\int f \, d\mu_{y,x} = L_{y,x}f = \langle f(A)y, x \rangle = \langle y, \overline{f}(A)x \rangle$$

$$= \overline{\langle \overline{f}(A)x, y \rangle} = \overline{L_{x,y}\overline{f}} = \overline{\int \overline{f} \, d\mu_{x,y}}$$

$$= \int f \, d\overline{\mu_{x,y}}.$$

Now use the uniqueness of the Riesz representation.

(3) For f continuous on $\sigma(A)$ we have

$$\left| \int f \, d\mu_{x,y} \right| = |L_{x,y}f| = |\langle f(A)x, y \rangle|$$

$$\le \|f(A)\| \, \|x\| \, \|y\| \le \gamma_f \|x\| \, \|y\|,$$

where $\gamma_f = \sup_{z \in \sigma(A)} |f(z)|$. Taking the supremum over $f \in \mathcal{C}(\sigma(A))$ with $\gamma_f \le 1$ and using Exercise 17.11 proves (3). $\quad\square$

If f is a bounded Borel measurable function on $\mathbb{C}$, then $L_{y,x}\overline{f}$ is linear in y. By the Riesz representation theorem for Hilbert spaces, Theorem 19.9, there exists $w_x \in H$ such that $L_{y,x}\overline{f} = \langle y, w_x \rangle$ for all $y \in H$. We then have that for all $y \in H$,

$$L_{x,y}f = \int_{\sigma(A)} f(z) \, \mu_{x,y}(dz) = \int_{\sigma(A)} f(z) \, \overline{\mu_{y,x}(dz)}$$

$$= \overline{\int_{\sigma(A)} \overline{f(z)} \, \mu_{y,x}(dz)} = \overline{L_{y,x}\overline{f}}$$

$$= \overline{\langle y, w_x \rangle} = \langle w_x, y \rangle.$$

Since

$$\langle y, w_{x_1+x_2} \rangle = L_{y,x_1+x_2}\overline{f} = L_{y,x_1}\overline{f} + L_{y,x_2}\overline{f} = \langle y, w_{x_1} \rangle + \langle y, w_{x_2} \rangle$$

for all y and

$$\langle y, w_{cx} \rangle = L_{y,cx}\overline{f} = \overline{c}L_{y,x}\overline{f} = \overline{c}\langle y, w_x \rangle = \langle y, cw_x \rangle$$

for all y, we see that w_x is linear in x. We define $f(A)$ to be the linear operator on H such that $f(A)x = w_x$.

If C is a Borel measurable subset of $\mathbb{C}$, we let

$$E(C) = \chi_C(A). \tag{25.13}$$

Remark 25.29 Later on we will write the equation

$$f(A) = \int_{\sigma(A)} f(z)\, E(dz). \tag{25.14}$$

Let us give the interpretation of this equation. If $x, y \in H$, then

$$\langle E(C)x, y \rangle = \langle \chi_C(A)x, y \rangle = \int_{\sigma(A)} \chi_C(z)\, \mu_{x,y}(dz).$$

Therefore we identify $\langle E(dz)x, y \rangle$ with $\mu_{x,y}(dz)$. With this in mind, (25.14) is to be interpreted to mean that for all x and y,

$$\langle f(A)x, y \rangle = \int_{\sigma(A)} f(z)\, \mu_{x,y}(dz).$$

Theorem 25.30 *(1) $E(C)$ is symmetric.*
(2) $\|E(C)\| \leq 1$.
(3) $E(\emptyset) = 0, E(\sigma(A)) = I$.
(4) If C, D are disjoint, $E(C \cup D) = E(C) + E(D)$.
(5) $E(C \cap D) = E(C)E(D)$.
(6) $E(C)$ and A commute.
(7) $E(C)^2 = E(C)$, so $E(C)$ is a projection. If C, D are disjoint, then $E(C)E(D) = 0$.
(8) $E(C)$ and $E(D)$ commute.

Proof. (1) This follows from

$$\langle x, E(C)y \rangle = \overline{\langle E(C)y, x \rangle} = \overline{\int \chi_C(z)\, \mu_{y,x}(dz)}$$

$$= \int \chi_C(z)\, \mu_{x,y}(dz) = \langle E(C)x, y \rangle.$$

(2) Since the total variation of $\mu_{x,y}$ is bounded by $\|x\| \|y\|$, we obtain (2).

(3) $\mu_{x,y}(\emptyset) = 0$, so $E(\emptyset) = 0$. If f is identically equal to 1, then $f(A) = I$, and

$$\langle x, y \rangle = \int_{\sigma(A)} \mu_{x,y}(dz) = \langle E(\sigma(A))x, y \rangle.$$

This is true for all y, so $x = E(\sigma(A))x$ for all x.

(4) holds because $\mu_{x,y}$ is a measure, hence finitely additive.

(5) We first show

$$\langle (\chi_C g)(A)x, y \rangle = \langle \chi_C(A)g(A)x, y \rangle \tag{25.15}$$

if g is a real-valued continuous function. To see this, let f_n be a real-valued continuous function. Then

$$\langle f_n(A)g(A)x, y \rangle = \langle (f_n g)(A)x, y \rangle \tag{25.16}$$

by Proposition 25.27. The right hand side is

$$\int (f_n g)(z) \, \mu_{x,y}(dz).$$

If we take a sequence $\{f_n\}$ of real-valued continuous functions bounded by 1 such that $f_n \to \chi_C$ a.e. with respect to the measure $\mu_{x,y}$, then the right hand side of (25.16) converges to

$$\int (g\chi_C)(z) \, \mu_{x,y}(dz) = \langle (g\chi_C)(A)x, y \rangle.$$

If in addition $f_n \to \chi_C$ a.e. with respect to the measure $\mu_{y,g(A)x}$, we have

$$\langle f_n(A)g(A)x, y \rangle = \langle g(A)x, f_n(A)y \rangle = \overline{\langle f_n(A)y, g(A)x \rangle}$$

$$= \overline{\int f_n(z) \, \mu_{y,g(A)x}(dz)}$$

$$\to \overline{\int \chi_C(z) \, \mu_{y,g(A)x}(dz)} = \overline{\langle \chi_C(A)y, g(A)x \rangle}$$

$$= \langle y, \chi_C(A)g(A)x \rangle = \langle \chi_C(A)g(A)x, y \rangle.$$

Therefore (25.15) holds.

From (25.15) we obtain

$$\langle(\chi_C g)(A)x, y\rangle = \langle g(A)x, \chi_C(A)y\rangle. \tag{25.17}$$

Now replace g by g_n, where $\{g_n\}$ is a sequence of real-valued continuous functions bounded by 1. If $g_n \to \chi_D$ a.e. both with respect to the measure $\mu_{x,y}$ and with respect to the measure $\mu_{x,\chi_C(A)y}$, then the left hand side of (25.17) is equal to $\int (\chi_C g_n)(z)\, \mu_{x,y}(dz)$, which converges to

$$\int (\chi_C \chi_D)(z)\, \mu_{x,y}(dz) = \langle(\chi_C \chi_D)(A)x, y\rangle = \langle \chi_{C \cap D}(A)x, y\rangle.$$

The right hand side of (25.17) is equal to $\int g_n(z)\, \mu_{x,\chi_C(A)y}(dz)$, which converges to

$$\int \chi_D(z)\, \mu_{x,\chi_C(A)y}(dz) = \langle \chi_D(A)x, \chi_C(A)y\rangle.$$

Therefore

$$\langle \chi_{C \cap D}(A)x, y\rangle = \langle \chi_C(A)\chi_D(A)x, y\rangle$$

for all x and y. (5) follows.

(6) By (25.15) we know that

$$\langle(\chi_D g)(A)x, y\rangle = \langle \chi_D(A)g(A)x, y\rangle$$

for all x and y when g is a real-valued continuous function on $\sigma(A)$. By a very similar argument we can show

$$\langle(\chi_D g)(A)x, y\rangle = \langle g(A)\chi_D(A)x, y\rangle.$$

If we combine these two equations, we see that

$$\langle g(A)\chi_D(A)x, y\rangle = \langle \chi_D(A)g(A)x, y\rangle$$

for all $x, y \in H$. Since this holds for all x and y, we have $g(A)\chi_D(A) = \chi_D(A)g(A)$. Taking $g(z)$ first equal to $\mathrm{Re}\, z$ and then equal to $\mathrm{Im}\, z$ and using linearity yields (6).

(7) Setting $C = D$ in (5) shows $E(C) = E(C)^2$, so $E(C)$ is a projection. If $C \cap D = \emptyset$, then $E(C)E(D) = E(\emptyset) = 0$, as required.

(8) Writing

$$E(C)E(D) = E(C \cap D) = E(D \cap C) = E(D)E(C)$$

proves (8). □

The family $\{E(C)\}$, where C ranges over the Borel subsets of $\mathbb{C}$ is called the *spectral resolution of the identity*. We explain the name in just a moment.

Here is the *spectral theorem* for bounded symmetric operators.

Theorem 25.31 *Let H be a separable Hilbert space over the complex numbers and A a bounded symmetric operator. There exists a operator-valued measure E satisfying (1)–(8) of Theorem 25.30 such that*

$$f(A) = \int_{\sigma(A)} f(z) E(dz), \qquad (25.18)$$

for bounded Borel measurable functions f. Moreover, the measure E is unique.

Remark 25.32 When we say that E is an operator-valued measure, here we mean that (1)–(8) of Theorem 25.30 hold. We use Remark 25.29 to give the interpretation of (25.18).

Remark 25.33 If f is identically one, then (25.18) becomes

$$I = \int_{\sigma(A)} E(d\lambda),$$

which shows that $\{E(C)\}$ is a decomposition of the identity. This is where the name "spectral resolution" comes from.

Proof of Theorem 25.31. Given Remark 25.32, the only part to prove is the uniqueness, and that follows from the uniqueness of the measure $\mu_{x,y}$. □

25.7 Exercises

Exercise 25.1 Prove that if A, B, and C are bounded operators from a Hilbert space to itself, then
(1) $A(BC) = (AB)C$;
(2) $A(B + C) = AB + AC$ and $(B + C)A = BA + CA$.

Exercise 25.2 Prove that if A is a bounded symmetric operator, then so is A^n for each $n \geq 1$.

Exercise 25.3 Suppose $H = \mathbb{C}^n$ and Ax is multiplication of the vector $x \in H$ by a $n \times n$ matrix M. Prove that A^*x is multiplication of x by the conjugate transpose of M.

Exercise 25.4 Let $(X, \mathcal{A}, \mu)$ be a σ-finite measure space and $F : X \times X \to \mathbb{C}$ a jointly measurable function such that $F(y, x) = \overline{F(x, y)}$ and (25.4) holds. Prove that if A is defined by (25.5), then A is a bounded symmetric operator.

Exercise 25.5 If C_1, C_2 are subsets of a Hilbert space whose closures are compact, prove that the closure of

$$C_1 + C_2 = \{x + y : x \in C_1, y \in C_2\}$$

is also compact.

Exercise 25.6 Prove that if H is a Hilbert space, K is a compact symmetric operator on H, and Y is a closed subspace of X, then the map $K|_Y$ is compact.

Exercise 25.7 Suppose K is a bounded compact symmetric operator with non-negative eigenvalues $\lambda_1 \geq \lambda_2 \geq \ldots$ and corresponding eigenvectors $z_1, \ldots, z_n$. Prove that for each n,

$$\lambda_n = \max_{x \perp z_1, \cdots, z_{n-1}} \frac{(Kx, x)}{\|x\|^2}.$$

This is known as the *Rayleigh principle*.

Exercise 25.8 Let K be a compact bounded symmetric operator and let $z_1, \ldots, z_n$ be eigenvectors with corresponding eigenvalues $\lambda_1 \geq \lambda_2 \geq \cdots \geq \lambda_n$. Let X be the linear subspace spanned by $\{z_1, \ldots, z_N\}$. Prove that if $y \in X$, we have $\langle Ky, y \rangle \geq \lambda_n \langle y, y \rangle$.

Exercise 25.9 Prove that the nth largest eigenvalue for a compact bounded symmetric operator satisfies

$$\lambda_n = \max \Big\{ \min_{x \in S_n} \frac{\langle Kx, x \rangle}{\|x\|^2} : S_n \text{ is a linear subspace}$$

$$\text{of dimension } n \Big\}.$$

This is known as *Fisher's principle*.

Exercise 25.10 Prove that the nth largest eigenvalue for a compact bounded symmetric operator satisfies

$$\lambda_n = \min\left\{ \max_{x \in S_{n-1}^\perp} \frac{\langle Kx, x\rangle}{\|x\|^2} : S_{n-1} \text{ is a linear subspace} \right.$$

$$\left. \text{of dimension } n-1 \right\}.$$

This is known as *Courant's principle.*

Exercise 25.11 We say A is a *positive operator* if $\langle Ax, x\rangle \geq 0$ for all x. (In the case of matrices, the term used is *positive definite.*) Suppose A and B are compact positive symmetric operators and that $B - A$ is also a positive operator. Suppose A and B have eigenvalues α_k, β_k, resp., each arranged in decreasing order, i.e., $\alpha_1 \geq \alpha_2 \geq \cdots$ and similarly for the β_k. Prove that $\alpha_k \leq \beta_k$ for all k.

Exercise 25.12 Let A be a compact symmetric operator. Find necessary and sufficient conditions on a continuous function f such that $f(A)$ is a compact symmetric operator.

Exercise 25.13 Let A be a bounded symmetric operator, suppose $z, w \in \sigma(A)^c$, $R_z = (z - A)^{-1}$, and $R_w = (w - A)^{-1}$. Prove the *resolvent identity*

$$R_w - R_z = (z - w)R_w R_z.$$

Exercise 25.14 Suppose A is a bounded symmetric operator and f is a continuous function on $\sigma(A)$. Let P_n be polynomials which converge uniformly to f on $\sigma(A)$. Suppose $\lambda \in \mathbb{C}$ and suppose that there exists $\varepsilon > 0$ such that $d(\lambda, \sigma(P_n(A))) \geq \varepsilon$ for each n. Prove that $\lambda \notin \sigma(f(A))$.

Exercise 25.15 Prove that K is a compact symmetric positive operator if and only if all the eigenvalues of K are non-negative.

Exercise 25.16 Let A be a bounded symmetric operator, not necessarily compact. Prove that if $A = B^2$ for some bounded symmetric operator B, then A is a positive operator..

Exercise 25.17 Let A be a bounded symmetric operator whose spectrum is contained in $[0, \infty)$. Prove that A has a square root, that is, there exists a bounded symmetric operator B such that $A = B^2$.

Exercise 25.18 Let A be a bounded symmetric operator, not necessarily compact. Prove that A is a positive operator if and only if $\sigma(A) \subset [0, \infty)$.

Exercise 25.19 Let A be a bounded symmetric operator. Prove that $\mu_{x,x}$ is a real non-negative measure.

Exercise 25.20 Suppose that A is a bounded symmetric operator, $C_1, \ldots, C_n$, are disjoint Borel measurable subsets of $\mathbb{C}$, and $a_1, \ldots, a_n$ are complex numbers. Prove that

$$\left\| \sum_{i=1}^{n} c_i E(C_i) \right\| = \max_{1 \le i \le n} |c_i|.$$

Exercise 25.21 Prove that if A is a bounded symmetric operator and f is a bounded Borel measurable function, then

$$\|f(A)x\|^2 = \int_{\sigma(A)} |f(z)|^2 \, \mu_{x,x}(dz).$$

Exercise 25.22 Prove that if A is a bounded symmetric operator and $\{s_n\}$ is a sequence of simple functions such that s_n converges uniformly to a bounded Borel measurable f on $\sigma(A)$, then

$$\|f(A) - s_n(A)\| \to 0.$$

Chapter 26

Distributions

Mathematical physicists often talk about the Dirac delta function, which is supposed to be a function that is equal to 0 away from 0, equal to infinity at 0, and which has integral equal to 1. Of course, no measurable function can have these properties. The delta function can be put on a solid mathematical footing through the use of the theory of distributions. The term *generalized function* is also used, although there are other notions of generalized functions besides that of distributions.

For simplicity of notation, in this chapter we restrict ourselves to dimension one, but everything we do can be extended to $\mathbb{R}^n$, $n > 1$, although in some cases a more complicated proof is necessary. See [5] for the n dimensional case.

26.1 Definitions and examples

We use C_K^∞ for the set of C^∞ functions on $\mathbb{R}$ with compact support. Let $Df = f'$, the derivative of f, $D^2 f = f''$, the second derivative, and so on, and we make the convention that $D^0 f = f$.

If f is a continuous function on $\mathbb{R}$, let $\operatorname{supp}(f)$ be the *support* of f, the closure of the set $\{x : f(x) \neq 0\}$. If $f_j, f \in C_K^\infty$, we say $f_j \to f$ in the C_K^∞ sense if there exists a compact subset K such that $\operatorname{supp}(f_j) \subset K$ for all j, f_j converges uniformly to f, and $D^m f_j$ converges uniformly to $D^m f$ for all m.

We have not claimed that C_K^∞ with this notion of convergence is a Banach space, so it doesn't make sense to talk about bounded linear functionals. But it does make sense to consider continuous linear functionals. A map $F : C_K^\infty \to \mathbb{C}$ is a *continuous linear functional on* C_K^∞ if $F(f + g) = F(f) + F(g)$ whenever $f, g \in C_K^\infty$, $F(cf) = cF(f)$ whenever $f \in C_K^\infty$ and $c \in \mathbb{C}$, and $F(f_j) \to F(f)$ whenever $f_j \to f$ in the C_K^∞ sense.

A *distribution* is defined to be a complex-valued continuous linear functional on C_K^∞.

Here are some examples of distributions.

Example 26.1 If g is a continuous function, define

$$G_g(f) = \int_{\mathbb{R}} f(x)g(x)\, dx, \qquad f \in C_K^\infty. \qquad (26.1)$$

It is routine to check that G_g is a distribution.

Note that knowing the values of $G_g(f)$ for all $f \in C_K^\infty$ determines g uniquely up to almost everywhere equivalence. Since g is continuous, g is uniquely determined at every point by the values of $G_g(f)$.

Example 26.2 Set $\delta(f) = f(0)$ for $f \in C_K^\infty$. This distribution is the *Dirac delta function.*

Example 26.3 If g is integrable and $k \geq 1$, define

$$F(f) = \int_{\mathbb{R}} D^k f(x)g(x)\, dx, \qquad f \in C_K^\infty.$$

Example 26.4 If $k \geq 1$, define $F(f) = D^k f(0)$ for $f \in C_K^\infty$.

There are a number of operations that one can perform on distributions to get other distributions. Here are some examples.

Example 26.5 Let h be a C^∞ function, not necessarily with compact support. If F is a distribution, define $M_h(F)$ by

$$M_h(F)(f) = F(fh), \qquad f \in C_K^\infty.$$

It is routine to check that $M_h(F)$ is a distribution.

Example 26.1 shows how to consider a continuous function g as a distribution. Defining G_g by (26.1),

$$M_h(G_g)(f) = G_g(fh) = \int (fh)g = \int f(hg) = G_{hg}(f).$$

Therefore we can consider the operator M_h we just defined as an extension of the operation of multiplying continuous functions by a C^∞ function h.

Example 26.6 If F is a distribution, define $D(F)$ by

$$D(F)(f) = F(-Df), \qquad f \in C_K^\infty.$$

Again it is routine to check that $D(F)$ is a distribution.

If g is a continuously differentiable function and we use (26.1) to identify the function g with the distribution G_g, then

$$D(G_g)(f) = G_g(-Df) = \int (-Df)(x)g(x)\, dx$$
$$= \int f(x)(Dg)(x)\, dx = G_{Dg}(f), \qquad f \in C_K^\infty,$$

by integration by parts. Therefore $D(G_g)$ is the distribution that corresponds to the function that is the derivative of g. However, $D(F)$ is defined for any distribution F. Hence the operator D on distributions gives an interpretation to the idea of taking the derivative of any continuous function.

Example 26.7 Let $a \in \mathbb{R}$ and define $T_a(F)$ by

$$T_a(F)(f) = F(f_{-a}), \qquad f \in C_K^\infty,$$

where $f_{-a}(x) = f(x + a)$. If G_g is given by (26.1), then

$$T_a(G_g)(f) = G_g(f_{-a}) = \int f_{-a}(x)g(x)\, dx$$
$$= \int f(x)g(x - a)\, dx = G_{g_a}(f), \qquad f \in C_K^\infty,$$

by a change of variables, and we can consider T_a as the operator that translates a distribution by a.

Example 26.8 Define R by

$$R(F)(f) = F(Rf), \qquad f \in C_K^\infty,$$

where $Rf(x) = f(-x)$. Similarly to the previous examples, we can see that R reflects a distribution through the origin.

Example 26.9 Finally, we give a definition of the convolution of a distribution with a continuous function h with compact support. Define $C_h(F)$ by

$$C_h(F)(f) = F(f * Rh), \qquad f \in C_K^\infty,$$

where $Rh(x) = h(-x)$. To justify that this extends the notion of convolution, note that

$$C_h(G_g)(f) = G_g(f * Rh) = \int g(x)(f * Rh)(x)\,dx$$

$$= \int \int g(x)f(y)h(y-x)\,dy\,dx = \int f(y)(g * h)(y)\,dy$$

$$= G_{g*h}(f),$$

or C_h takes the distribution corresponding to the continuous function g to the distribution corresponding to the function $g * h$.

One cannot, in general, define the product of two distributions or quantities like $\delta(x^2)$.

26.2   Distributions supported at a point

We first define the support of a distribution. We then show that a distribution supported at a point is a linear combination of derivatives of the delta function.

Let G be open. A distribution F is zero on G if $F(f) = 0$ for all C_K^∞ functions f for which $\operatorname{supp}(f) \subset G$.

Lemma 26.10 *If F is zero on G_1 and G_2, then F is zero on $G_1 \cup G_2$.*

Proof. This is just the usual partition of unity proof. Suppose f has support in $G_1 \cup G_2$. We will write $f = f_1 + f_2$ with $\operatorname{supp}(f_1) \subset$

G_1 and supp $(f_2) \subset G_2$. Then $F(f) = F(f_1) + F(f_2) = 0$, which will achieve the proof.

Fix $x \in \text{supp}\,(u)$. Since G_1, G_2 are open, we can find h_x such that h_x is non-negative, $h_x(x) > 0$, h_x is in C_K^∞, and the support of h_x is contained either in G_1 or in G_2. The set $B_x = \{y : h_x(y) > 0\}$ is open and contains x.

By compactness we can cover supp f by finitely many balls $\{B_{x_1}, \ldots, B_{x_m}\}$. Let h_1 be the sum of those h_{x_i} whose support is contained in G_1 and let h_2 be the sum of those h_{x_i} whose support is contained in G_2. Then let

$$f_1 = \frac{h_1}{h_1 + h_2} f, \qquad f_2 = \frac{h_2}{h_1 + h_2} f.$$

Clearly supp $(f_1) \subset G_1$, supp $(f_2) \subset G_2$, $f_1 + f_2 > 0$ on $G_1 \cup G_2$, and $f = f_1 + f_2$. $\qquad \square$

If we have an arbitrary collection of open sets $\{G_\alpha\}$, F is zero on each G_α, and supp (f) is contained in $\cup_\alpha G_\alpha$, then by compactness there exist finitely many of the G_α that cover supp (f). By Lemma 26.10, $F(f) = 0$.

The union of all open sets on which F is zero is an open set on which F is zero. The complement of this open set is called the *support* of F.

Example 26.11 The support of the Dirac delta function is $\{0\}$. Note that the support of $D^k \delta$ is also $\{0\}$.

Define

$$\|f\|_{C^N(K)} = \max_{0 \le k \le N} \sup_{x \in K} |D^k f(x)|.$$

Proposition 26.12 *Let F be a distribution and K a fixed compact set. There exist N and c depending on F and K such that if $f \in C_K^\infty$ has support in K, then*

$$|F(f)| \le c \|f\|_{C^N(K)}.$$

Proof. Suppose not. Then for each m there exists $f_m \in C_K^\infty$ with support contained in K such that $F(f_m) = 1$ and $\|f\|_{C^N(K)} \le 1/m$.

Therefore $f_m \to 0$ in the sense of C_K^∞. However $F(f_m) = 1$ while
$F(0) = 0$, a contradiction. □

Proposition 26.13 *Suppose F is a distribution and* $\mathrm{supp}\,(F) =$
$\{0\}$. *There exists N such that if $f \in C_K^\infty$ and $D^j f(0) = 0$ for*
$j \leq N$, *then $F(f) = 0$.*

Proof. Let $\varphi \in C^\infty$ be 0 on $[-1, 1]$ and 1 on $|x| > 2$. Let
$g = (1 - \varphi)f$. Note $\varphi f = 0$ on $[-1, 1]$, so $F(\varphi f) = 0$ because F is
supported on $\{0\}$. Then

$$F(g) = F(f) - F(\varphi f) = F(f).$$

Thus is suffices to show that $F(g) = 0$ whenever $g \in C_K^\infty$, $\mathrm{supp}\,(g)$
$\subset [-3, 3]$, and $D^j g(0) = 0$ for $0 \leq j \leq N$.

Let $K = [-3, 3]$. By Proposition 26.12 there exist N and
c depending only on F such that $|F(g)| \leq c\|g\|_{C^N(K)}$. Define
$g_m(x) = \varphi(mx)g(x)$. Note that $g_m(x) = g(x)$ if $|x| > 2/m$.

Suppose $|x| < 2/m$ and $g \in C_K^\infty$ with support in $[-3, 3]$ and
$D^j g(0) = 0$ for $j \leq N$. By Taylor's theorem, if $j \leq N$,

$$D^j g(x) = D^j g(0) + D^{j+1} g(0)x + \cdots + D^N g(0)\frac{x^{N-j}}{(N-j)!} + R$$

$$= R,$$

where the remainder R satisfies

$$|R| \leq \sup_{y \in \mathbb{R}} |D^{N+1} g(y)| \frac{|x|^{N+1-j}}{(N+1-j)!}.$$

Since $|x| < 2/m$, then

$$|D^j g(x)| = |R| \leq c_1 m^{j-1-N} \tag{26.2}$$

for some constant c_1.

By the definition of g_m and (26.2),

$$|g_m(x)| \leq c_2 |g(x)| \leq c_3 m^{-N-1},$$

where c_2 and c_3 are constants. Again using (26.2),

$$|Dg_m(x)| \leq |\varphi(mx)|\,|Dg(x)| + m|g(x)|\,|D\varphi(mx)| \leq c_4 m^{-N}.$$

Continuing, repeated applications of the product rule show that if $k \leq N$, then

$$|D^k g_m(x)| \leq c_5 m^{k-1-N}$$

for $k \leq N$ and $|x| \leq 2/m$, where c_5 is a constant.

Recalling that $g_m(x) = g(x)$ if $|x| > 2/m$, we see that $D^j g_m(x) \to D^j g(x)$ uniformly over $x \in [-3, 3]$ if $j \leq N$. We conclude

$$F(g_m - g) = F(g_m) - F(g) \to 0.$$

However, each g_m is 0 in a neighborhood of 0, so by the hypothesis, $F(g_m) = 0$; thus $F(g) = 0$. $\qquad \square$

By Example 26.6, $D^j \delta$ is the distribution such that $D^j \delta(f) = (-1)^j D^j f(0)$.

Theorem 26.14 *Suppose F is a distribution supported on $\{0\}$. Then there exist N and constants c_i such that*

$$F = \sum_{i=0}^{N} c_i D^i \delta.$$

Proof. Let $P_i(x)$ be a C_K^∞ function which agrees with the polynomial x^i in a neighborhood of 0. Taking derivatives shows that $D^j P_i(0) = 0$ if $i \neq j$ and equals $i!$ if $i = j$. Then $D^j \delta(P_i) = (-1)^i i!$ if $i = j$ and 0 otherwise.

Use Proposition 26.13 to determine the integer N. Suppose $f \in C_K^\infty$. By Taylor's theorem, f and the function

$$g(x) = \sum_{i=0}^{N} D^i f(0) P_i(x)/i!$$

agree at 0 and all the derivatives up to order N agree at 0. By the conclusion of Proposition 26.13 applied to $f - g$,

$$F\left(f - \sum_{i=0}^{N} \frac{D^i f(0)}{i!} P_i\right) = 0.$$

Therefore

$$F(f) = \sum_{i=0}^{N} \frac{D^i f(0)}{i!} F(P_i) = \sum_{i=0}^{N} (-1)^i \frac{D^i \delta(f)}{i!} F(P_i)$$

$$= \sum_{i=0}^{N} c_i D^i \delta(f)$$

if we set $c_i = (-1)^i F(P_i)/i!$ Since f was arbitrary and the c_i do not depend on f, this proves the theorem. $\square$

26.3   Distributions with compact support

In this section we consider distributions whose supports are compact sets.

Theorem 26.15 *If F has compact support, there exist L and continuous functions g_j such that*

$$F = \sum_{j \le L} D^j G_{g_j}, \tag{26.3}$$

where G_{g_j} is defined by Example 26.1.

Example 26.16 The delta function is the derivative of h, where h is 0 for $x < 0$ and 1 for $x \ge 0$. In turn h is the derivative of g, where g is 0 for $x < 0$ and $g(x) = x$ for $x \ge 0$. Therefore $\delta = D^2 G_g$.

Proof. Let $h \in C_K^\infty$ and suppose h is equal to 1 on the support of F. Then $F((1-h)f) = 0$, or $F(f) = F(hf)$. Therefore there exist N and c_1 such that

$$|F(hf)| \le c_1 \|hf\|_{C^N(K)}.$$

By the product rule,

$$|D(hf)| \le |h(Df)| + |(Dh)f| \le c_2 \|f\|_{C^N(K)},$$

and by repeated applications of the product rule,

$$\|hf\|_{C^N(K)} \le c_3\|f\|_{C^N(K)}.$$

Hence

$$|F(f)| = |F(hf)| \le c_4\|f\|_{C^N(K)}.$$

Let $K = [-x_0, x_0]$ be a closed interval containing the support of F. Let $C^N(K)$ be the N times continuously differentiable functions whose support is contained in K. We will use the fact that $C^N(K)$ is a complete metric space with respect to the metric $\|f-g\|_{C^N(K)}$.

Define

$$\|f\|_{H^M} = \Big(\sum_{k\le M} \int |D^k f|^2 \, dx \Big)^{1/2}, \qquad f \in C_K^\infty,$$

and let H^M be the completion of $\{f \in C_K^\infty : \operatorname{supp}(f) \subset K\}$ with respect to this norm. It is routine to check that H^M is a Hilbert space.

Suppose $M = N + 1$ and $x \in K$. Then using the Cauchy-Schwarz inequality and the fact that $K = [-x_0, x_0]$,

$$|D^j f(x)| = |D^f(x) - D^j f(-x_0)| = \Big| \int_{-x_0}^{x} D^{j+1} f(y) \, dy \Big|$$

$$\le |2x_0|^{1/2} \Big(\int_{\mathbb{R}} |D^{j+1} f(y)|^2 \, dy \Big)^{1/2}$$

$$\le c_5 \Big(\int_{\mathbb{R}} |D^{j+1} f(y)|^2 \, dy \Big)^{1/2}.$$

This holds for all $j \le N$, hence

$$\|u\|_{C^N(K)} \le c_6 \|u\|_{H^M}. \tag{26.4}$$

Recall the definition of completion from Section 20.4. If $g \in H^M$, there exists $g_m \in C^N(K)$ such that $\|g_m - g\|_{H^M} \to 0$. In view of (26.4), we see that $\{g_m\}$ is a Cauchy sequence with respect to the norm $\|\cdot\|_{C^N(K)}$. Since $C^N(K)$ is complete, then g_m converges with respect to this norm. The only possible limit is g. Therefore $g \in C^N(K)$ whenever $g \in H^M$.

Since $|F(f)| \le c_4\|f\|_{C^N(K)} \le c_4 c_6 \|f\|_{H^M}$, then F can be viewed as a bounded linear functional on H^M. By the Riesz representation theorem for Hilbert spaces (Theorem 19.9), there exists $g \in H^M$ such that

$$F(f) = \langle f, g \rangle_{H^M} = \sum_{k \le M} \langle D^k f, D^k g \rangle, \qquad f \in H^M.$$

Now if $g_m \to g$ with respect to the H^M norm and each $g_m \in C^N(K)$, then

$$\langle D^k f, D^k g \rangle = \lim_{m \to \infty} \langle D^k f, D^k g_m \rangle = \lim_{m \to \infty} (-1)^k \langle D^{2k} f, g_m \rangle$$
$$= (-1)^k \langle D^{2k} f, g \rangle = (-1)^k G_g(D^{2k} f)$$
$$= D^{2k} G_g(f)$$

if $f \in C_K^\infty$, using integration by parts and the definition of the derivative of a distribution. Therefore

$$F = \sum_{k \le M} (-1)^k D^{2k} G_{g_k},$$

which gives our result if we let $L = 2M$, set $g_j = 0$ if j is odd, and set $g_{2k} = (-1)^k g$. $\qquad \square$

26.4   Tempered distributions

Let $\mathcal{S}$ be the class of complex-valued C^∞ functions u such that $|x^j D^k u(x)| \to 0$ as $|x| \to \infty$ for all $k \ge 0$ and all $j \ge 1$. $\mathcal{S}$ is called the *Schwartz class*. An example of an element in the Schwartz class that is not in C_K^∞ is e^{-x^2}.

Define

$$\|u\|_{j,k} = \sup_{x \in \mathbb{R}} |x|^j |D^k u(x)|.$$

We say $u_n \in \mathcal{S}$ converges to $u \in \mathcal{S}$ in the sense of the Schwartz class if $\|u_n - u\|_{j,k} \to 0$ for all j, k.

A continuous linear functional on $\mathcal{S}$ is a function $F : \mathcal{S} \to \mathbb{C}$ such that $F(f + g) = F(f) + F(g)$ if $f, g \in \mathcal{S}$, $F(cf) = cF(f)$ if $f \in \mathcal{S}$ and $c \in \mathbb{C}$, and $F(f_m) \to F(f)$ whenever $f_m \to f$ in the sense of the Schwartz class. A *tempered distribution* is a continuous linear functional on $\mathcal{S}$.

Since $C_K^\infty \subset \mathcal{S}$ and $f_n \to f$ in the sense of the Schwartz class whenever $f_n \to f$ in the sense of C_K^∞, then any continuous linear functional on $\mathcal{S}$ is also a continuous linear functional on C_K^∞. Therefore every tempered distribution is a distribution.

Any distribution with compact support is a tempered distribution. If g grows slower than some power of $|x|$ as $|x| \to \infty$, then G_g is a tempered distribution, where $G_g(f) = \int f(x)g(x)\,dx$.

For $f \in \mathcal{S}$, recall that we defined the Fourier transform $\mathcal{F}f = \hat{f}$ by

$$\hat{f}(u) = \int f(x)e^{ixu}\,dx.$$

Theorem 26.17 *$\mathcal{F}$ is a continuous map from $\mathcal{S}$ into $\mathcal{S}$.*

Proof. For elements of $\mathcal{S}$, $D^k(\mathcal{F}f) = \mathcal{F}((ix)^k)f)$. If $f \in \mathcal{S}$, $|x^k f(x)|$ tends to zero faster than any power of $|x|^{-1}$, so $x^k f(x) \in L^1$. This implies $D^k \mathcal{F}u$ is a continuous function, and hence $\mathcal{F}u \in C^\infty$.

We see by Exercise 26.11 that

$$u^j D^k(\mathcal{F}f)(u) = i^{k+j}\mathcal{F}(D^j(x^k f))(u). \qquad (26.5)$$

Using the product rule, $D^j(x^k f)$ is in L^1. Hence $u^j D^k \mathcal{F}f(u)$ is continuous and bounded. This implies that every derivative of $\mathcal{F}f(u)$ goes to zero faster than any power of $|u|^{-1}$. Therefore $\mathcal{F}f \in \mathcal{S}$.

Finally, if $f_m \to f$ in the sense of the Schwartz class, it follows by the dominated convergence theorem that $\mathcal{F}(f_m)(u) \to \mathcal{F}(f)(u)$ uniformly over $u \in \mathbb{R}$ and moreover $|u|^k D^j(\mathcal{F}(f_m)) \to |u|^k D^j(\mathcal{F}(f))$ uniformly over $\mathbb{R}$ for each j and k. $\qquad \square$

If F is a tempered distribution, define $\mathcal{F}F$ by

$$\mathcal{F}F(f) = F(\hat{f})$$

for all $f \in \mathcal{S}$. We verify that $\mathcal{F}G_g = G_{\hat{g}}$ if $g \in \mathcal{S}$ as follows:

$$\mathcal{F}(G_g)(f) = G_g(\hat{f}) = \int \hat{f}(x)g(x)\,dx$$

$$= \int\int e^{iyx}f(y)g(x)\,dy\,dx = \int f(y)\hat{g}(y)\,dy$$

$$= G_{\hat{g}}(f)$$

if $f \in \mathcal{S}$.

Note that for the above equations to work, we used the fact that $\mathcal{F}$ maps $\mathcal{S}$ into $\mathcal{S}$. Of course, $\mathcal{F}$ does not map C_K^∞ into C_K^∞. That is why we define the Fourier transform only for tempered distributions rather than all distributions.

Theorem 26.18 $\mathcal{F}$ *is an invertible map on the class of tempered distributions and* $\mathcal{F}^{-1} = (2\pi)^{1/2}\mathcal{F}R$. *Moreover* $\mathcal{F}$ *and* R *commute.*

Proof. We know

$$f(x) = (2\pi)^{-1/2} \int \widehat{f}(-u)e^{ixu}\,du, \qquad f \in \mathcal{S},$$

so $f = (2\pi)^{-1/2}\mathcal{F}R\mathcal{F}f$, and hence $\mathcal{F}R\mathcal{F} = (2\pi)^{1/2}I$, where I is the identity. Then if H is a tempered distribution,

$$(2\pi)^{-1/2}\mathcal{F}R\mathcal{F}H(f) = R\mathcal{F}H((2\pi)^{-1/2}\mathcal{F}f) = \mathcal{F}H((2\pi)^{-1/2}R\mathcal{F}f)$$
$$= H((2\pi)^{-1/2}\mathcal{F}R\mathcal{F}f) = H(f).$$

Thus

$$(2\pi)^{-1/2}\mathcal{F}R\mathcal{F}H = H,$$

or

$$(2\pi)^{-1/2}\mathcal{F}R\mathcal{F} = I.$$

We conclude $A = (2\pi)^{-1/2}\mathcal{F}R$ is a left inverse of $\mathcal{F}$ and $B = (2\pi)^{-1/2}R\mathcal{F}$ is a right inverse of $\mathcal{F}$. Hence $B = (A\mathcal{F})B = A(\mathcal{F}B) = A$, or $\mathcal{F}$ has an inverse, namely, $(2\pi)^{-1/2}\mathcal{F}R$, and moreover $R\mathcal{F} = \mathcal{F}R$. $\qquad\square$

26.5   Exercises

Exercise 26.1 Can C_K^∞ be made into a metric space such that convergence in the sense of C_K^∞ is equivalent to convergence with respect to the metric? If so, is this metric space complete? Can it be made into a Banach space?

Exercise 26.2 We define a metric for $\mathcal{S}$ by setting

$$d(f,g) = \sum_{k,j} \frac{1}{2^{j+k}} \cdot \frac{\|f - g\|_{j,k}}{1 + \|f - g\|_{j,k}}.$$

Prove that this is a metric for $\mathcal{S}$. Prove that a sequence converges in the sense of the Schwartz class if and only if it converges with respect to the metric. Is $\mathcal{S}$ with this metric complete?

Exercise 26.3 Prove that if $f \in C_K^\infty$, then

$$F(f) = \lim_{\varepsilon \to 0} \int_{|x| > \varepsilon} \frac{f(x)}{x}\,dx$$

exists. Prove that F is a distribution.

Exercise 26.4 Suppose U is a continuous linear map from C_K^∞ into C_K^∞. If F is a distribution, define TF by

$$TF(f) = F(Uf), \qquad f \in C_K^\infty.$$

(1) Prove that TF is a distribution.
(2) Suppose V is a continuous linear map from C_K^∞ into itself such that $\int g(Uf) = \int (Vg)f$ for every $f, g \in C_K^\infty$. Prove that if $g \in C_K^\infty$, then

$$TG_g = G_{Vg}.$$

Exercise 26.5 If μ is a finite measure defined on the Borel σ-algebra, prove that F given by $F(f) = \int f\,d\mu$ is a distribution.

Exercise 26.6 Suppose g is a continuously differentiable function and h is its derivative in the classical sense. Prove that $DG_g = G_h$.

Exercise 26.7 A *positive distribution* F is one such that $F(f) \geq 0$ whenever $f \geq 0$. Prove that if K is a compact set and F is a positive distribution, then there exists a constant c such that

$$|F(f)| \leq c \sup_{x \in K} |f(x)|$$

for all f supported in K.

Exercise 26.8 Prove that if F is a positive distribution with compact support, then there exists a measure μ such that $F(f) = \int f \, d\mu$ for $f \in C_K^\infty$.

Exercise 26.9 Suppose

$$F(f) = \lim_{\varepsilon \to 0} \int_{1 \geq |x| \geq \varepsilon} \frac{f(x)}{x} \, dx.$$

Show that F is a distribution with compact support. Find explicitly L and the functions g_j guaranteed by Theorem 26.15, and prove that F has the representation given by (26.3).

Exercise 26.10 Let $g_1(x) = e^x$ and $g_2(x) = e^x \cos(e^x)$. Prove that G_{g_2} is a tempered distribution but G_{g_1} is not.

Exercise 26.11 Prove (26.5).

Exercise 26.12 Determine $\mathcal{F} G_1$, $\mathcal{F}\delta$, and $\mathcal{F} D^j \delta$ for $j \geq 1$.

Bibliography

[1] L. Ahlfors. *Complex Analysis.* McGraw-Hill, New York, 1979.

[2] S. Axler, P. Bourdon, and R. Wade. *Harmonic Function Theory*, 2nd ed. Springer, New York, 2010.

[3] R.F. Bass. *Probabilistic Techniques in Analysis.* Springer, New York, 1975.

[4] G.B. Folland. *Real Analysis: Modern Techniques and their Applications*, 2nd ed. Wiley, New York, 1999.

[5] P.D. Lax. *Functional Analysis.* Wiley, New York, 2002.

[6] H. Royden. *Real Analysis*, 3rd ed. Macmillan, New York, 1988.

[7] W. Rudin. *Principles of Mathematical Analysis*, 3rd ed. McGraw-Hill, New York, 1976.

[8] W. Rudin. *Real and Complex Analysis*, 3rd ed. McGraw-Hill, New York, 1987.

[9] E. M. Stein. *Singular Integrals and Differentiability Properties of Functions.* Princeton Univ. Press, Princeton, 1970.

Index

Made in the USA
San Bernardino, CA
30 August 2016